KU-328-623

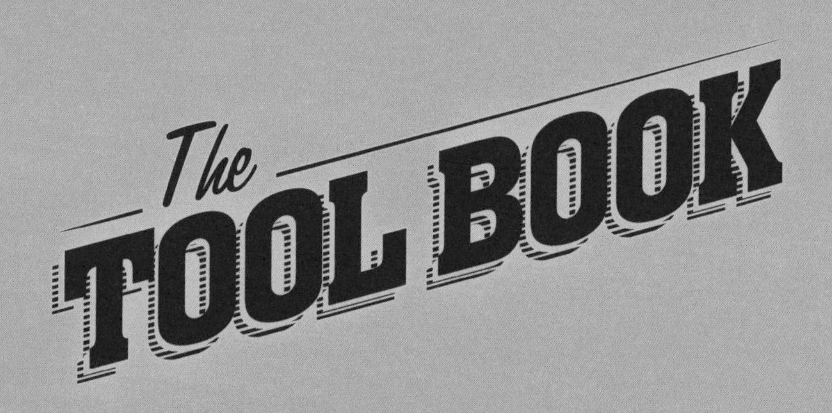
The
TOOL BOOK

The
TOOL BOOK
A TOOL-LOVER'S GUIDE TO OVER 200 HAND TOOLS
DK

Writers
Phil Davy, Luke Edwardes-Evans,
Jo Behari, Matthew Jackson

Senior Editor Ruth O'Rourke-Jones

Editor Jamie Ambrose

Project Art Editor Vicky Read

Managing Editor Dawn Henderson

Managing Art Editor Marianne Markham

Asset Management Emily Reid

Jacket Designer Steven Marsden

Producer Rebecca Fallowfield

Creative Technical Support
Sonia Charbonnier, Tom Morse

Photography John Spence
at MMS Marketing Services

Illustrator Andrew Torrens

Art Director Maxine Pedliham

Publishing Director Mary-Clare Jerram

First published in Great Britain in 2018
by Dorling Kindersley Limited
80 Strand, London, WC2R 0RL

10 9 8 7 6 5 4 3 2 1
001–305962–May/2018

A CIP catalogue record for this book is available from the British Library.
ISBN: 978-0-2413-0211-8

Printed and bound in China

A WORLD OF IDEAS:
SEE ALL THERE IS TO KNOW
www.dk.com

CONTENTS

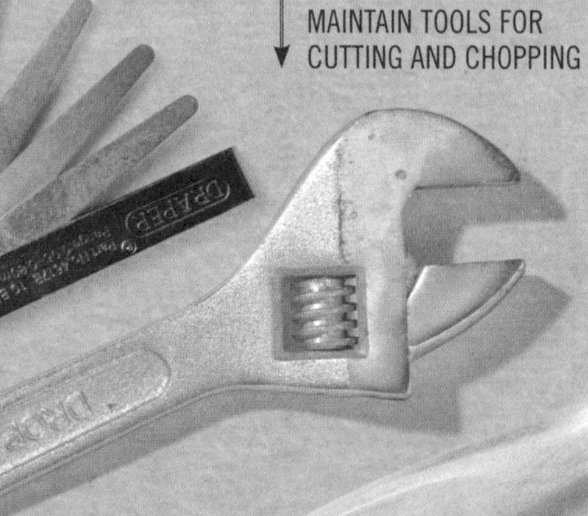

98 THE TOOLS FOR FIXING & FASTENING

142 THE TOOLS FOR STRIKING & BREAKING

166 THE TOOLS FOR DIGGING & GROUNDWORK

190 THE TOOLS FOR SHAPING & SHARPENING

218 THE TOOLS FOR FINISHING & DECORATING

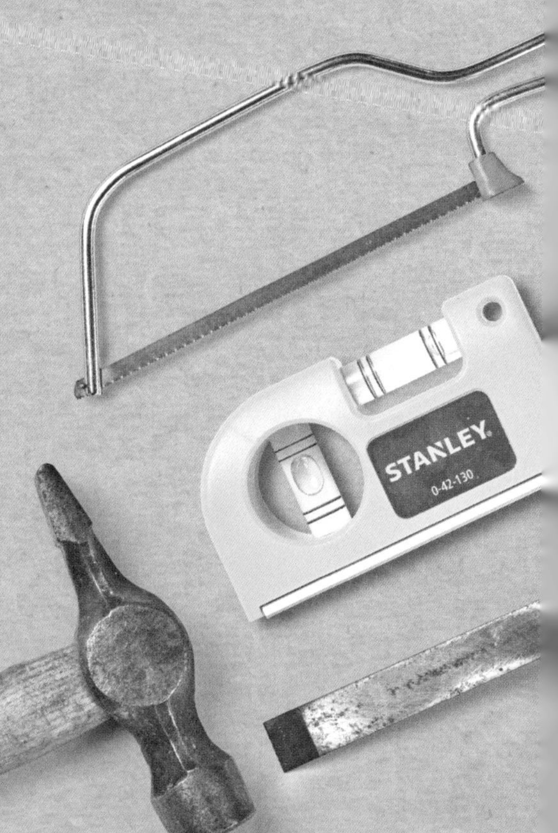

THE TOOL BOOK

FOREWORD

BY NICK OFFERMAN

Photo Emily Shur

Tools. Yes!

The mere mention of the word thrills my blood, much as I have been given to understand the words "chocolate", or "gin", or "fireman" can sometimes titillate others. Another similarly rousing but unrelated term for me is "wizardry". Like many, I have always been strongly attracted to the magicks described in stories of enchantment; the eldritch knowledge and "old ways" passed down over generations. I am a complete sucker for the sorts of charms and glamours to be found in your ***Chronicles of Narnia***, your ***Lord of the Rings***, your ***Dungeons & Dragons***, and your ***Harry Potters***, particularly those containing the stripe of conjuring that requires a mystical object, like a wand, a staff, or a ring. And now that I think about it, perhaps my infatuation with "tools" and "wizardry" are not so unrelated after all.

As a youth in rural Illinois (the American state that contains Chicago and corn, among other things), much of my spare time was spent learning to gainfully employ many of the implements represented in this magnificent catalogue you are about to read. My many teachers included my mother and my father, along with their own siblings and parents, most of whom maintained a family farm operation, raising pigs, soya beans, and corn. Every household in the family cultivated a robust garden and attractive yard (some more so than others, as in any family) whilst perennially investing in the beauty and tidiness of their houses and outbuildings.

These extremely practical humans were always in the process of getting good work done, and that usually with tools. Their sphere of influence was not limited to their yards, either, because neighbours in our community would readily share tools, as well as their kind attentions and the labour of their hands. In my 30 or so years living as an "adult" I have inhabited and visited a pretty full spectrum of living arrangements, from abject poverty (working in theatre) to the lap of luxury (working in television), and I can attest from my experience that the homes in which health and love and economy most readily flourished were the ones that knew the use of some of the noble equipment in this here book. I daresay that the older I grow, the more the wrenches and the hammers begin to resemble the tackle of wizards.

I think that had anyone suggested to me in my teenage years that my love of tools might be remotely comparable to my reverence for sorcery, I would have muttered something terribly pithy from the film ***Highlander*** under my breath, perhaps the epic "It's better to burn out than to fade away…", to which someone might have said "isn't that a Neil Young lyric?", to which I would have defiantly stated "You don't know dick", and lurched off in a very Emilio Estevez fashion to admire my cassette tapes. In my crotchety middle age, however, I am finding it easier to comprehend the reason that I find an adze every bit as compelling as Gandalf's magic sword Glamdring. It's because both carry the deep wisdom of true handcraft. Through the painstaking centuries of trial and error on the part of man and womankind (or elves, in the case of the sword), these clever sticks have been imbued with the cumulative, arcane intelligence of the generations of hands that perfected their forms, and thereby, their uses.

One of my most poignant brand of tool memories is when my dad would take me out in the yard to help him plant trees, or do anything with a shovel or a spade. The honest mastery with which he brandished said item to sculpt an accurate and practical hole out of the gravelly soil might just as easily have been found inhabiting a virtuoso soloist's bow, as she sawed away on some Tchaikovsky (***Valse Sentimentale*** please and thank you). To mine, a child's perception, his skills could easily be accused of witchcraft, especially when the spell did not end at the digging. My dear dad was not only accomplishing his task of landscaping, he was also teaching me to use the spade. To this day, when I grip a shovel's staff in my hands and place my foot upon the shoulder of the blade in preparation for its initial plunge into the ground, my dad is there with me, his hands upon mine, his gentle tutelage in my ears.

Think about that. I know that he learned the ways of the shovel from his dad, as well as other workers in the neighbourhood from his own youth, but where did his dad get it? And what of the myriad other tools in which he schooled me, not to mention the other methods I gleaned from my aunts and uncles? It stands to reason that my very own erudition in the art of the hole can be traced ancestrally all the way back to the first inventor of the shovel. Since we humorously cognizant monkeys first began to swing a stone at the end of a stick, we have been quietly and confidently passing along our collective knowledge to the next generation, again and again over the millennia, each apprentice acquiring an eventual mastery, occasionally adding slight improvements, until we humans can now print a book like this one, full of colours and handsomely worded pictures, containing among its juicy offerings categories of tools like "Digging and Groundwork" and "Striking and Breaking". It's downright Homeric.

The tools in this book are not those of the specialist. It is unlikely that you will build a serviceable rocket ship or a nickelodeon with the stuff in here. On the contrary, the authors have chosen to honour the venerable tools of the generalist. What you can create with these charismatic utensils is a limitless sense of adventure and accomplishment – a life rife with practical industry. The list does not necessarily contain the items that spring to mind when one considers creating "art", as it were, but (in *my* book) they are precisely the devices necessary for the art of living well, and I think that is about as magical as it can get.

PLAN YOUR

Work Area

Whether you're fixing a bike, decorating the house, or making furniture you need somewhere to work. Depending on space and budget this could be a folding workbench that stores in a cupboard, or a fully equipped workshop in a shed. For occasional jobs a temporary solution may do, but for long-term projects a more permanent work area is necessary.

Work surface

For many tasks a solid work surface makes a big difference. A workbench doesn't need to be huge, but it should be as solid as possible. You can fit a vice to the front or one end for working with timber, metals, or other materials. A portable Workmate or similar bench is a cheaper, lightweight alternative and can be used outdoors or moved from one room to another.

Tool storage

Tools should be stored so that you can find them easily. Displaying them on a wall or in a dedicated cupboard means you can spot if a tool is missing and is safer than riffling through a toolbox. Keeping a work area tidy and well organized helps you work efficiently.

Heat, light, and power

For lighting and mains sockets, you will need a power supply, which should be installed by a qualified electrician. You may also need heating as an unheated work area in cold weather can be miserable. Think about the type of heating: a mobile gas heater is efficient, but a naked flame in a dusty workshop is not wise. An electric heater may be safer but costly to warm a large space.

Environment

Some hand tools, such as hammers, can be noisy in use. For frequent work, consider insulating a workshop or restricting work times: think about the neighbours! Fire safety is a must, especially if storing flammable materials. Fire extinguishers are inexpensive, but make sure you get the correct type to suit the situation.

FOCUS ON...

Security

Keeping your tools safe should be a priority. Don't rely on a flimsy padlock on the shed door. Thieves will aim for the easiest method of breaking and entering, so think about how you would gain entry to your workshop if you lost the keys. Adding internal shutters to windows, security bars across doors, and battery-operated alarms can help. Track tools by using a security-marking kit to label them invisibly with your postcode – it will show up under ultraviolet light.

Woodworkers often store tools *in purpose-made cabinets with hinged doors. This a good use of wall space and protects the contents from workshop dust. Fitting cabinet locks or padlocks makes tools more secure, and keeps them out of the reach of children.*

Far left: *In a garage or shed hang tools on hooks or pegs above a bench. This makes them easy to reach and you can spot if a tool is missing.*

Left: *Decorating means turning a room into a temporary work area. Empty the space so you have clear access and can move a pasting table or stepladder around as needed.*

FOCUS ON...

SAFETY

Even hand tools can be hazardous if abused or maintained poorly. For example, a blunt chisel is arguably more risky than a sharp one as it's more likely to slip. Check the condition of your tools regularly and replace any that are damaged and beyond repair. Make sure that you wear suitable personal protective equipment and that this is stored carefully when not in use. Label drawers or cupboards so you know just where to find the relevant item.

A simple first aid kit should include plasters, eyewash, and dressings. It's not necessary to have a large, comprehensive box as individual items can be replaced if they're used or become out of date.

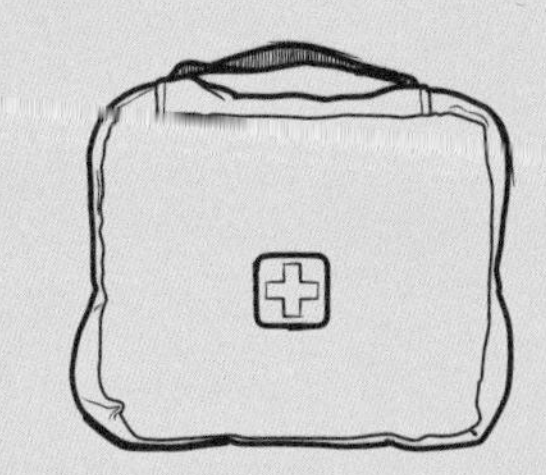

Work gloves prevent splinters and cuts when handling rough-sawn timber or heavy, metal items. Heavy leather or fabric gloves can be cumbersome – lighter vinyl gloves are easier to wear and still give protection.

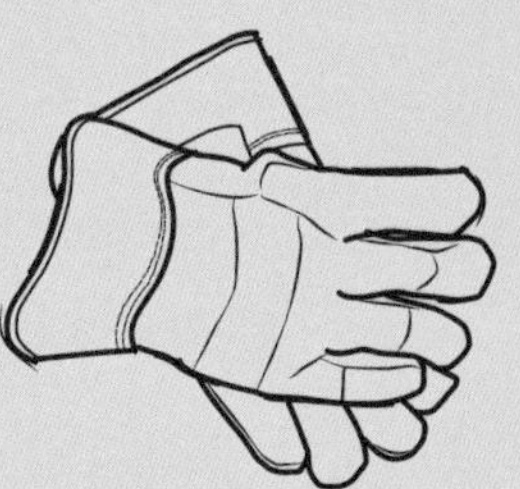

Eye protection is vital. Clear safety glasses are comfortable and some may be worn over spectacles. If welding, you'll need proper full-face protection.

Choosing a

Tool Belt

If you're working from a ladder it can be frustrating when you don't have the right tools to hand. To overcome this, wear a tool belt and pouch around your waist so you don't need to climb up and down to fetch an item you've forgotten. Outer pockets are handy for storing small items like nails or screws.

Spanners *should be stored together when using several sizes.*

Claw hammer *is too big for pockets – store in its own loop.*

Tool loop *is good for storing large items, but check they won't fall through.*

> “Don’t OVERLOAD your belt with tools as this could become TIRING to wear over a LONG PERIOD”

FOCUS ON…

Pockets & Holsters

Tool belts are traditionally made from sturdy leather, with pockets stitched and riveted. Durable fabric belts offer a similar number of pockets and hoops, sometimes detachable, with polypropylene webbing belts. Belts have steel or plastic buckles and are adjustable for size. Shaped holsters for bulkier tools are usually steel. When buying, check the fit and consider how many tools you will need. Too many pockets could tempt you to load up the belt unnecessarily.

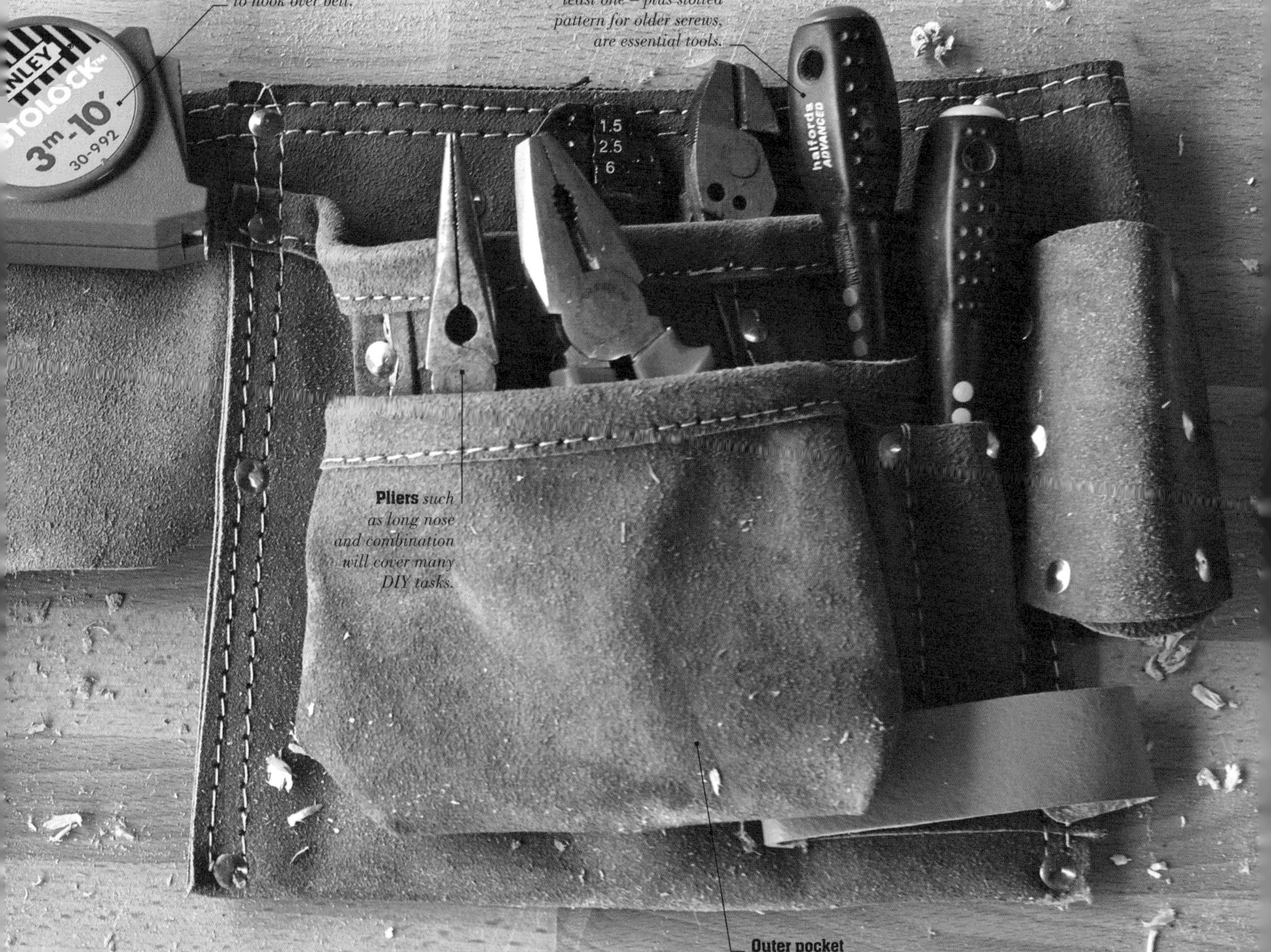
Retractable tape measure *has clip on back designed to hook over belt.*
Pozi screwdriver *– at least one – plus slotted pattern for older screws, are essential tools.*
Pliers *such as long nose and combination will cover many DIY tasks.*
Outer pocket *is good for storing screws and nails.*
3m-10′
30-992
1.5
2.5
6
halfords
ADVANCED

Choosing a Tool Box

When there is no dedicated work space it makes sense to keep hand tools in a tool box. This means items can be stowed in a cupboard, moved easily from one job to another, or transported in the back of a car. Tool boxes are usually heavy duty and come in a variety of sizes and styles. Lids are often hinged and good boxes allow a padlock to be fitted.

Steel tool box

A steel box is a sturdy option, ideal for bike maintenance or mechanic's tools. Cantilevered versions reveal lower compartments as the top of the box opens outwards. A full-length handle makes them easy to carry, but steel boxes tend to be heavier than plastic ones. Filled with tools, they are quite weighty. It's a good idea to line compartments with anti-slip matting or bubblewrap to protect valuable tools. If keeping a steel tool box in an unheated area check the contents regularly. Condensation will lead to rusting of metal tools, so you may want to treat these with an anti-corrosive spray.

Plastic tool box

A plastic tool box will not rust and is better at protecting precision marking and measuring tools, as well as metal woodworking tools. Some lightweight plastic models are more susceptible to damage. Heavy-duty boxes (made from structural foam) may have rubber seals around the lid, making them water resistant.

Tool bags

The traditional tool bag is made from canvas with reinforced rope handles and brass eyelets around the rim to secure the contents. These have been largely replaced by reinforced, synthetic fabric bags with numerous internal and external pockets.

Small plastic tool boxes *are ideal for hobbyist tools. Larger, more durable ones suit heavier tools and can usually be padlocked.*

Measuring and marking *tools should be stored on soft, anti-slip matting cut to fit compartments.*

Steel tool box

Tape measures *are used frequently and should be readily accessible when you open the lid.*

Mallet *and other heavy tools should be placed at the bottom of the box in the largest area.*
Fixings *and other loose items should be placed in clear plastic bags or smaller boxes.*
Clarke
DIN ISO 8764
Hammers *are stored with heavier kit at base of box.*
Sanding block *is neatly packed to make most of available space.*

Choosing a

Tool Shed

A shed is a safe place to store your tools but it can also be where you carry out DIY, bike maintenance, or woodwork. A suitable shed should be positioned on a firm, level surface, so you may need to do some preparation work. Start by laying paving slabs to create a base, or use concrete, which is more work but will be more solid and last indefinitely.

Metal shed

Made from a series of corrugated steel panels bolted or clipped together, this type of shed is suitable for garden tools, stepladders, and large items such as folding benches or sawhorses. The steel is normally galvanized and pre-painted, making it fire and rot resistant. Arguably more durable than timber, metal sheds are virtually maintenance-free, with no roofing felt to replace or panels that require painting. Roof panels are also corrugated for strength. A potential problem with such sheds is condensation, which can lead to rusting of tools if not kept in check. Metal sheds are unlikely to have windows, though this makes them secure. Doors may be sliding or hinged and can be padlocked.

Timber shed

Traditional sheds are built from softwood, which should be treated with preservative to prevent rot and insect attack. Consisting of timber sections bolted together, this framework is covered in shiplap or tongue-and groove boards nailed horizontally. Roofs are pitched to shed rain and snow. Once fixed to the walls, the roof material is covered in mineral felt and nailed in place. Flooring can be chipboard or plywood. Windows may be glass or clear plastic and are likely to be fixed in budget sheds. The outside should be maintained with an exterior wood finish or paint. A timber shed is easy to insulate and walls can be lined with MDF or plywood. With lighting and electrical sockets this can provide an excellent small workshop.

A traditional timber *shed provides a secure work area to pursue interests, like bike maintenance, woodwork, DIY, and gardening.*

1 *Work platform* 2 *Tool belt* 3 *Claw hammers* 4 *Shelving for small tools* 5 *Clamps* 6 *Saw horses and saws* 7 *Heavy outdoor tools on floor* 8 *Lighter garden tools on hooks*

Folding saw horse *is lightweight and can be hung on wall.*

Portable work bench *has folding legs and adjustable jaws, making it ideal for DIY jobs.*

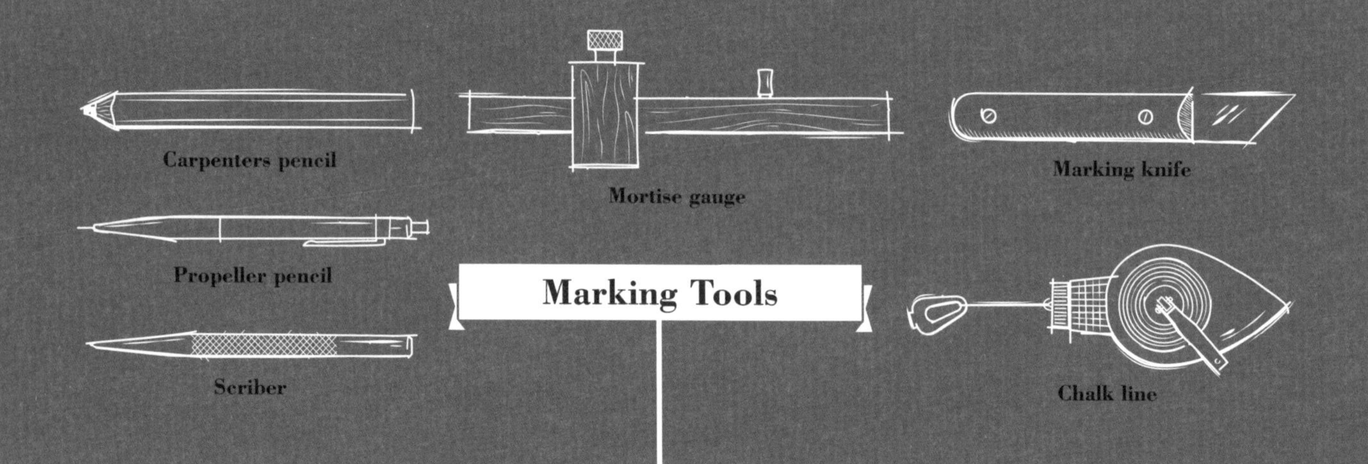

THE TOOLS for MEASURING & MARKING

From simple scribers and rulers to more complex digital levels and calipers, accurate measuring and marking tools are essential to getting your project off to a good start.

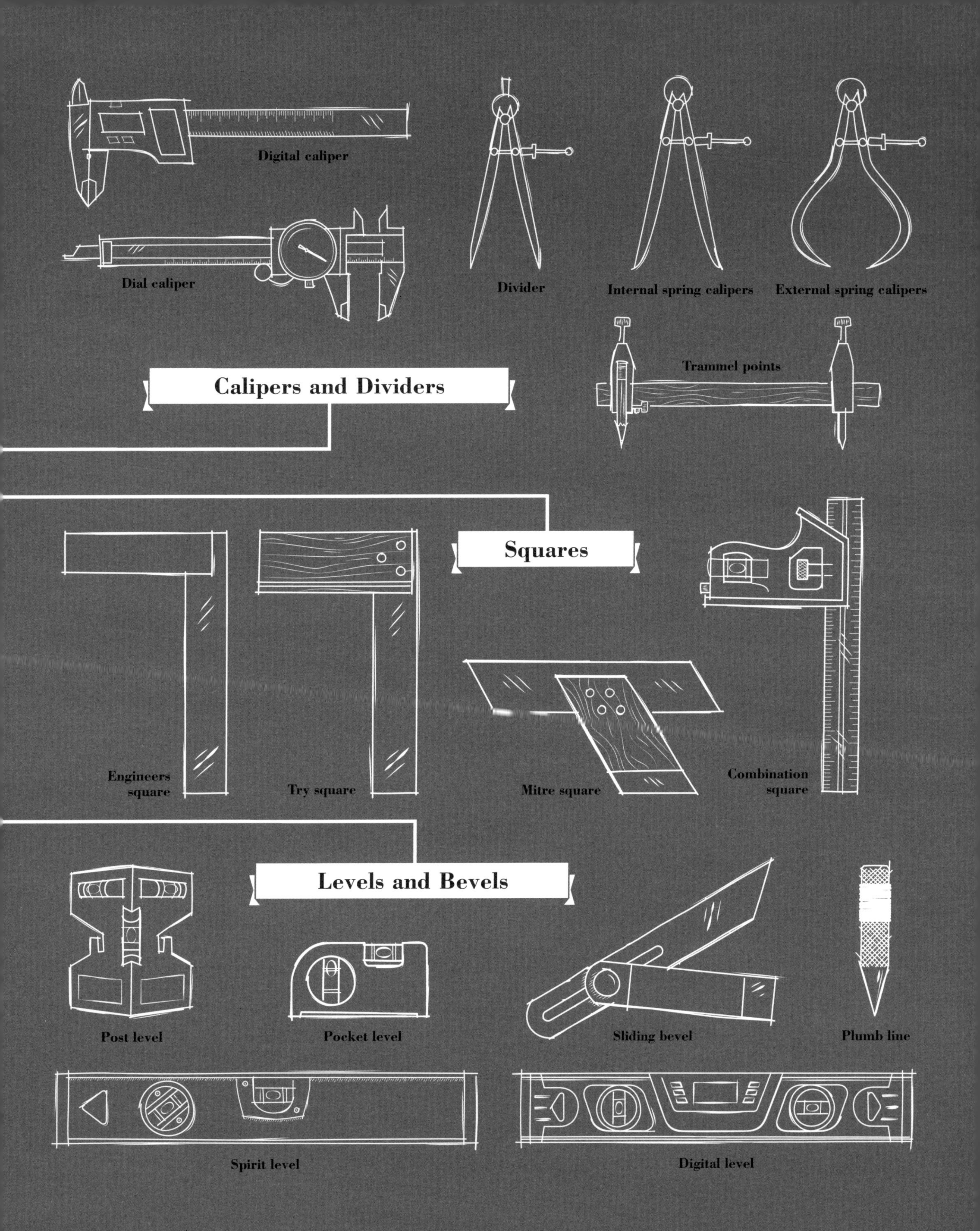

Digital caliper
Dial caliper
Divider
Internal spring calipers
External spring calipers
Trammel points
Calipers and Dividers
Squares
Engineers square
Try square
Mitre square
Combination square
Levels and Bevels
Post level
Pocket level
Sliding bevel
Plumb line
Spirit level
Digital level

HISTORY OF MEASURING & MARKING

c.3000 BCE — FIRST CHALK LINE

Ancient Egyptian builders used an early form of the chalk line. A cord, coated with wet red or yellow ochre, was held taut between two points and then snapped onto a surface to leave a straight line. This technique is still used in modern construction today, with chalk used instead of ochre.

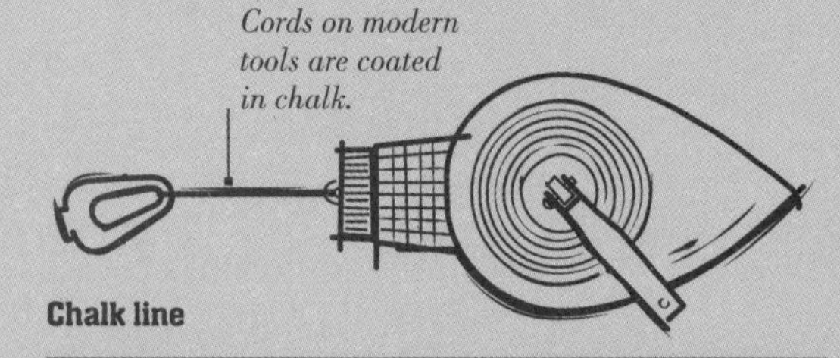

Chalk line

c.2650 BCE — MEASURING RODS

Copper-alloy bars were used as measuring rods in Mesopotamia. In 1916, a well-preserved, demarcated bar was found at excavations at Nippur (present-day Iraq). It is thought to be a Sumerian cubit, equal to around 51.85cm (20.4in). Cubits varied across the Middle East. The Egyptian royal cubit of around the same period was 52.3cm (20.6in).

c.2600 BCE — EARLY PLUMB LINE

Buildings such as the Great Pyramid at Giza led to the development of levels, including an early plumb line to check walls were vertical. The tool was E-shaped and had a weighted line suspended from the upper edge of the "E". The tool was held against the wall to check that the string touched the lower outside edge of the E shape.

2.3 MILLION
blocks were used in the building of the Great Pyramid of Giza.

"MAN IS THE MEASURE OF ALL THINGS."
PROTAGORAS
481–411 BCE

c.2600 BCE — EGYPTIAN A-FRAME LEVEL

Egyptians used an A-frame tool to check horizontal levels. The frame was set on the surface to be checked and a plumb line was suspended from the centre. This method was still being used across Europe well into the 19th century.

Plumb line hangs from apex of frame.

Weight attached to end of cord.

A-frame

c.1290 BCE — FIRST SQUARE

The square was also developed in ancient Egypt and it may have been used it to cut stone accurately for building temples, pyramids, and other monuments. Two pieces of wood were joined at a right angle, sometimes with a diagonal rod acting as a brace. Such artefacts have been discovered in tombs such as that of an artisan named Sennedjem, at Deir el-Medina, Egypt.

THE YARD

1305 – In England, Edward I defined the yard as the distance from the tip of his nose to his outstretched thumb:

0.9M (3FT)

c.1070 BCE

EGYPTIAN RULERS

The Egyptians used various rulers, from ceremonial stone cubit-rods found in temples to wooden rulers used by carpenters. The standard was a royal cubit, defined as seven palm widths measured across the fingers, totalling roughly 52.3cm (20.6in). Egyptian masons used wooden rulers with a bevelled edge.

Ancient Egyptian ruler

c.600 BCE

BASIC DIVIDERS AND CALIPERS

Dividers – many similar to modern compasses – and calipers were used by both the ancient Greeks and Romans; as ancient calipers were wooden, however, most have not survived. One rare example from the 7th century BCE, with one fixed and one movable jaw, was discovered during the excavation of a Greek shipwreck off the coast of Tuscany.

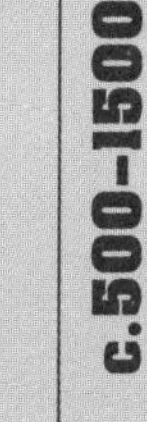

Roman dividers had curved or straight legs.

Divider

c.500–1500

MEDIEVAL DIVIDERS

By the Middle Ages, the caliper was used for woodwork, but huge dividers were used by architects planning large stone constructions, such as cathedrals. These tools were often half a man's height!

1452–1519

RENAISSANCE DIVIDER

Leonardo da Vinci refined the divider by adding a knuckle-joint hinge to increase the instrument's rigidity. His notes include a compass with interchangeable points, including a clamp for graphite or chalk, and a beam compass with a screw adjustment used to draw large circles.

Leonardo wanted to make the divider more stable when the legs were open by increasing the contact area of the hinge points.

Leonardo's divider

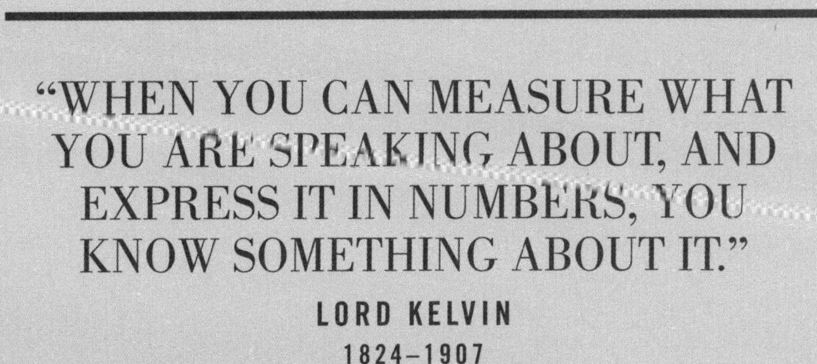

"WHEN YOU CAN MEASURE WHAT YOU ARE SPEAKING ABOUT, AND EXPRESS IT IN NUMBERS, YOU KNOW SOMETHING ABOUT IT."

LORD KELVIN
1824–1907

1600s

BEVEL SQUARE

Adjustable bevel squares were first used to measure and create angles other than 90 degrees during the mid-17th century. Some of these were fixed to commonly used angles such as 45 degrees, while the angle bevel could be adjusted to any angle desired.

EARLY SPIRIT LEVELS

consisted of a sealed glass tube containing alcohol and an air bubble. Before being used as surveying instruments, they were used on telescopes.

Choosing a
Marking Tool

Accurate marking out is a basic principle when working with wood, metal, plastic, or any other surface. Without dependable tools you cannot achieve accuracy later on in a project. Marking tools should be sturdy and made from quality materials, rather than flimsy and unreliable. Simple is often better than complex.

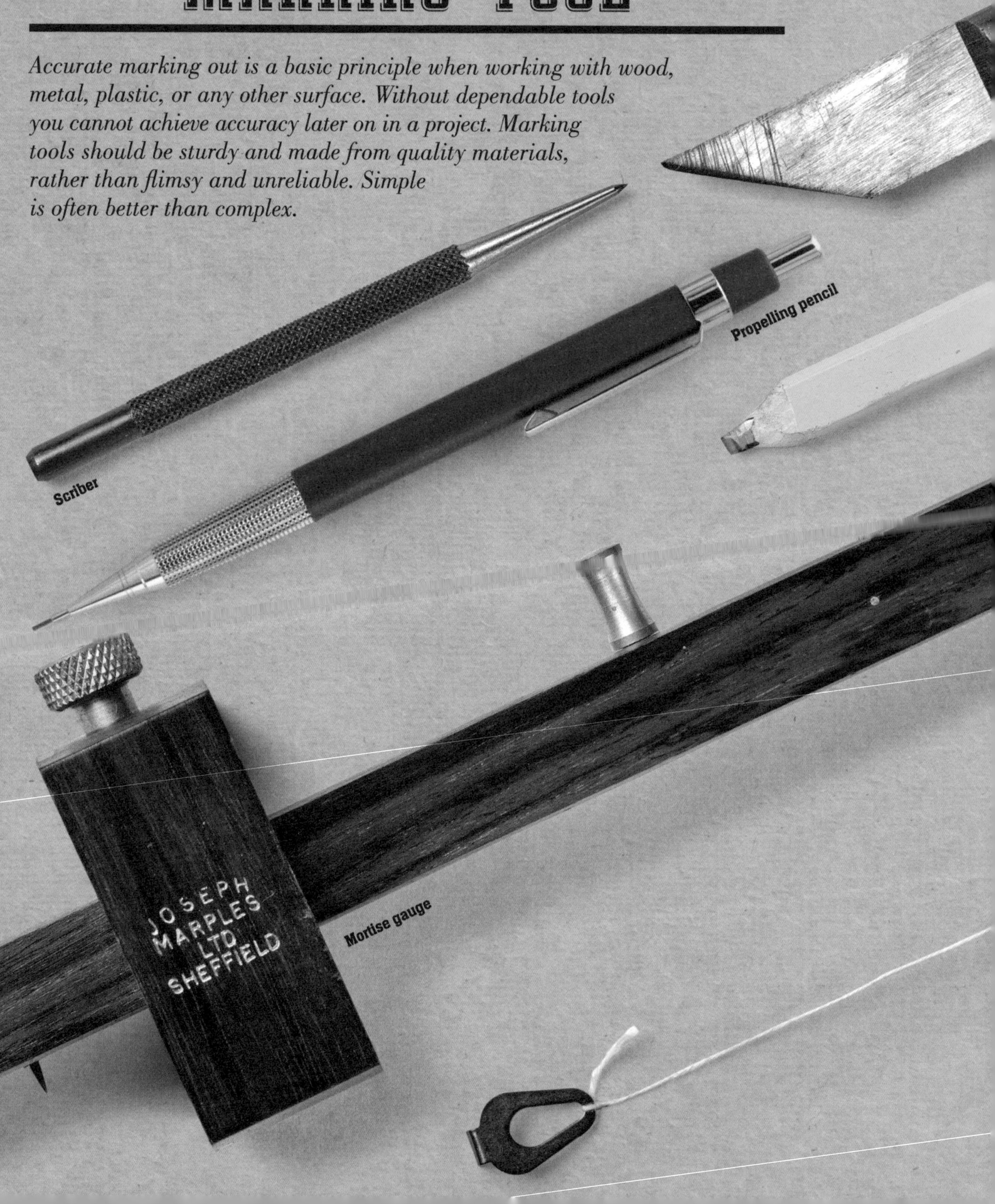

Propelling pencil

Scriber

Mortise gauge

JOSEPH MARPLES LTD SHEFFIELD

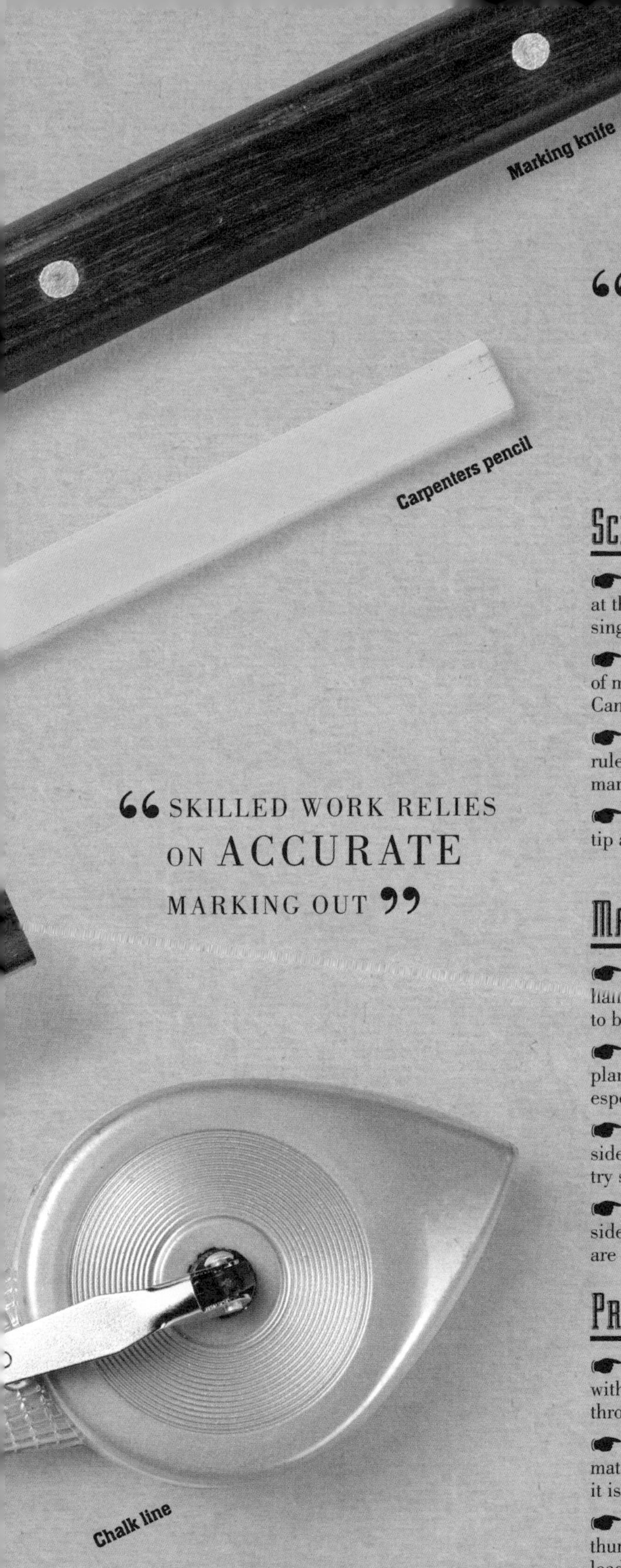

“SKILLED WORK RELIES ON ACCURATE MARKING OUT”

“MARKING TOOLS SHOULD BE STURDY AND MADE FROM QUALITY MATERIALS”

Scriber

What it is Hardened steel point at the end of a slim handle. May be single- or double-ended.

Use it for Mainly scoring surfaces of metal prior to cutting or machining. Can also be used on other materials.

How to use Run it along a steel rule or an engineer's square for accurate marking, including at 90 degrees.

Look for A precision-engineered tip and check that the grip is non-slip.

Marking Knife

What it is A hardwood or metal handle with skewed steel blade, ground to bevel on one side.

Use it for Ensuring clean lines on planed timber by severing wood fibres, especially if marking joints before sawing.

How to use Hold and pull the flat side of the blade against a steel rule or try square when marking a line.

Look for The bevel on left or right side of blade to suit user. Japanese tools are made of laminated steel.

Propelling Pencil

What it is A lead or graphite core with mechanical jaws that propel it through the outer casing as the tip wears.

Use it for Marking wood and other materials. With consistent point size, it is sturdier than a common HB pencil

How to use Activated by pressing thumb button at the opposite end. The lead can be retracted to prevent breakage.

Look for New refills that are correct diameter and hardness grade. A pocket clip and eraser tip are useful.

Carpenters Pencil

What it is A rectangular wood body with graphite core. Sturdier than standard pencils and unlikely to break.

Use it for Approximate marking of timber and other materials. Not really fine enough for general woodwork.

How to use Sharpen to a chisel point with a knife and use like an ordinary standard pencil.

Look for Interchangeable coloured leads in some plastic versions, which work like propelling pencils.

Mortise Gauge

What it is Twin steel pins that create parallel lines in wood. Hardwood stock slides along stem, locking with a screw.

Use it for Marking exact position of mortises parallel to the edge of planed timber. Also for marking matching tenons.

How to use Set pins to chisel width and overall distance from stock with steel rule. Hold against wood and push tool.

Look for Flush brass strips set into the face of the stock that reduce wear and increase the life of the tool.

Chalk Line

What it is A long, retractable string contained within a metal or plastic box filled with coloured chalk.

Use it for Marking a long, straight cutting line on rough-sawn timber, particularly where the edge is uneven.

How to use Pull out the string and hook clip over timber end. Lift string taut and snap it against surface to create a line.

Look for Easy string-rewind action and a self-sealing grommet for cleaner refilling with chalk.

SIDE VIEW
Twin pins for marking mortise. End pin is fixed, inner one adjusts.
Brass thumbscrew locks sliding pin and stock position on stem.
Pins, angled top view
JOSEPH
MARPLES
LTD
SHEFFIELD
Hardwood stock with flat face that runs against timber.
TOP VIEW
Twin brass strips inlaid into face of stock to reduce wear.
Fixed brass section screwed into
cut in stem.
Thumbscrew, angled side view
Knurled edge on thumbscrew to aid ease of use.

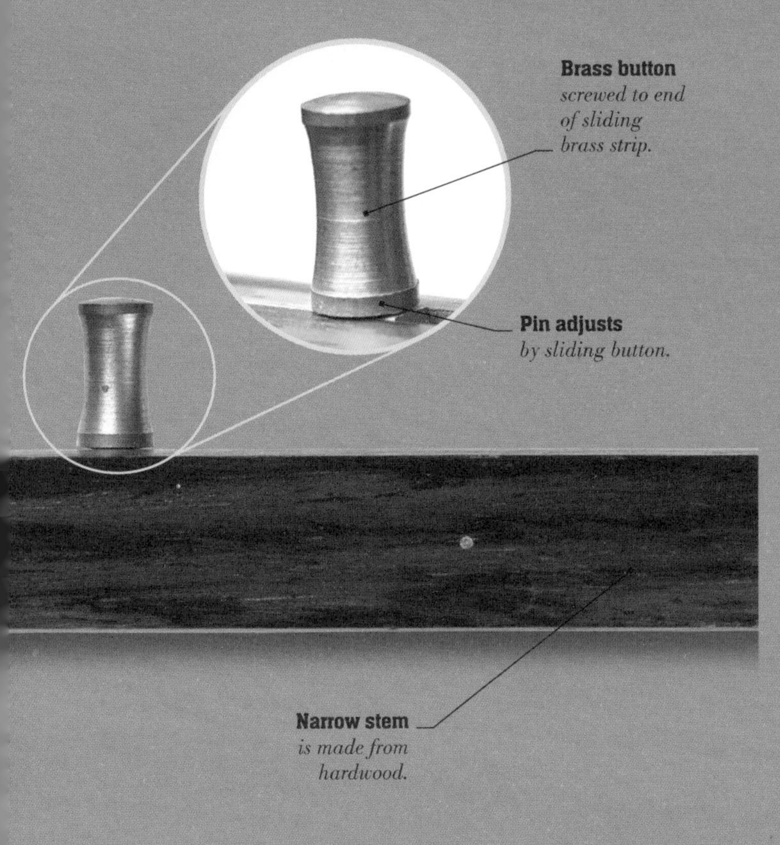

STRUCTURE OF A MORTISE GAUGE

A mortise gauge is traditionally made of rosewood or a similar dense hardwood, with brass facings and adjusters, so it is one of the most attractive hand tools. Although it may have a very specialized function, it would be difficult to manage without it if you're doing a job that involves mortise and tenon joints.

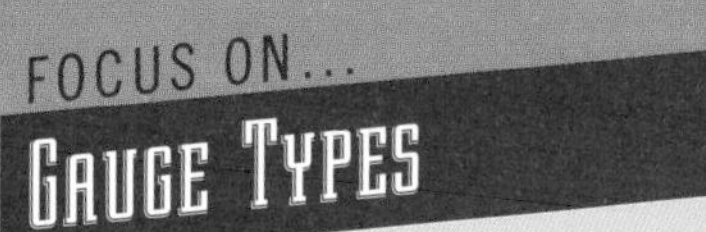

FOCUS ON...

GAUGE TYPES

Both marking and mortise gauges have tiny pins that scribe fine lines along wood. Although visually similar, the cutting gauge uses a small knife blade instead. This is normally sharpened to a V-point. All three gauges are used in the same way.

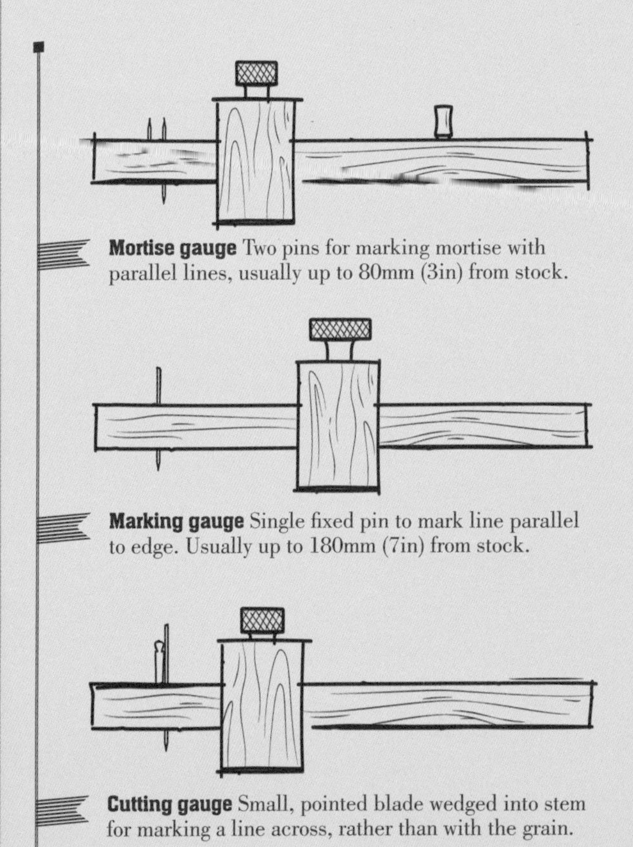

Mortise gauge Two pins for marking mortise with parallel lines, usually up to 80mm (3in) from stock.

Marking gauge Single fixed pin to mark line parallel to edge. Usually up to 180mm (7in) from stock.

Cutting gauge Small, pointed blade wedged into stem for marking a line across, rather than with the grain.

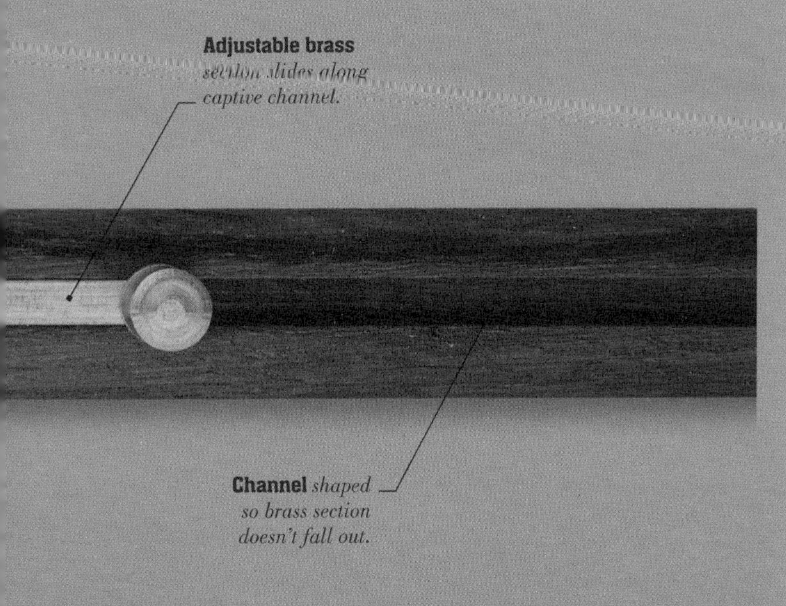

" TO PROTECT FROM MOISTURE AND PREVENT STOCK STICKING, STORE GAUGES IN PLASTIC BAGS "

USING A

Mortise Gauge

A more sophisticated version of a marking gauge (which has just a single pin), this tool is used specifically when setting out the position of a mortise, or rectangular hole, in timber. Twin pins are adjusted to the exact chisel width required, and the stock is then set at the selected distance from the timber's edge.

The Process

Before you start

- **Check the dimensions** Make sure that the timber is planed to the correct width and thickness before using the gauge.
- **Choose the chisel** Select the chisel (mortise or firmer) closest to the finished mortise size and set the gauge accordingly.

1 Set the gauge

Holding the chisel against the gauge, adjust the sliding pin so that both pins just nip the outside of the blade. Lock this setting with the thumbscrew on the stock or adjuster on the stem. Recheck the setting against the chisel.

2 Lock the stock

Use a steel rule to measure the stock distance. This will give you the precise dimension of the mortise from the edge of the timber. Lock the stock using its thumbscrew and check the setting against the rule.

Rotate *the thumbscrew to lock the stock.*

FOCUS ON...

Scoring Action

The pins of a mortise gauge are incredibly sharp and score the surface of wood with minimal pressure. This means that the wood is marked clearly but not split and damaged. The gauge is rotated so it rests on the wood with little weight behind it. As the gauge is pushed along, the pin concentrates the pressure to score the surface lightly. This only works for marking along the grain, not across it.

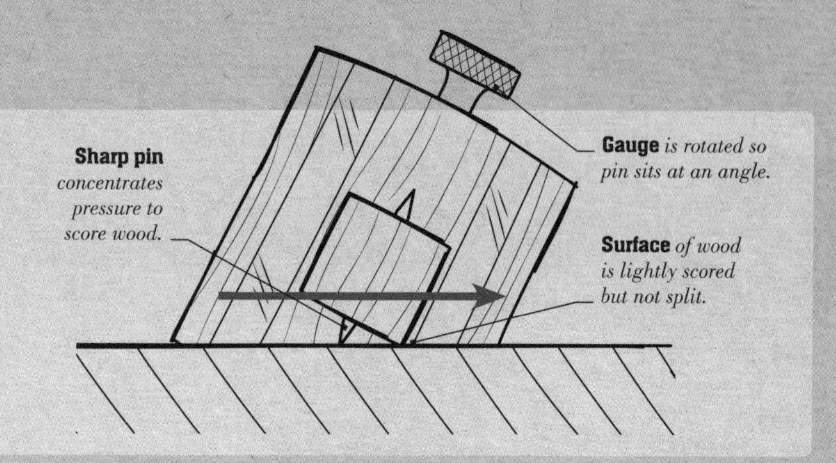

4 Shade the area

Once the parallel lines have been marked out, square off the ends with a square or steel rule. Then mark out the area to be cut away between the parallel lines. Cross-hatching with a carpenters pencil works well.

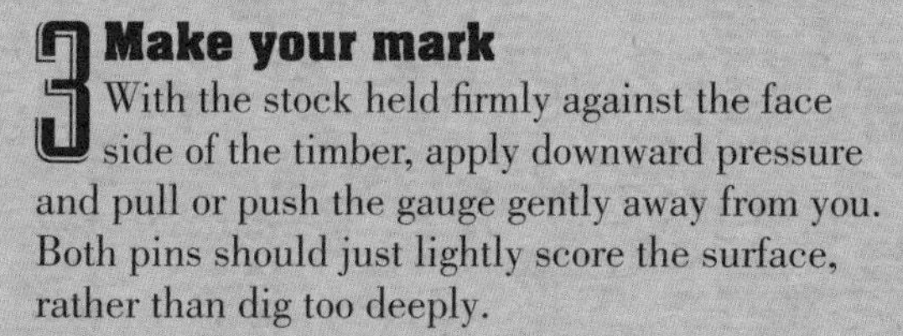

3 Make your mark

With the stock held firmly against the face side of the timber, apply downward pressure and pull or push the gauge gently away from you. Both pins should just lightly score the surface, rather than dig too deeply.

After you finish

- **Mark again** Mark out matching tenons without readjusting the mortise gauge pins.
- **Cut the mortise** Use a mortise or firmer chisel to chop out the mortise from both sides of timber with the appropriate mallet.

Choosing a Measure

Whether you are marking out the boundary for a new workshop or checking the tiniest clearance at a lathe, the correct measuring tool is crucial. You may be working in fractions of a millimetre or your project may span several metres, but without precise measurements you will struggle to achieve accurate work.

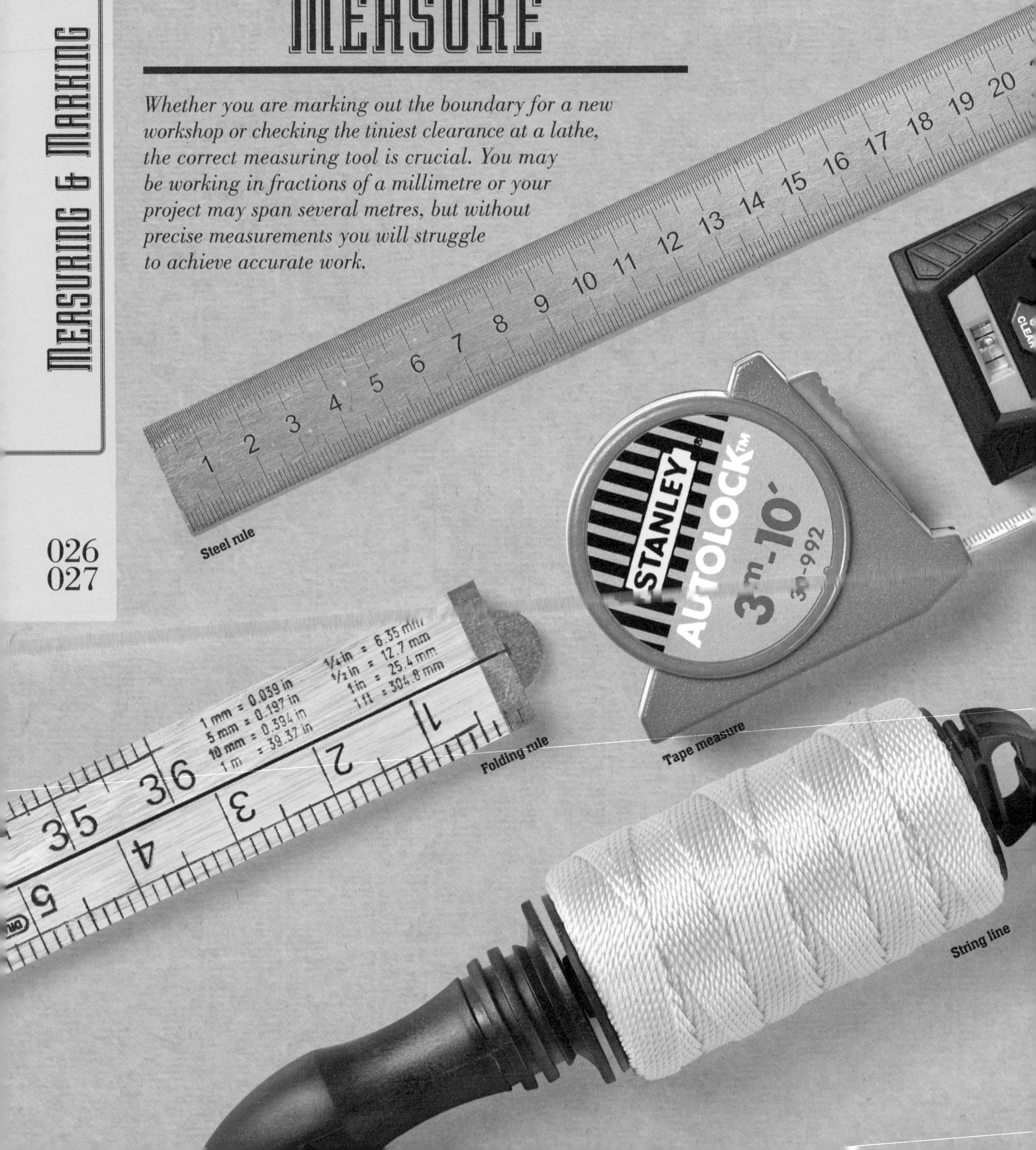

Steel rule

Folding rule

Tape measure

String line

Laser measure

Feeler gauge

> “DON’T BE TEMPTED TO MIX METRIC AND IMPERIAL MEASUREMENTS ON THE SAME PROJECT”

Steel Rule

What it is A sturdier version of a school ruler, usually made of stainless steel to prevent corrosion.

Use it for Small measurements and layout work. Often metric on one side, imperial on reverse.

How to use Hold down firmly, then use it as a guide for drawing lines with a pencil or marking knife.

Look for Lengths of 150mm and 300mm (6in and 12in), clear graduations and a brushed finish for easier reading.

Folding Rule

What it is Hinged boxwood or plastic ruler with 10 sections. Extends to 1m or 2m (3–6.5 ft), folds flat for storage.

Use it for Building projects where a measuring tape is too flexible. Useful for confined openings such as doorways.

How to use Unfold as many sections as necessary and measure from the square end of tool.

Look for Rigidity when opened. Both metric and imperial graduations are useful.

Tape Measure

What it is Flexible steel blade from 2m to 10m (6.5–33ft) overall length, contained in metal or plastic case.

Use it for General measuring over distance. Longer tapes have wider blades for better rigidity.

How to use Clip the end hook to edge of object, or hold against wall or framework for internal measurements.

Look for A thumb button to lock the blade open, a belt clip, and recoil action that is not too fierce.

Laser Measure

What it is Battery-powered electronic device, with digital display, that uses a laser to measure accurate distance.

Use it for Measuring rooms and buildings; used mainly for indoor measures or in poor light conditions.

How to use Place device against wall, switch on and read display. Some models calculate areas and volumes.

Look for Models with both metric and imperial displays, good battery life, and a protective bag or case.

String Line

What it is Tough, weatherproof cord up to 100mm (4in) long, usually wound around a plastic spool for storage.

Use it for Setting out reference lines over a distance for brickwork, walling, or fencing work.

How to use Tie one end around a nail or pin placed in ground. Undo line and secure tautly at far end with another nail.

Look for Bright-coloured lines, which are easier to work with. Cut off frayed ends when necessary.

Feeler Gauge

What it is Set of extremely thin, hardened steel blades of precise thickness, each marked with size.

Use it for Making adjustments to car, motorbike, and petrol lawnmower engines. Blades fold into case for storage.

How to use Insert the tapered end of blade into gap. Correct size is reached when blade contacts both surfaces.

Look for Check whether you need metric or imperial gauge sizes, although combination sets are available.

> "IT IS NOT WRONG TO DO AS CERTAIN POOR AND SIMPLE MEN ARE WONT TO SAY, WHO TELL US WE MUST MARK SEVEN TIMES AND CUT ONCE."
>
> BENVENUTO CELLINI

CHOOSING A

Caliper or Divider

It is tricky to measure external diameters of cylinders or internal diameters of bowls accurately using standard rules or tape measures. Mechanical calipers have adjustable legs and are used particularly by woodturners. Dial and digital calipers are tools used more in engineering tasks that give precise dimensions via their relevant displays.

Dial caliper

Digital caliper

External spring caliper

Internal spring caliper

“ DIVIDERS AND COMPASSES HAVE SHARP POINTS SO TAKE CARE WHEN HANDLING THEM ”

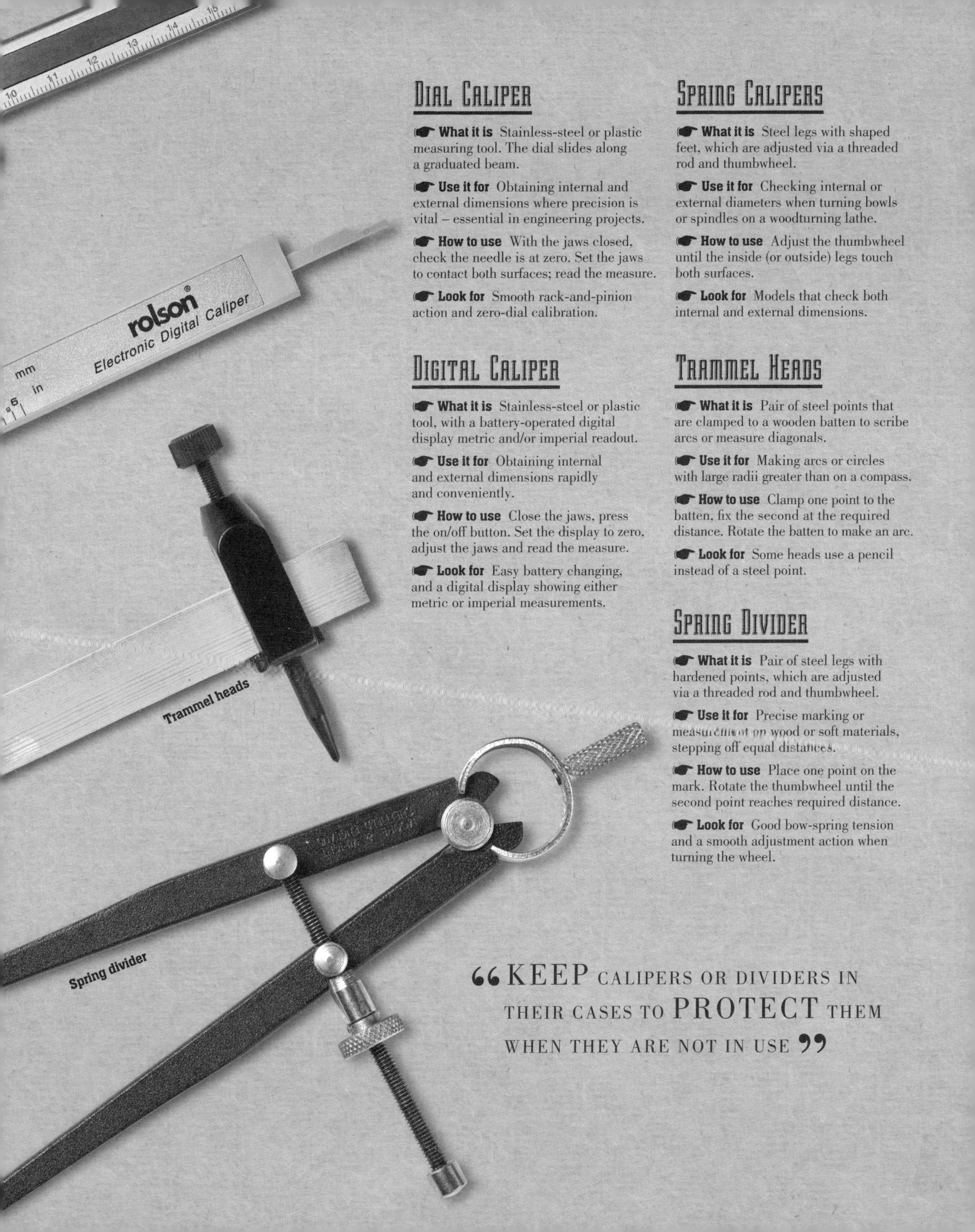

Dial Caliper

☛ **What it is** Stainless-steel or plastic measuring tool. The dial slides along a graduated beam.

☛ **Use it for** Obtaining internal and external dimensions where precision is vital – essential in engineering projects.

☛ **How to use** With the jaws closed, check the needle is at zero. Set the jaws to contact both surfaces; read the measure.

☛ **Look for** Smooth rack-and-pinion action and zero-dial calibration.

Digital Caliper

☛ **What it is** Stainless-steel or plastic tool, with a battery-operated digital display metric and/or imperial readout.

☛ **Use it for** Obtaining internal and external dimensions rapidly and conveniently.

☛ **How to use** Close the jaws, press the on/off button. Set the display to zero, adjust the jaws and read the measure.

☛ **Look for** Easy battery changing, and a digital display showing either metric or imperial measurements.

Spring Calipers

☛ **What it is** Steel legs with shaped feet, which are adjusted via a threaded rod and thumbwheel.

☛ **Use it for** Checking internal or external diameters when turning bowls or spindles on a woodturning lathe.

☛ **How to use** Adjust the thumbwheel until the inside (or outside) legs touch both surfaces.

☛ **Look for** Models that check both internal and external dimensions.

Trammel Heads

☛ **What it is** Pair of steel points that are clamped to a wooden batten to scribe arcs or measure diagonals.

☛ **Use it for** Making arcs or circles with large radii greater than on a compass.

☛ **How to use** Clamp one point to the batten, fix the second at the required distance. Rotate the batten to make an arc.

☛ **Look for** Some heads use a pencil instead of a steel point.

Spring Divider

☛ **What it is** Pair of steel legs with hardened points, which are adjusted via a threaded rod and thumbwheel.

☛ **Use it for** Precise marking or measurement on wood or soft materials, stepping off equal distances.

☛ **How to use** Place one point on the mark. Rotate the thumbwheel until the second point reaches required distance.

☛ **Look for** Good bow-spring tension and a smooth adjustment action when turning the wheel.

“KEEP CALIPERS OR DIVIDERS IN THEIR CASES TO PROTECT THEM WHEN THEY ARE NOT IN USE”

Structure of a Digital Caliper

A contemporary version of the traditional dial caliper, the digital caliper is much easier to read and faster to use. The display may be solar-powered or it may rely on a button battery. Better-quality tools are made from stainless steel, while cheaper versions may be of plastic or carbon fibre.

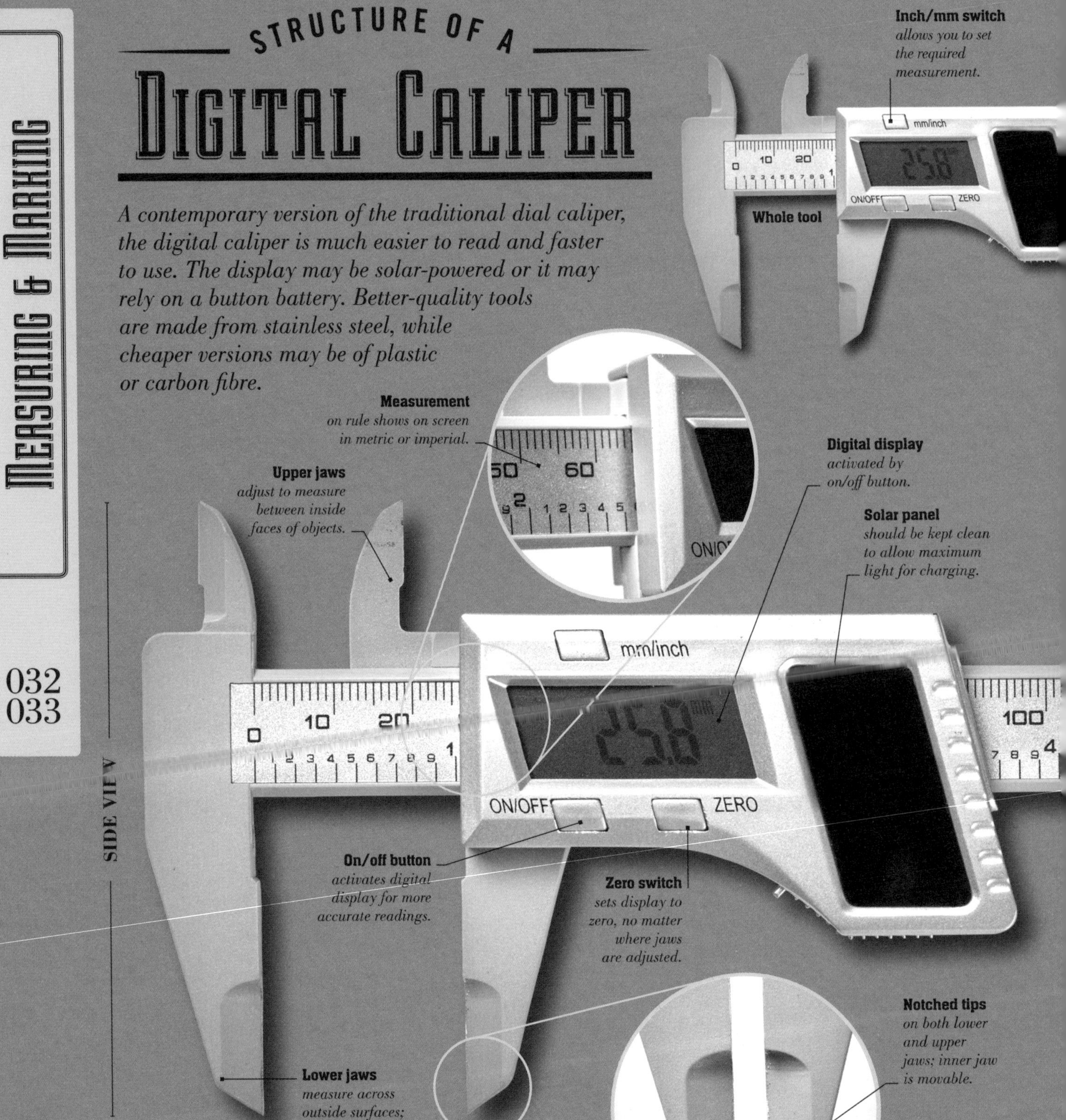

USING A DIGITAL CALIPER

One of the easiest measuring tools to use, the digital caliper is also one of the most precise. As well as internal and external measurements, a pin at the end of the beam can also be used for checking depth, which is revealed as you slide the jaws apart.

Depth gauge *uses sliding pin to measure depth.*

Main scale *marked out in either metric or imperial for accuracy and ease of use.*

Beam or blade *marked with graduations up to 150mm (6in).*

“ALWAYS BUY STAINLESS-STEEL TOOLS FOR MEASURING AND MARKING IF YOU CAN AFFORD THEM”

The Process

Before you start

Clean the caliper Wipe clean surfaces to ensure they are free of grease and grime before using the caliper.

Check the battery If no readout is visible, check the battery and replace it if necessary. If using a solar-powered model, make sure it has been properly charged.

1 Slide the jaws

Press the on/off button. Select the unit needed by pressing the metric/imperial button, then slide the jaws closed. Check that the display is set to zero by using the appropriate button.

2 Take the readings

For internal measurements, slide the upper jaws open on the reading head so that they make contact with the two inside faces. Read the display. For external measurements, slide the lower jaws around the outside surfaces of the object to be measured, then read the display.

After you finish

Remove the battery If you're unlikely to use the tool for several months, take out the battery. This prevents the possibility of corrosion destroying the battery connection.

Store it safely Replace the digital caliper in its storage case or drawer to keep it clean and dry.

FOCUS ON...

CAPACITANCE

Electronic sensors embedded along the beam detect changes in electrical charge, known as capacitance, as the distance changes between the jaws. The back of the display head contains a network of lines etched onto a printed circuit board that interact with a similar pattern of copper tracks on the beam, forming a variable capacitor. As the head travels along the beam, it sends a signal to a chip within the caliper, generating the readings shown on the LCD display.

Choosing a Square

Regardless of whether you're working with metal, wood, or sheet materials of various types, at some point you will need to use a square. You need it not only for marking lines that are perpendicular to an edge, but also for checking angles, or that a workpiece is square before additional work can go ahead on it.

Engineers try square

Carpenters try square

Mitre square

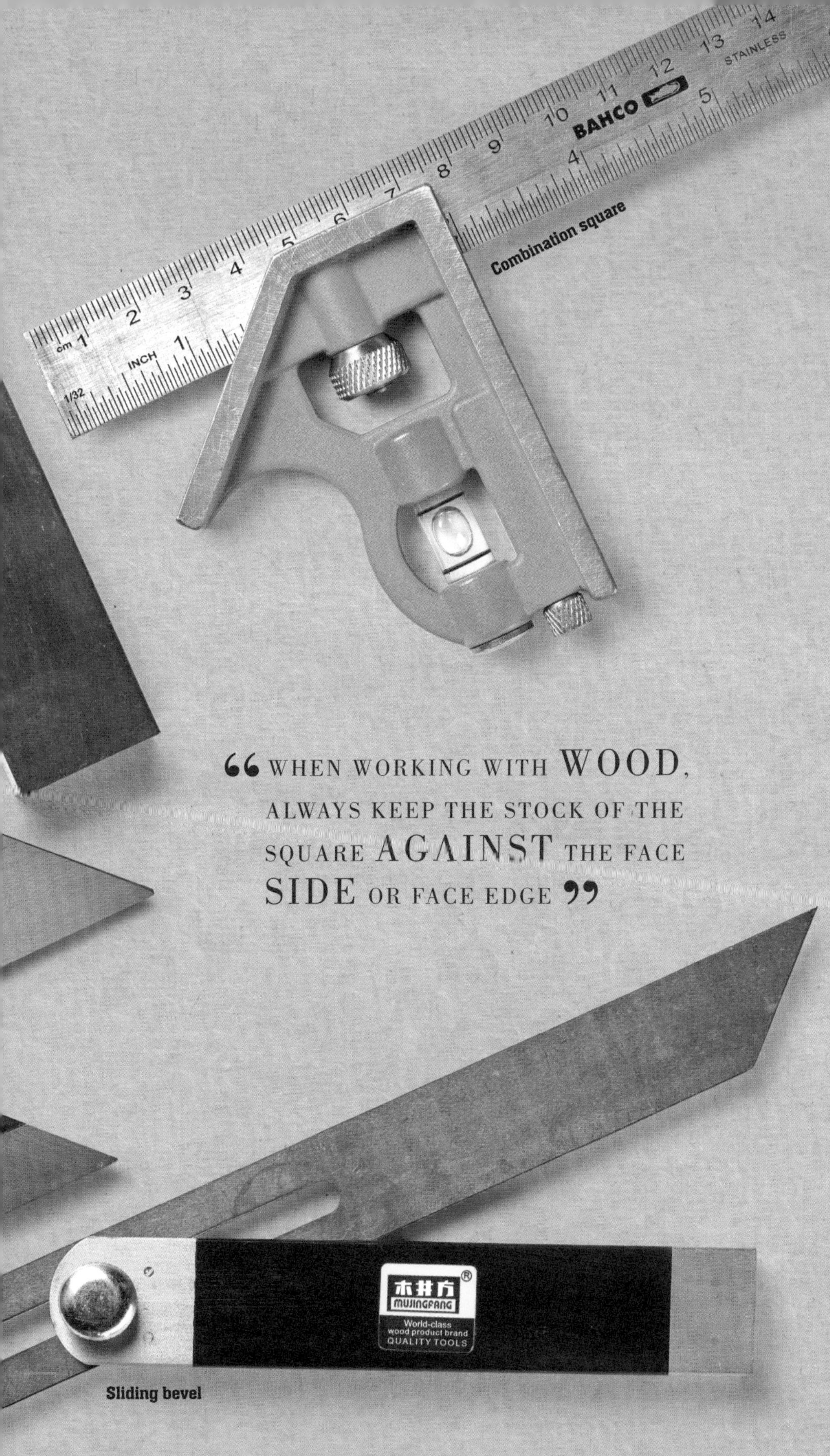

Combination square

Sliding bevel

> “WHEN WORKING WITH **WOOD**, ALWAYS KEEP THE STOCK OF THE SQUARE **AGAINST** THE FACE **SIDE** OR FACE EDGE”

Try Squares

What it is A hardened steel blade riveted to a wood or plastic stock (carpenter's) or an all-steel stock (engineer's) at precisely 90 degrees.

Use it for General wood- and metalworking; marking items before sawing or further work. Testing for squareness.

How to use Hold stock firmly against the workpiece edge. Mark a line along the outside of the blade.

Look for A brass facing to hardwood stocks for greater endurance.

Mitre Square

What it is A hardened steel blade riveted to a hardwood or metal stock at precisely 45 degrees.

Use it for Checking and marking out 45-degree angles on materials.

How to use Hold the stock firmly against the workpiece. Mark a line along the outside of the blade.

Look for On new tools, sharp edges may need filing slightly before use.

Combination Square

What it is An adjustable stock that slides along a rule, locking with a thumbscrew.

Use it for Common for marking 45-degree angles; as a rule or level, for checking depth; also as a try square.

How to use Loosen the thumbscrew, slide the stock along the rule, retighten.

Look for A heavy cast-iron stock for reliability and accuracy. Most models include a spirit level and scriber in the stock.

Sliding Bevel

What it is A hardwood, plastic, or aluminium stock with a steel blade that can be locked at any angle.

Use it for Checking existing angles, adjusting blades on machines, and marking out material.

How to use Hold the stock firmly against the edge of the workpiece, swing the blade to the required angle, and tighten.

Look for An easy but firm locking action via a lever or thumbscrew.

TOP VIEW

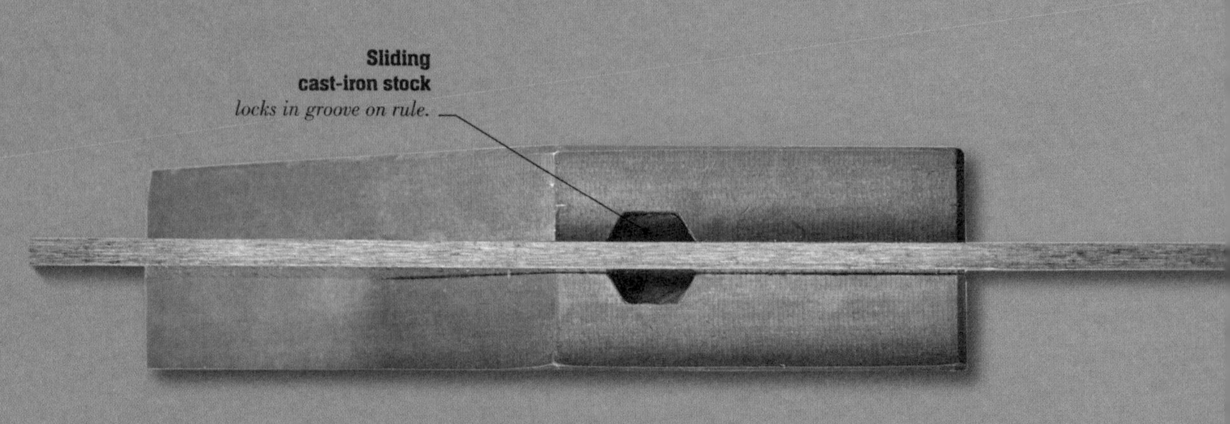

SIDE VIEW

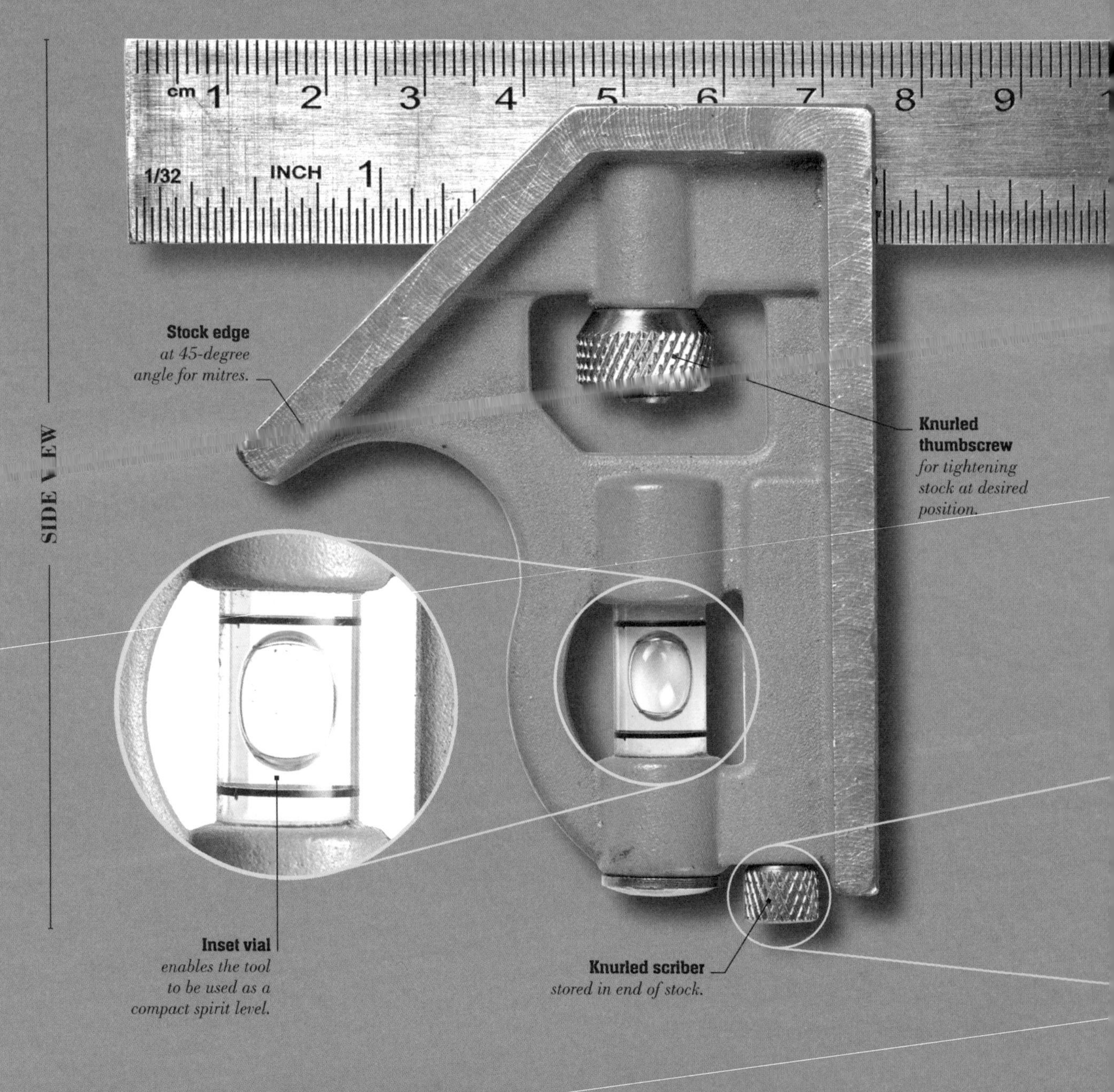

STRUCTURE OF A

COMBINATION SQUARE

A combination square is a multifunctional tool that is used to mark out both timber and metal at the start of a project, as well as to check mitres, corners, and right angles. Unlike a conventional square, a combination stock incorporates a spirit level and scriber, and it can also be used for measuring depth.

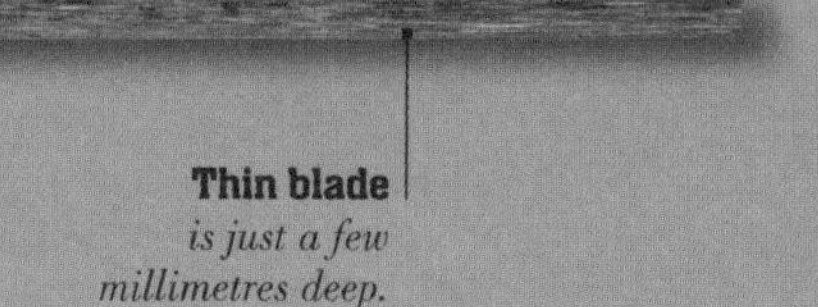

Thin blade *is just a few millimetres deep.*

Stainless-steel rule *with metric or imperial graduations. Length from 150–400mm (6–16in).*

“MORE **VERSATILE** THAN A TRY OR ENGINEER’S SQUARE, THIS **TOOL** HAS NUMEROUS USES”

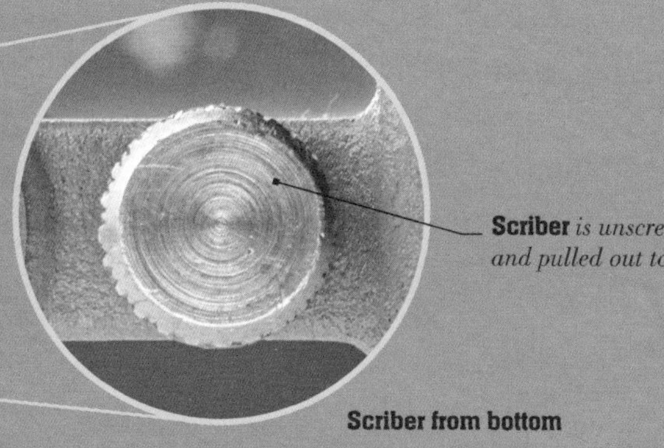

Scriber *is unscrewed and pulled out to use.*

Scriber from bottom

FOCUS ON...

COMBI SQUARE HEADS

Combination squares can include a variety of interchangable heads that can be fixed on to the blade. The standard head is used to check angles of 90- or 45-degrees, and is probably all you need for most woodworking and DIY jobs. More complex combination sets, often used by metal-workers and engineers, may include additional heads, such as a protractor.

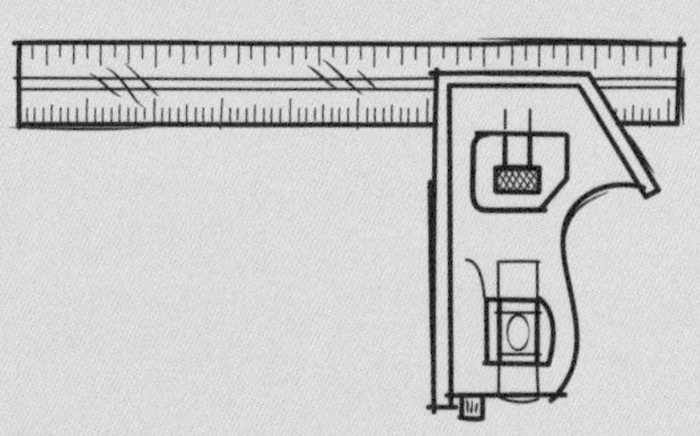

Combination square A basic combination square has a standard, or square, head fitted on to the blade.

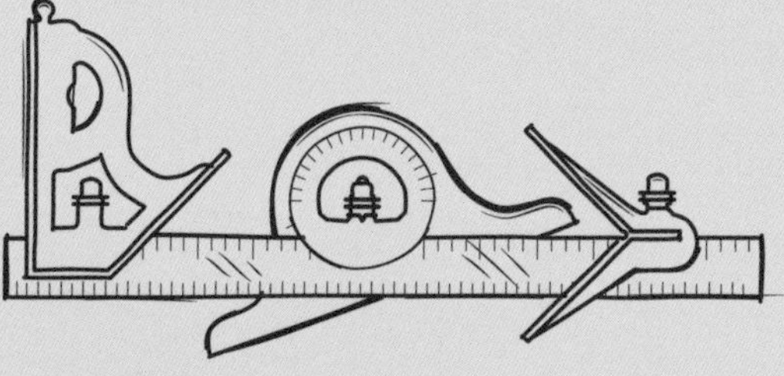

Combination sets A combination set includes more specialized heads, such as a protractor, centre head, and 45-degree holder, as shown here.

USING A
Combination Square

Having a small (150mm/6in) combination square can come in handy, particularly when marking out woodworking joints or checking internal corners in wood or metal where space may be too tight to allow the use of larger squares. The small version takes up very little room in the toolbox and is a useful addition to any do-it-yourselfer's toolkit.

The Process

Before you start

Check the size Ensure that the square's ruler width and length are adequate for the workpiece.

Test the square If using it for the first time, and especially if you've bought a used tool, check your square is true. Lay the stock against the edge of a straight piece of ply or MDF and draw a perpendicular line. Flip the square and draw a second line over the first. If they line up, the square is true.

Check the workpiece Reference surfaces should be completely straight. In woodwork this is usually the face side or face edge.

Cutting guide *drawn by running pencil along blade.*

2 Mark a right angle for cutting

To mark timber to length before sawing, measure from one end with a tape measure or steel rule. Mark with a sharp pencil and square a line across the face side and face edge. You may find it helpful to add a third line down the back edge to guide the saw.

1 Draw a sliding guide line

Slide the stock until the required length is visible on ruler. Lock the thumbscrew, then hold stock firmly against workpiece edge. Hold pencil at the end of blade and slide the square along workpiece to create a parallel line.

"IF YOU CANNOT MEASURE IT, YOU CANNOT CONTROL IT." LORD KELVIN

FOCUS ON...

Right Angles

There are three corners on a square that should be perfect right angles. But don't assume that blade and stock on your square are set precisely at 90 degrees. To check, hold the stock firmly against the dead straight edge of a board. Draw a pencil line along the outside of the blade. Flip the tool over and then repeat the process. Both lines will coincide if the square is accurate. Check this when buying new tools.

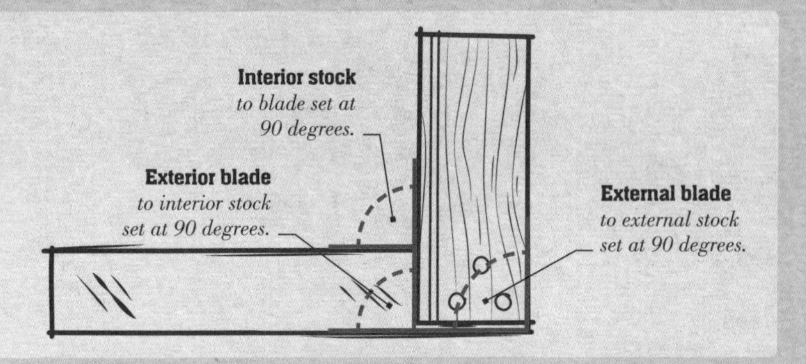

Interior stock *to blade set at 90 degrees.*

Exterior blade *to interior stock set at 90 degrees.*

External blade *to external stock set at 90 degrees.*

With shoulder *of square on edge of timber, rule lies at 45 degrees.*

4 Check an interior angle

To make sure internal corners are square, slide the stock to the end of the blade and tighten the thumbscrew. Hold the square against both interior surfaces and check the angle for accuracy. This is useful when gluing together a box or drawer.

3 Mark 45 degrees

To mark a 45-degree cutting angle, hold the shoulder of the square against the edge of the workpiece with one hand. Draw a line across using the rule to mark the angle of cut.

Built-in scriber *at end of stock can be used in place of pencil.*

After you finish

Keep it clean The rules on combination squares can rust if exposed to moisture. Wipe the blade with a cloth and a drop of oil or paraffin to protect it after use.

Store it safely Like all measuring tools, combination squares should be treated with respect, as they are relatively fragile tools. If your square came in a purpose-made case, replace it there for storage. If not, store it so that it lies as flat as possible.

CHOOSING A
LEVEL

A spirit level is an essential tool for general building work, renovation, or landscaping projects. Besides checking that horizontal and vertical surfaces are true, a long level can also be used as a straight edge when cutting plasterboard or marking out sheet materials. A short level may be a more useful choice in a confined space.

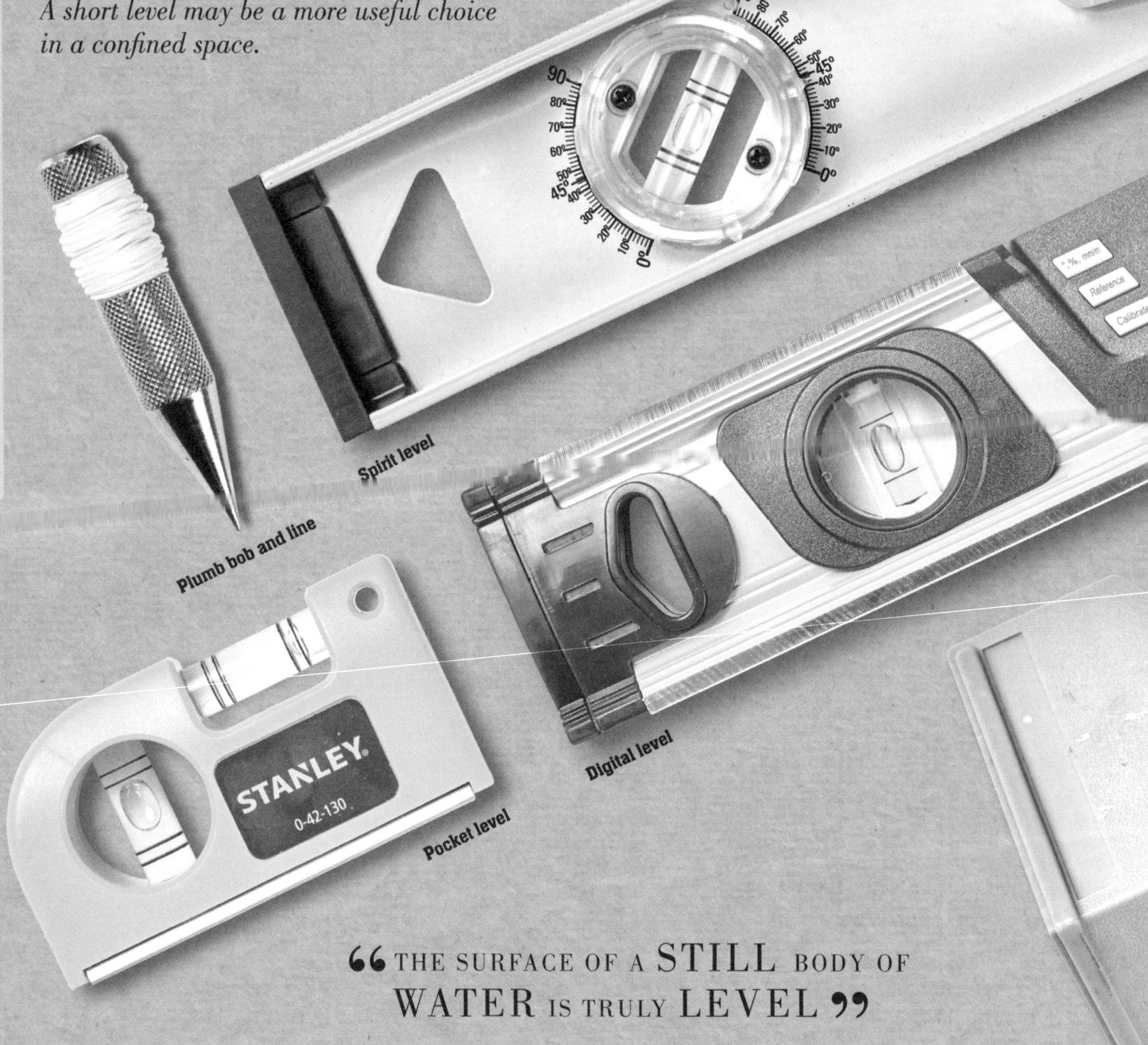

Spirit level

Plumb bob and line

Digital level

Pocket level

“THE SURFACE OF A STILL BODY OF WATER IS TRULY LEVEL”

Post level

PLUMB BOB

☛ **What it is** A tapered brass or steel weight (bob) suspended on a fine nylon or cotton string tied to its opposite end.

☛ **Use it for** Checking walls and studs are plumb. Transferring marks from floor to ceiling. Hanging wallpaper.

☛ **How to use** Drive a nail into a surface perpendicular to the item being checked, such as a ceiling joist beside a wall stud. Tie the string to the nail. Once the bob is still, measure the distance from the item's surface to the top of the string and to the bob tip. If they match, the surface is plumb.

☛ **Look for** As the tool relies on gravity, the line must hang free for accuracy.

SPIRIT LEVEL

☛ **What it is** A long, rectangular aluminium box with liquid-filled vials embedded in the ends and in the centre.

☛ **Use it for** Checking horizontal surfaces are level (parallel) or vertical surfaces are perpendicular (plumb).

☛ **How to use** Place the tool on surface or against edge. Check the appropriate bubble is centred between the lines: the centre vial finds the true horizontal, the end vials find the true vertical.

☛ **Look for** Shock-absorbing rubber end caps provide protection. Vials with a magnifying lens are easier to read.

POCKET LEVEL

☛ **What it is** A compact tool for checking the levels of smaller items or working in confined spaces.

☛ **Use it for** Levelling pictures and paintings, shelves, light switches, and wall tiles.

☛ **How to use** Hold tool against vertical or horizontal surface. The surface is level or plumb when the bubble is centred.

☛ **Look for** A magnetic strip enables easy use on metal surfaces. A belt clip is also handy.

DIGITAL LEVEL

☛ **What it is** A tool similar to a spirit level, but with an LCD screen displaying angles in both degrees and percentages.

☛ **Use it for** Checking precise angles (degrees) on roof timbers or inclinations of sloping surfaces (mm per metre).

☛ **How to use** Place the tool on the surface to be checked and turn on the level. Use hold button to retain display.

☛ **Look for** Backlighting makes the LCD easier to read. Audible bleeps indicate when surfaces are perfectly level or plumb.

POST LEVEL

☛ **What it is** A compact and angled level with three vials that allow for use around corners.

☛ **Use it for** Checking that fence posts or pipework are plumb in all their respective vertical planes.

☛ **How to use** Hold the level against two surfaces of the object, such as a post corner, and check that all bubbles are centred in their vials.

☛ **Look for** Models with built-in magnets are useful for checking metal surfaces.

“A VERTICAL SURFACE IS THE LINE DEFINED BY A PLUMB LINE”

Structure of a Spirit Level

The vial is a crucial component of a spirit level, and any tool will have at least two: one for horizontal and one for vertical work. High-tech digital models will also measure angles and inclination, which are displayed on a clear LCD screen. Although spirit level bodies are usually extruded aluminium, wooden levels are still available. Lengths range from around 250 –2,440mm (10–96in).

Whole tool

End cap *is made of soft material to absorb shock if level is dropped. Makes tool more durable.*

Edges of tool *have been machined flat for absolute accuracy. They also form a useful hand grip.*

Hanging hole *for storage.*

SIDE VIEW

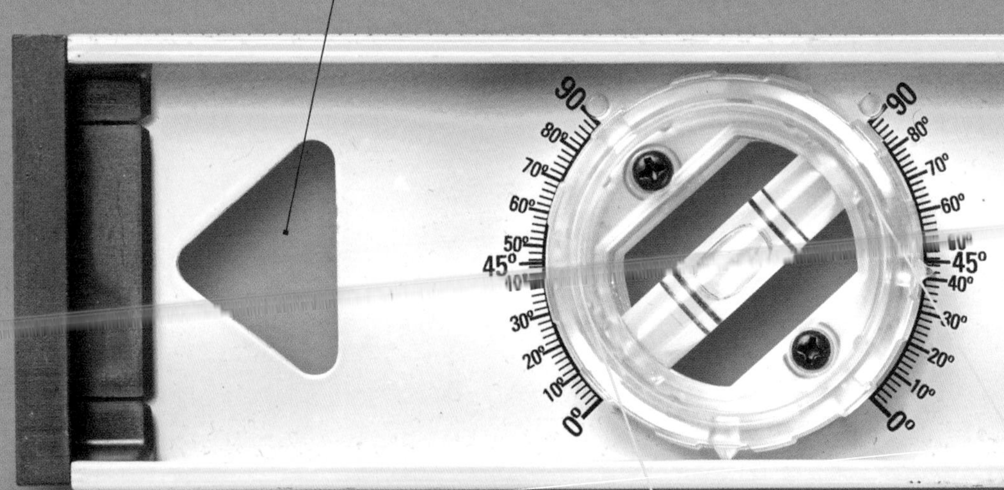

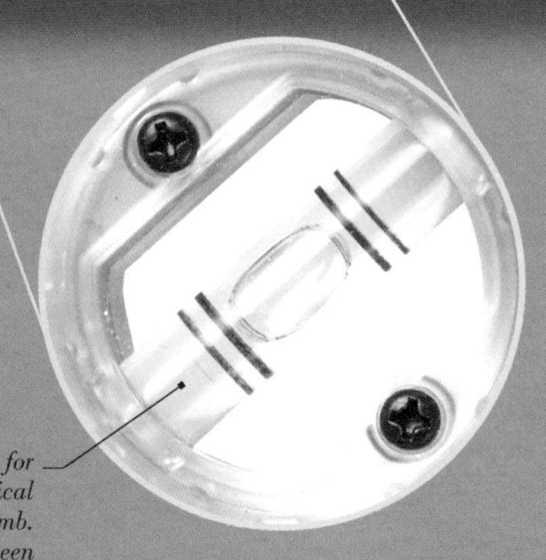

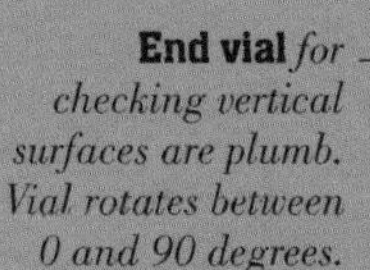

End vial *for checking vertical surfaces are plumb. Vial rotates between 0 and 90 degrees.*

FOCUS ON...

The Bubble

The vial of a spirit level is not completely filled with fluid in order to create an air bubble. Because it is less dense than the coloured fluid, the bubble rises to the top of the vial if unobstructed. When the tool is level, the high point is in the vial's centre, so the bubble settles there. If the tool is not level, the bubble heads for higher ground, or the far end – right or left – of the vial.

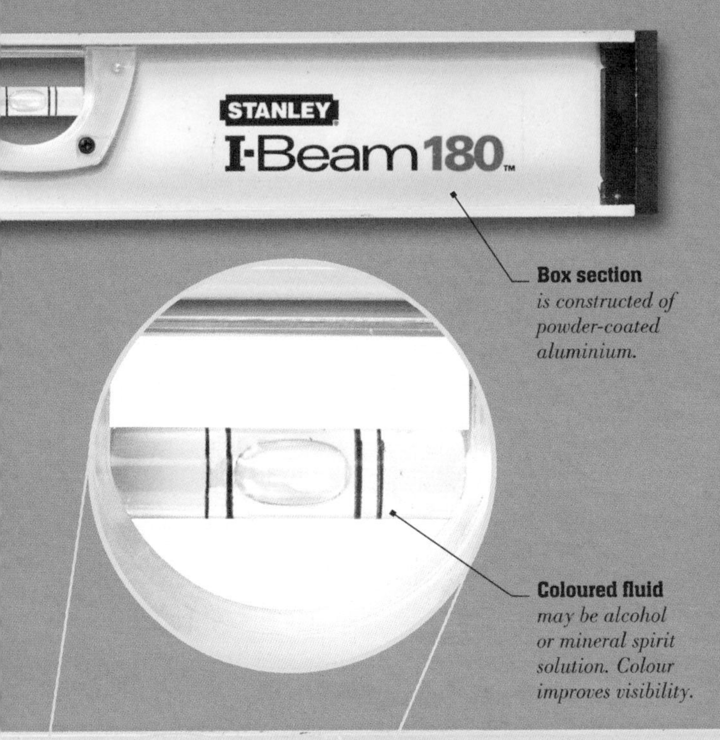

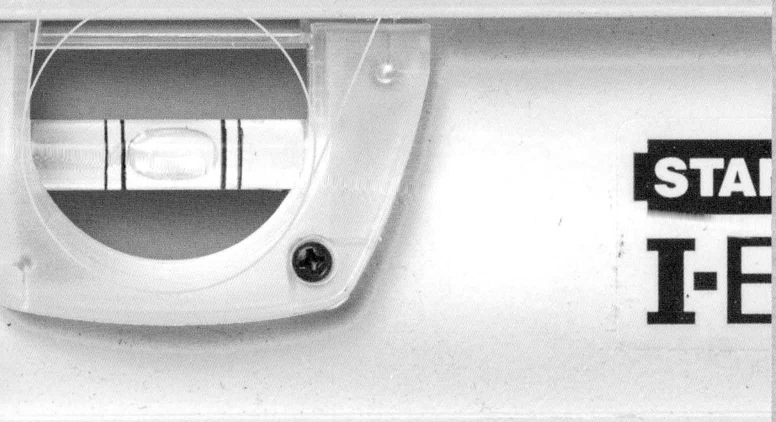

“CALIBRATE YOUR LEVEL BY PLACING IT ON A FLAT SURFACE, NOTE THE READING, FLIP THE LEVEL AND NOTE THE BUBBLE’S POSITION. IF BOTH READINGS MATCH THE LEVEL IS BALANCED”

USING A

SPIRIT LEVEL

A longer level means greater accuracy. If you need to check that a surface is horizontal across a greater span than the tool length, substitute a piece of straight, parallel timber between the two points with the level on top.

The Process

Before you start

Inspect the level Make sure the level is completely clean. Remove any dirt or debris from tool edges.

Prepare your ground If using the level in outdoor work (such as concrete or building decking), hammer pointed pegs into uneven ground. Level tops of pegs first to obtain reference points and measure down from these.

1 To check a horizontal surface

Place the spirit level on the horizontal surface and allow bubble to come to rest. If the bubble does not appear exactly between divisions on vial, adjust one end of object until the bubble is centralized.

2 To mark a wall

If using the level when marking a wall, place a pencil mark at one end. Flip the level around and mark at the opposite end, continuing this method until sufficient distance is covered.

3 To check a surface is plumb

To check the plumb of a vertical surface, hold the level against the edge. If the bubble is not centred in the vial, then the vertical surface is not perpendicular or plumb and needs adjusting.

After you finish

Clean it up Always wipe down or wash off a level if it is muddy or has been used around concrete. An aluminium body will not rust, but any debris attached to its surface can throw its measurement out of true.

Put it away Store your level in a dry, secure place. Some levels have hoels in the end to hang them up.

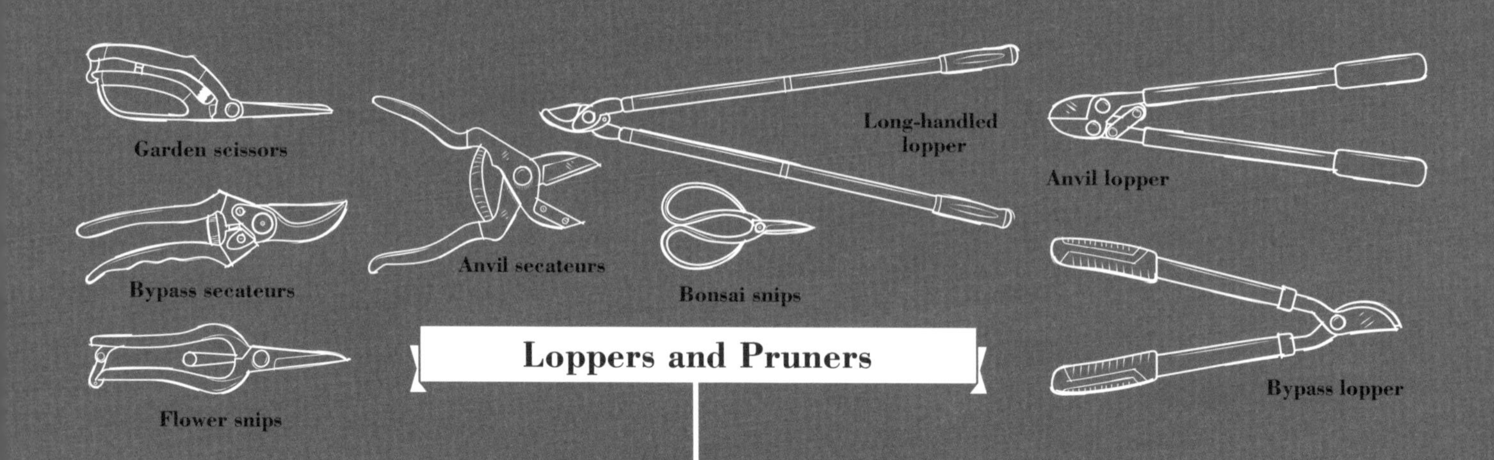

THE TOOLS *for* CUTTING & CHOPPING

Most tasks in the workshop or the garden involve some cutting or chopping. Whether you need to split logs, prune shrubs, clear weeds, or saw through a pipe, there's a specific tool to help you.

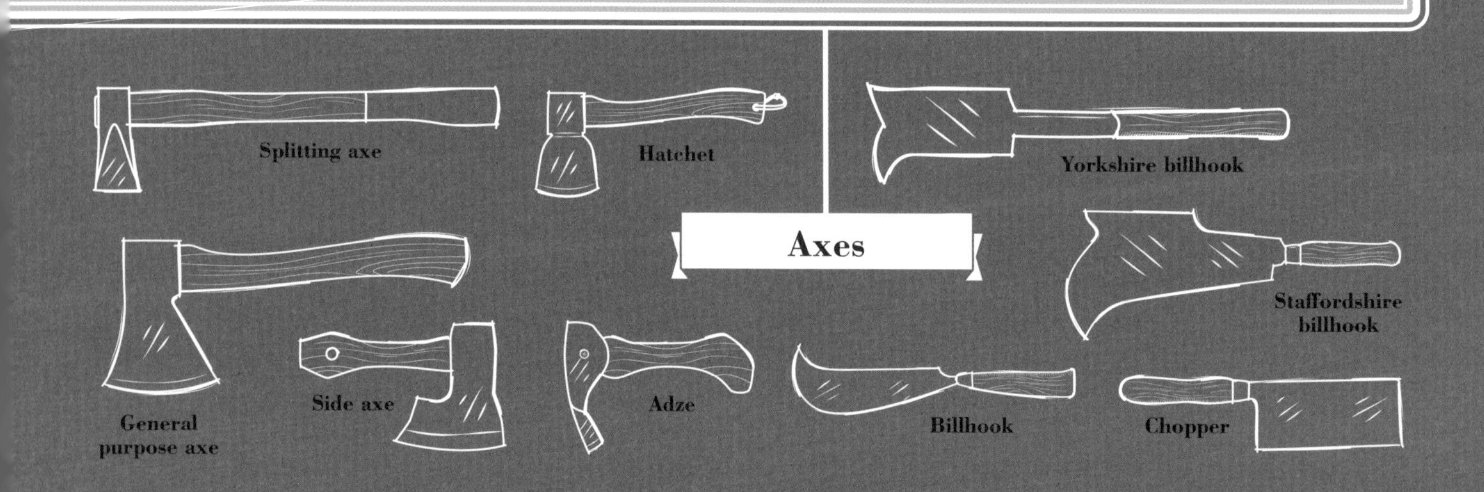

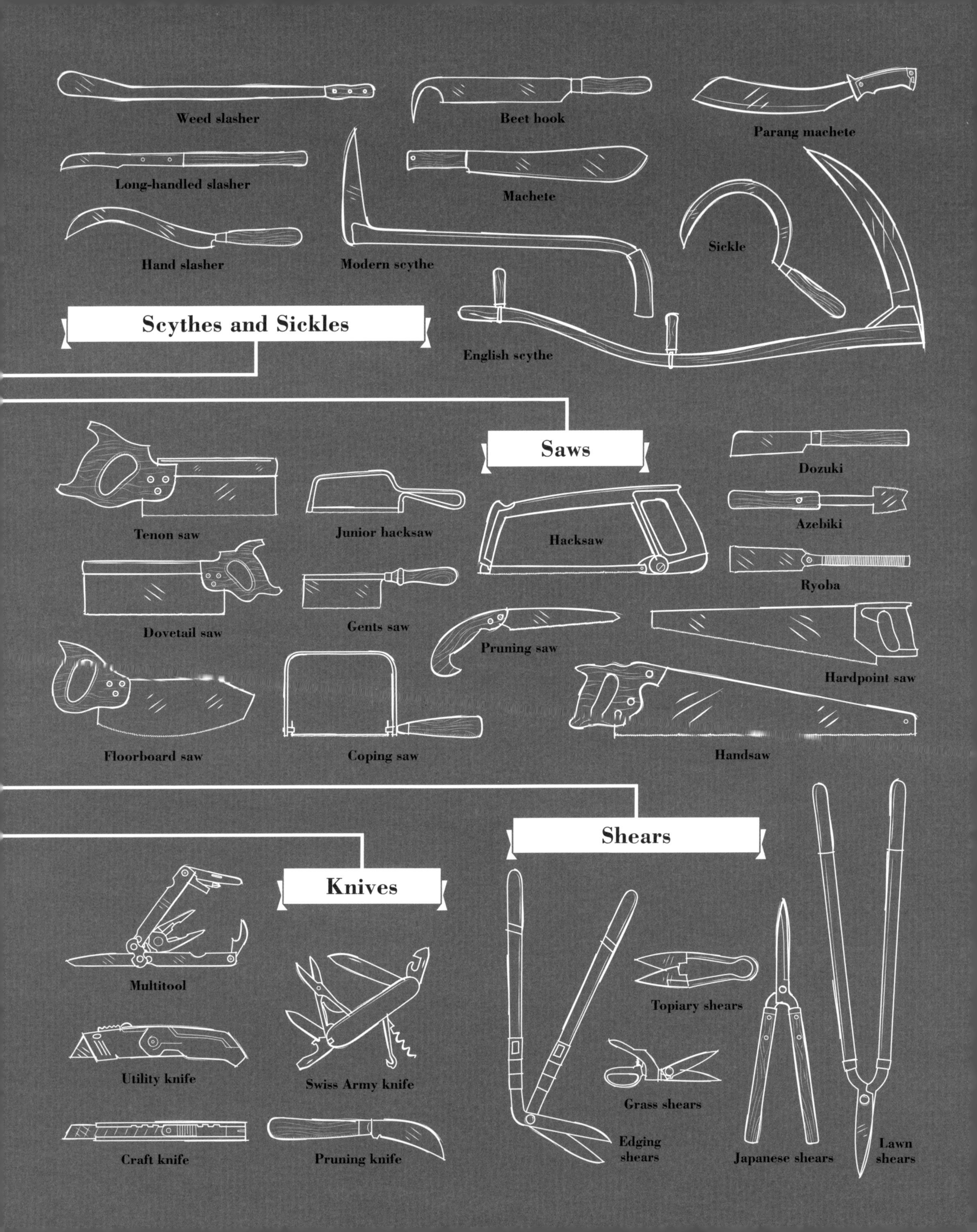
Weed slasher
Beet hook
Parang machete
Long-handled slasher
Machete
Hand slasher
Modern scythe
Sickle
Scythes and Sickles
English scythe
Saws
Dozuki
Tenon saw
Junior hacksaw
Hacksaw
Azebiki
Ryoba
Dovetail saw
Gents saw
Pruning saw
Hardpoint saw
Floorboard saw
Coping saw
Handsaw
Shears
Knives
Multitool
Topiary shears
Utility knife
Swiss Army knife
Grass shears
Craft knife
Pruning knife
Edging shears
Japanese shears
Lawn shears

HISTORY OF

CUTTING & CHOPPING

3.3–1.7 MYA FIRST TOOLS

The oldest-known stone tools come from Kenya. Stone cutting tools, made by striking ("knapping") rocks against each other, have been found at numerous archaeological sites. Naturally shaped rocks were used for cutting throughout this period, before stone was worked into scrapers and hand-axes with serrated edges.

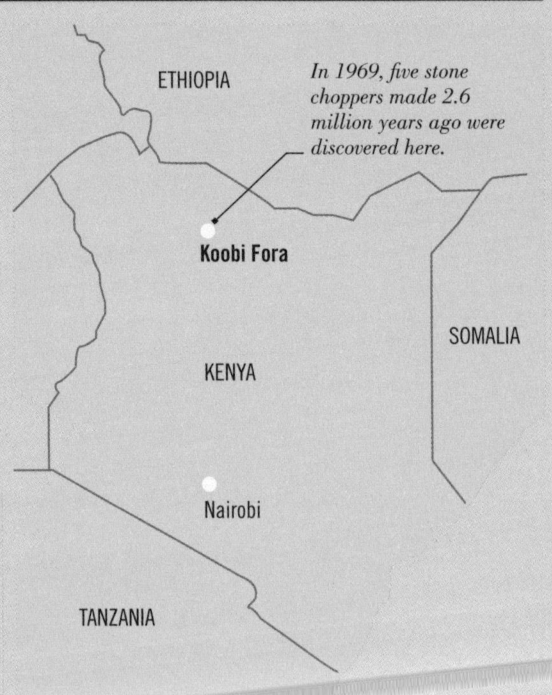

1.5 MYA EARLY AXE

Used for chopping, scraping, or as hunting knives, axe-like tools appeared in the Acheulean period. They were sharp-edged, tapered at one end, and rounded at the other.

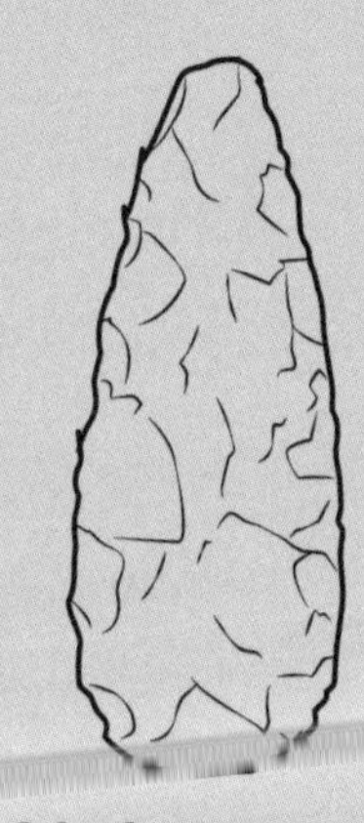

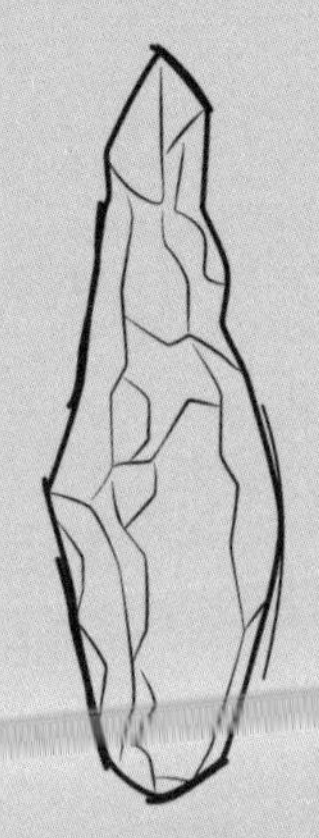

Acheulean axe

35,000 YA HANDLED KNIVES

Cro-Magnon man used stone blades, but also transformed other materials, such as bone, horn, and ivory from mammoths into tools. A graver was a narrow flint blade used to scrape out slivers of bone to make pins or needles. Handles, or hafts, were first attached to blades around this time, creating the first knives.

18,000–8000 BCE EARLY SICKLES

Sickles were first used in the Mesolithic Era. These tools were probably developed in Mesopotamia, and played an important role in the farming revolution by improving the efficiency of gathering crops.

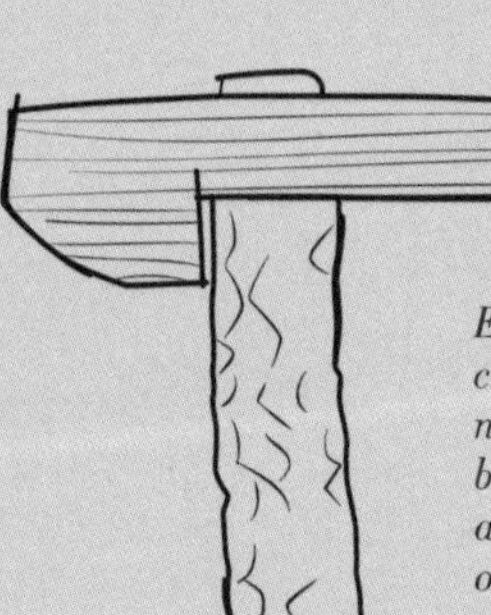

Neolithic sickle

"OFT DID THE HARVEST TO THEIR SICKLE YIELD…"

THOMAS GRAY

6500 BCE USING METALS

Before the advent of smelting, copper and meteoric iron were hammered into sheets to create tools with sharper and harder edges, including knives, choppers, and axe heads. Some were shaped with handles; others were set into handles of wood or bone.

1,083°C

The melting point of copper.

When alloyed with tin, this drops to

950°C

"WE SHAPE OUR TOOLS, AND THEN OUR TOOLS SHAPE US."

JOHN M CULKIN
THE SATURDAY REVIEW,
18 MARCH, 1967

c.3000–1900 BCE THE FIRST TRUE SAWS

In the Bronze Age, metal was smelted and cast, processes that improved many tools and weapons. Bladed saws were created from smelted and cast copper, with teeth that cut through wood, rather than hacking at its surface. Saws began to be used for woodworking, heralding the rise of many modern saws.

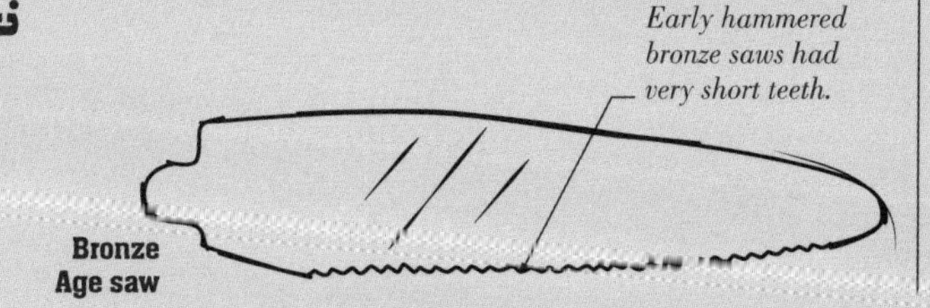

Bronze Age saw

2700 BCE AXES AND ADZES

Metal axes and adzes were used in ancient Egypt and in Mesopotamia. The Egyptians lashed metal blades to wooden handles, but the Mesopotamian tools created shaft holes to fix blade and handle together. Shaft-hole axes and adzes were also used in Crete around 700 years later.

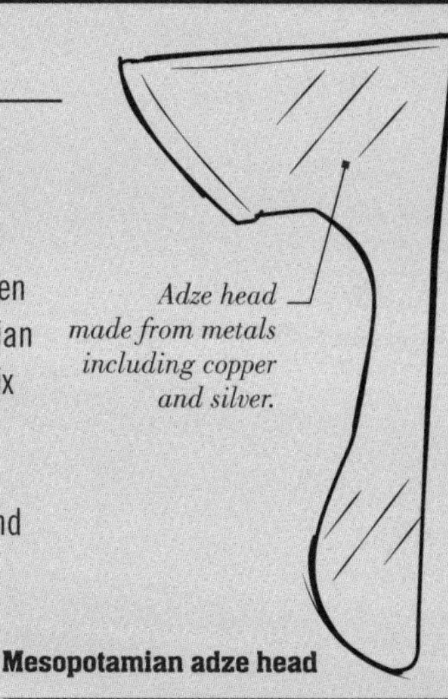

Mesopotamian adze head

IRON ALLOYS

More durable tools were formed in the Iron Age. Tools such as axes were made of wrought and cast iron.

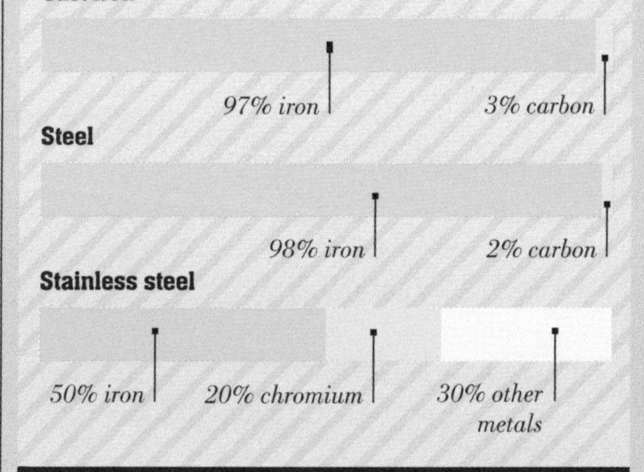

35 BCE–500 CE SAW DEVELOPMENT

The Romans improved saws by adding handles and frames. In the 1st century CE, historian Pliny the Elder noted that the setting of the teeth to create a kerf wider than the blade's thickness helped to minimize sawdust.

A SHARP KNIFE IS NOTHING WITHOUT A SHARP EYE.

c.500–1500 CROSSCUT SAWS

In the Middle Ages, long two-handled saws were developed for crosscutting green wood. Two men operated the saw, one pulling as the other pushed. The teeth of these saws were set in the same way as modern crosscut saws.

1819 FIRST SECATEURS

French aristocrat Antoine de Molleville invented the first hand pruners or *secateurs* (from the French word for "cutter").

Choosing a Saw

Some saws cut across woodgrain, others rip along it; look for teeth per inch (tpi) to know the cut type. A high tpi (10–12) means finer cuts but slower sawing than with a 4–5 tpi. Blade length and handle shape are also important. Saws for cutting metal have finer teeth than woodcutting saws.

Handsaw

Tenon saw

Hardpoint saw

Floorboard Saw

What it is A specialist tool with a convex row of teeth, a plastic or hardwood handle, and traditional or hardpoint teeth.

Use it for Cutting across floorboards without the need to lever them up first for conventional sawing.

How to use Start the cut with curved teeth across the board's centre. Flip the saw and continue with the straight teeth.

Look for Teeth that can be sharpened, although rarely available on new models.

Hardpoint Saw

What it is A plastic-handled saw with heat-treated teeth that stay sharper longer than traditional saws.

Use it for General-purpose sawing of timber and boards. Can't be resharpened.

How to use Place teeth at rear edge of wood, pull saw back to make groove. Hold saw at angle; cut on push/pull strokes.

Look for A soft-grip handle, a 550mm (22in) blade, and 7–8 tpi. Handle can be used to mark wood at 45 or 90 degrees.

Handsaw

What it is Hardwood-handled saw with teeth able to be sharpened. Blade length is around 500–660mm (20–26in).

Use it for Use a crosscut saw for cuts across grain; a rip saw for coarse cuts with the grain; a panel saw for sheet material.

How to use See hardpoint saw (above). Cut with push/pull strokes.

Look for A handle attached with screws, so it can be tightened as needed.

Tenon Saw

What it is Saw with reinforced brass or steel back on the blade and a hardwood or plastic handle. Blade length: 250–455mm (10–18in); 12–16 tpi.

Use it for Sawing joints in timber, and finer crosscut work than a handsaw.

How to use Start cut at the timber's back edge. Pull saw back to make groove, lower the blade, and cut horizontally.

Look for A heavy brass back to blade makes joint-cutting easier. Look down the tooth-line to check the blade is straight.

Gents Saw

What it is Smaller version of the dovetail saw, with even finer teeth. Blade length: 100–200mm (4–8in); up to 30 tpi.

Use it for Very fine cuts and precision work: making musical instruments, model-making, fine furniture.

How to use Align teeth at rear of wood. Pull backwards and gradually lower saw as you cut with push stroke.

Look for A blade length of 150mm (6in) is a useful size.

Dovetail Saw

What it is Like a small tenon saw; fine teeth and hardwood handle. Blade length: 200–250mm (8–10in); 16–22 tpi.

Use it for Small joints, especially dovetails; model-making, cabinetmaking.

How to use Start the cut at the wood's rear edge. Pull back to make a groove, then lower the blade and cut horizontally.

Look for A brass back to increase the weight for greater control.

CONTINUED

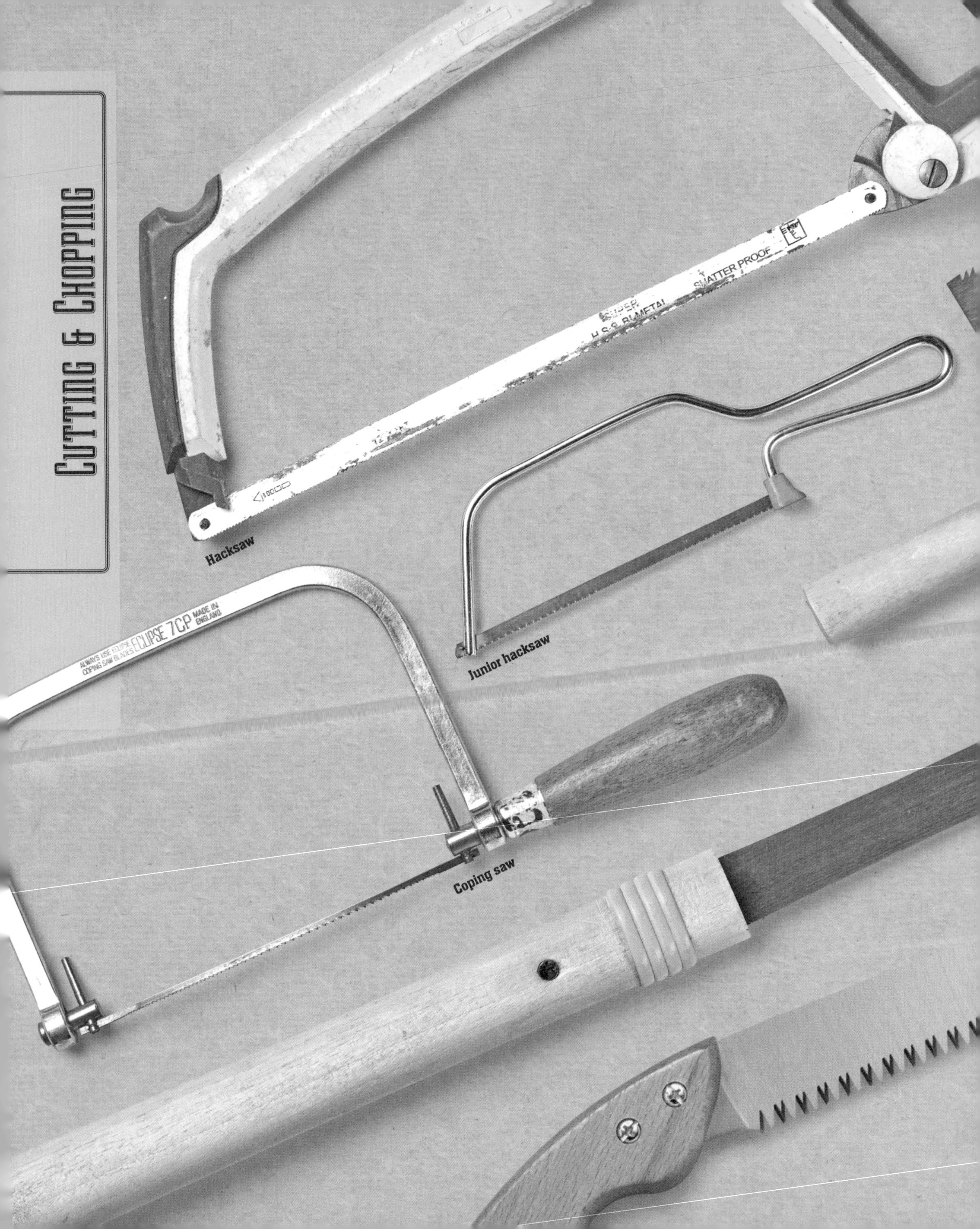

Hacksaw

Junior hacksaw

Coping saw

"THE TEETH OF JAPANESE SAWS, OR NOKOGIRI, POINT BACKWARDS AND CUT THE WOOD ON THE PULL STROKE"

Hacksaw

What it is A metal-framed, fine-toothed saw with the blade under tension. Replaceable blade is 300mm (12in).

Use it for Cutting metals, plastic pipes, ceramic tiles. Some saws cut flush.

How to use Place teeth on surface, pull saw back to make a groove. Cut with alternate push/pull strokes.

Look for Quick-release blade tensioning makes replacing blades easy.

Coping Saw

What it is Deep, metal-framed saw with a 150mm (6in) replaceable blade held under tension. Blade can be rotated.

Use it for Cutting curves or profiles in timber, sheet materials, ceramic tiles.

How to use Drill a clearance hole and insert blade end. Refit blade to frame, adjust tension and angle to make the cut.

Look for Make sure the blade is easy to tighten and release.

Junior Hacksaw

What it is Small hacksaw, with bent-steel frame and 150mm (6in) blade. Pins at blade ends slot in frame.

Use it for Cutting metal or plastic, bolts, general smaller-scale metalwork. Blades for wood are also available.

How to use Pull saw back to start the cut. Continue with push/pull strokes.

Look for Saws with adjustable tension offer greater control.

Ryoba

What it is Japanese combination saw with two rows of teeth (typically 10–16 tpi). Handle is wrapped in split bamboo.

Use it for Crosscutting timber with finer teeth; ripping along grain with larger teeth. Joints and general joinery work.

How to use Hold the saw at a low angle when cutting. Swap between teeth depending on timber density and thickness.

Look for A release lever, which makes blade replacement easier. Blade length should be 240mm (9in).

Dozuki

What it is Japanese saw with a folded steel back to provide support for the thin blade. Very fine teeth (18–20 tpi).

Use it for Fine crosscutting: joints, cabinetmaking, mouldings, precision work.

How to use Start cut at the rear of wood, lower the blade and keep it parallel to surface as you pull the saw back.

Look for Models with replaceable blades; otherwise saws can be expensive as teeth may be too small to sharpen.

Azebiki

What it is Japanese saw with two short rows of curved teeth on a hardened steel blade, with a hardwood handle.

Use it for Plunge cutting in sheet material and thinner timber without starting from an edge.

How to use Place teeth on pencil line, gently pull back to start the cut. Use fine teeth for crosscutting, coarse for ripping.

Look for A replaceable blade. A protective cover is also a good idea.

Pruning Saw

What it is A rigid, backless blade (fixed or folding, straight or curved), often with triple-ground teeth to cut both ways.

Use it for Removing small branches, pruning trees, shrubs, and general garden work where secateurs are too small.

How to use Support branch with your free hand while sawing to prevent a sudden drop.

Look for A folding handle to protect the teeth when the saw isn't in use. Ensure this locks securely when blade is open.

"A variety of BLADES for different materials makes a hacksaw a VERSATILE tool"

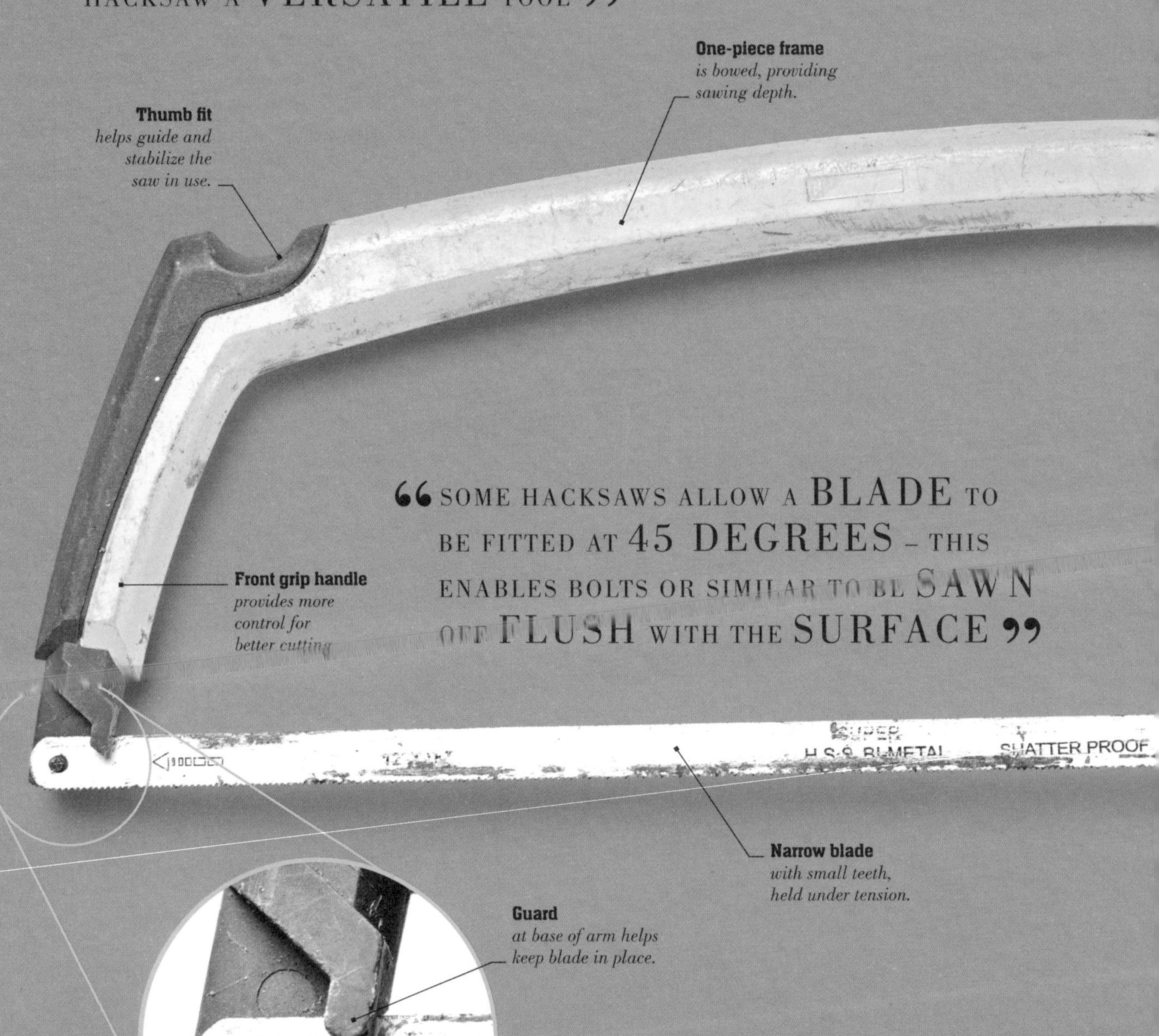

"Some hacksaws allow a BLADE to be fitted at 45 DEGREES – this enables bolts or similar to be SAWN off FLUSH with the SURFACE"

STRUCTURE OF A HACKSAW

Unlike saws for cutting timber, a hacksaw has a sturdy metal framework that holds a blade under tension. Used primarily for sawing soft and hard metals, its small teeth also make it suitable for plastic pipes and fittings. Blades are replaceable and have a standard length of 300mm (12in). Fit a tungsten carbide grit blade to cut tiles and glass.

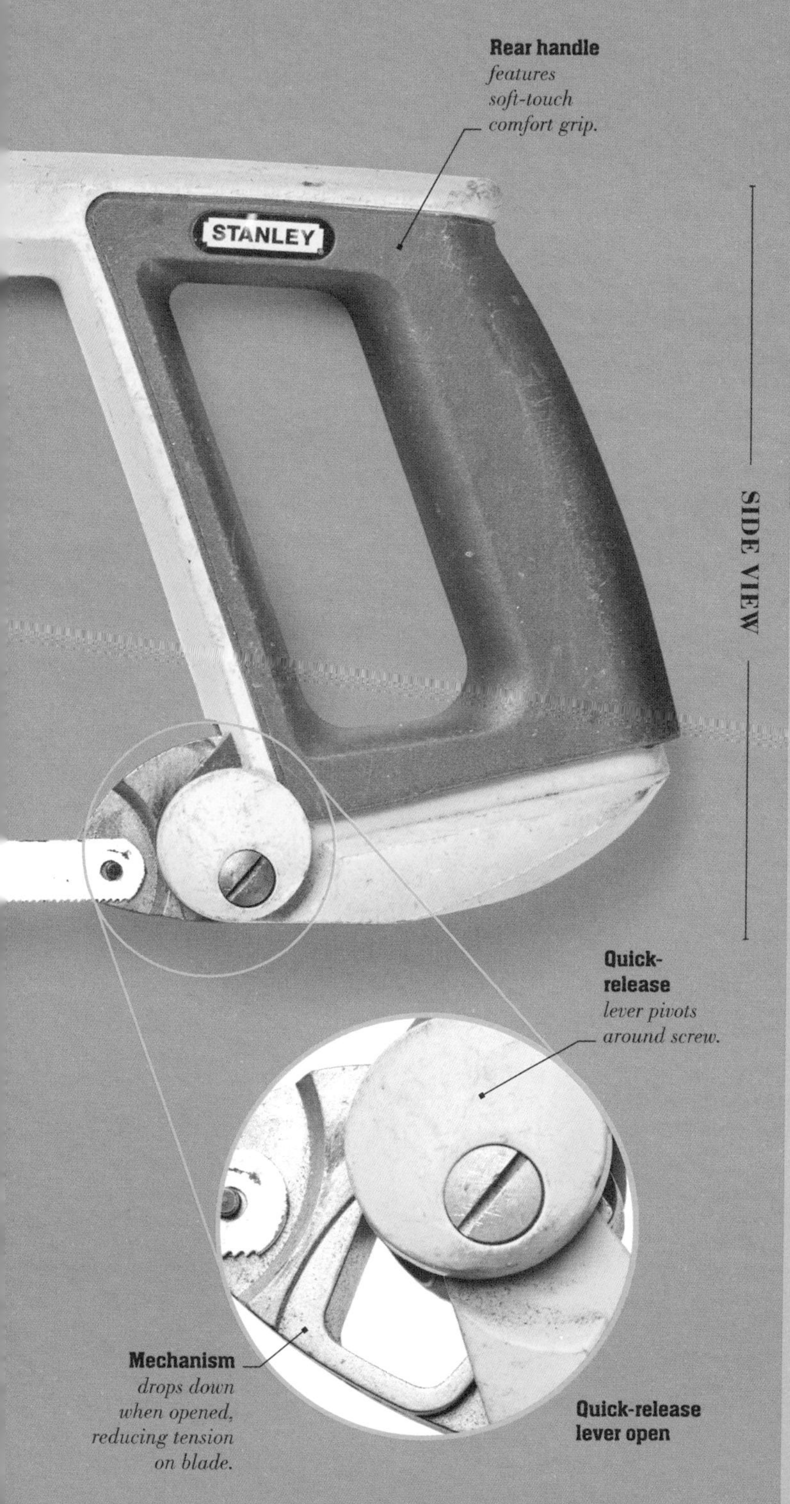

Rear handle *features soft-touch comfort grip.*

Quick-release *lever pivots around screw.*

Mechanism *drops down when opened, reducing tension on blade.*

Quick-release lever open

FOCUS ON...

Frames

The rigid steel or aluminium frame of a hacksaw supports a narrow, small-toothed blade, which is secured at each end by posts or studs, sometimes known as spigots. Once fitted, tension is applied in order to keep the blade taut enough for sawing without breaking it. On modern saws the frame incorporates an enclosed rear handle, which may be either bare metal or have a textured rubber grip.

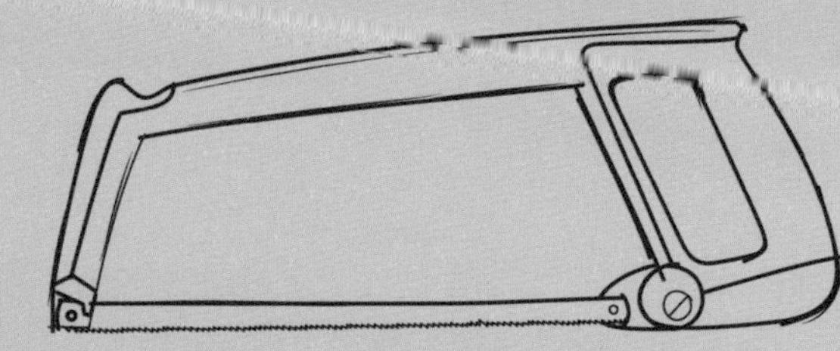

New hacksaw Has a tubular steel frame with a greater depth for sawing. Modern saws also feature a textured grip and a quick-release lever for easier blade-changing.

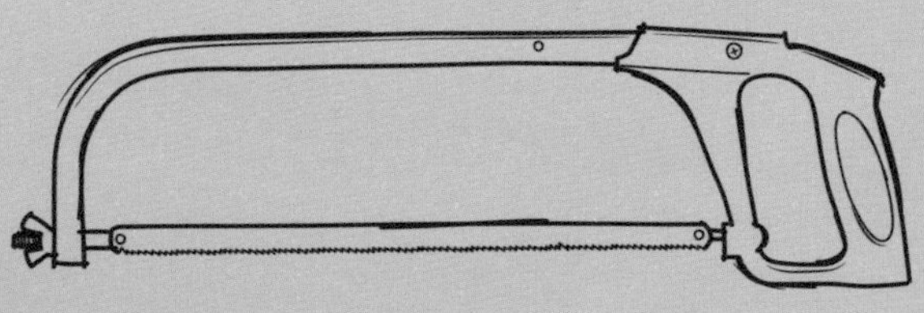

Old hacksaw The basic steel frame of an older model of hacksaw provides more limited clearance when sawing. Studs or pins hold the blade in place, and it is tensioned by turning a wing nut adjuster.

Using a

Hacksaw

Because its teeth are so small, sawing with a hacksaw is slow compared with using a handsaw – especially when cutting metal, as sawing too fast creates heat by friction. When sawing cylindrical objects like pipes, it can be difficult to achieve an even surface, but put a piece of tape around the pipe and you'll have a guide all the way around the diameter.

The Process

Before you start

- **Fit the blade** With the teeth facing forwards (pointing away from the handle), fit the blade by hooking each end over the frame pins or spigots.
- **Tighten the blade** Turn the wing nut or thumbscrew until the blade is taut. Newer tools may tighten automatically via a locking lever.
- **Secure workpiece** Never hold an item to be sawn in your hand while cutting. Always secure it to your bench or work surface using a clamp or vice.
- **Saw it safely** Wear gloves when working with metal objects with sharp edges, such as pipes or sheet metal.

1 Make your mark
Ensure that the object is gripped tightly in a vice or clamped to a workbench, and position your cut as close to the jaws as possible to prevent vibration. Use a piece of tape as a guide around pipes, or create a mark with a file.

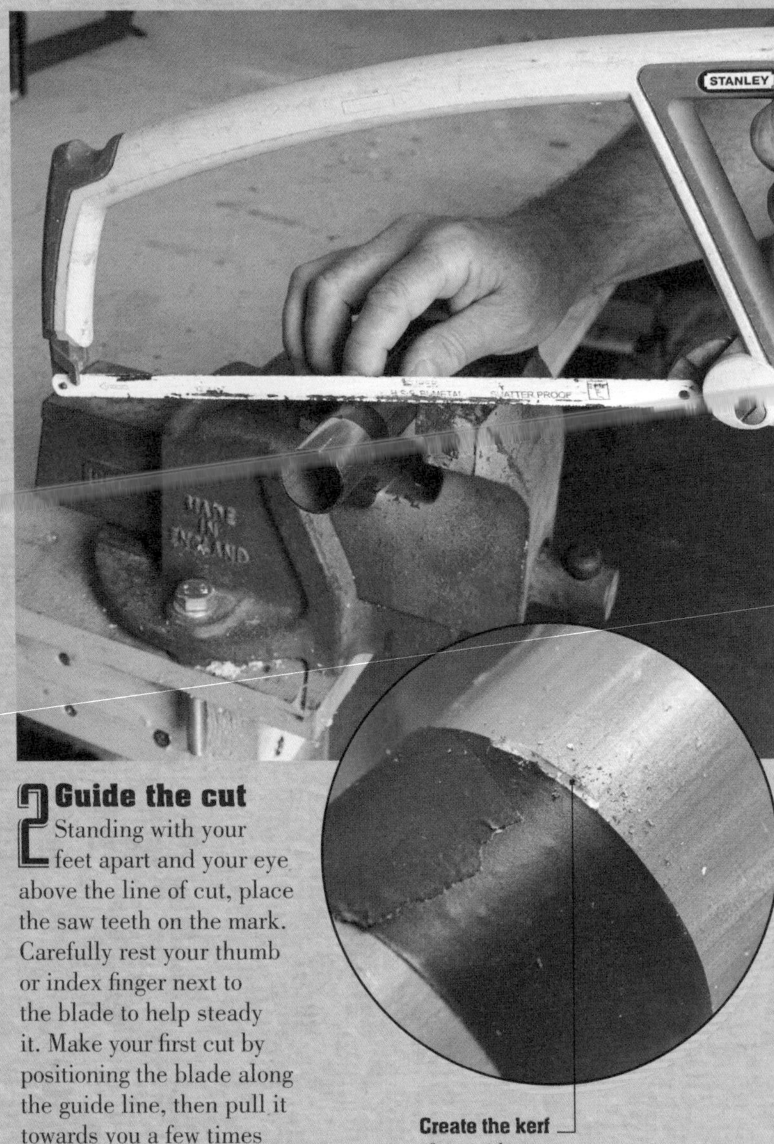

2 Guide the cut
Standing with your feet apart and your eye above the line of cut, place the saw teeth on the mark. Carefully rest your thumb or index finger next to the blade to help steady it. Make your first cut by positioning the blade along the guide line, then pull it towards you a few times to create a shallow groove.

Create the kerf *or first guide cuts for the blade by using gentle strokes of the saw.*

FOCUS ON...

The Blade

A general purpose carbon steel blade has 18 –32 tpi and will cut mild steel, soft metals, and rigid plastics. For hard metals a bi-metal universal blade with hardened teeth is more efficient and durable. The high speed steel edge and softer spring steel body welded together lets the blade flex without breaking. The blade is fixed with teeth facing forwards so it cuts on the push stroke, letting the user exert pressure when cutting tough material.

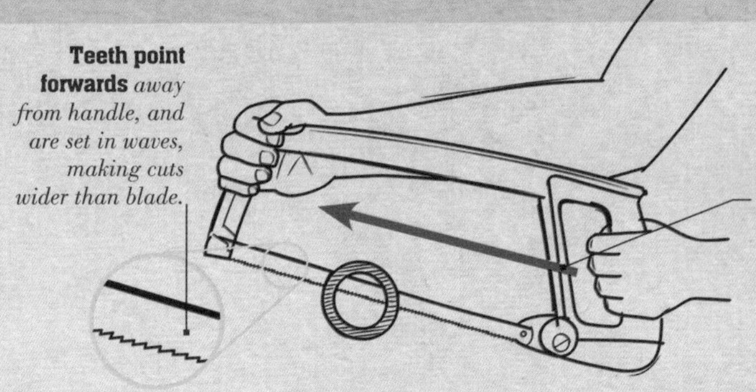

Teeth point forwards *away from handle, and are set in waves, making cuts wider than blade.*

Cut on push stroke *allows more pressure on the saw for harder-to-cut material.*

Stand *with your right leg forward if you are left-handed or left leg forward if you are right-handed.*

3 Sawing on

Grip the tool with both hands (unlike a woodcutting saw), with one hand on the handle and the other at the front of the frame. Push the blade forwards and backwards a few times, using as much blade length as possible on each stroke. Continue sawing back and forth, keeping the blade horizontal. Catch the offcut with your front hand as you complete the cut.

4 Finish the cut

Once you've cut through the object to your satisfaction, check that the cut edge is even all the way around. Metals in particular will have sharp, burred edges after sawing, so always use a file to remove these before releasing your object from the vice or clamp.

After you finish

- **Wipe off the blade** Wipe any debris off the blade after use. Apply a little oil with a soft cloth to prevent rusting.
- **Slacken off the tension** Loosen the wing nut a few turns before storing the saw in order to prolong the life of the blade.
- **Hang it up** Store the hacksaw by hanging it to keep the blade as sharp as possible for as long as possible.

Structure of a Handsaw

A traditional handsaw can be a highly rewarding tool to use, but its teeth must be sharp and they should be set correctly to tackle the job you have in mind. If you are cutting timber frequently, you may need both a crosscut and a rip saw in your tool kit. A panel saw has smaller teeth designed for cutting thinner pieces of wood, although these will blunt quickly when used on man-made boards.

Whole tool

SIDE VIEW

Toe of saw blade *is farthest from handle. Some have a hole here for hanging the tool.*

Top edge of blade *is skew-backed for better balance when sawing.*

Blade *is slightly flexible, and normally made from carbon steel.*

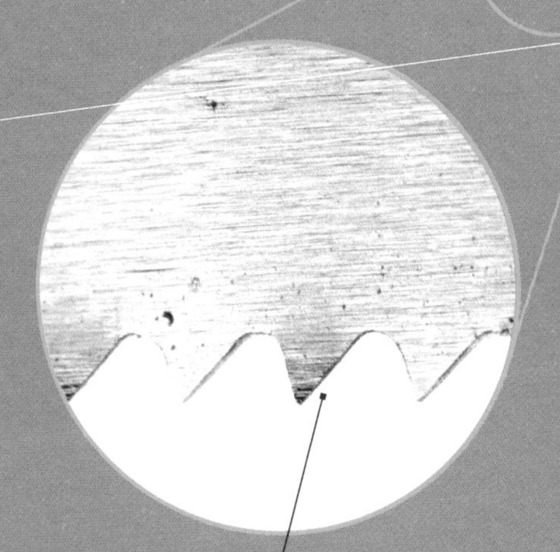

Teeth *are filed at angle on crosscut saw.*

Focus on... The Teeth

Traditional handsaws and backsaws, such as tenon or dovetail saws, have teeth that can be sharpened with a triangular saw file. Teeth are shaped individually and lean alternately to the left or right; this is known as the set. The set provides clearance as the saw slices through timber and cuts a channel, called a kerf, that is wider than the blade. Hardpoint saw teeth are electronically treated and cannot be sharpened, although they do stay sharper for longer.

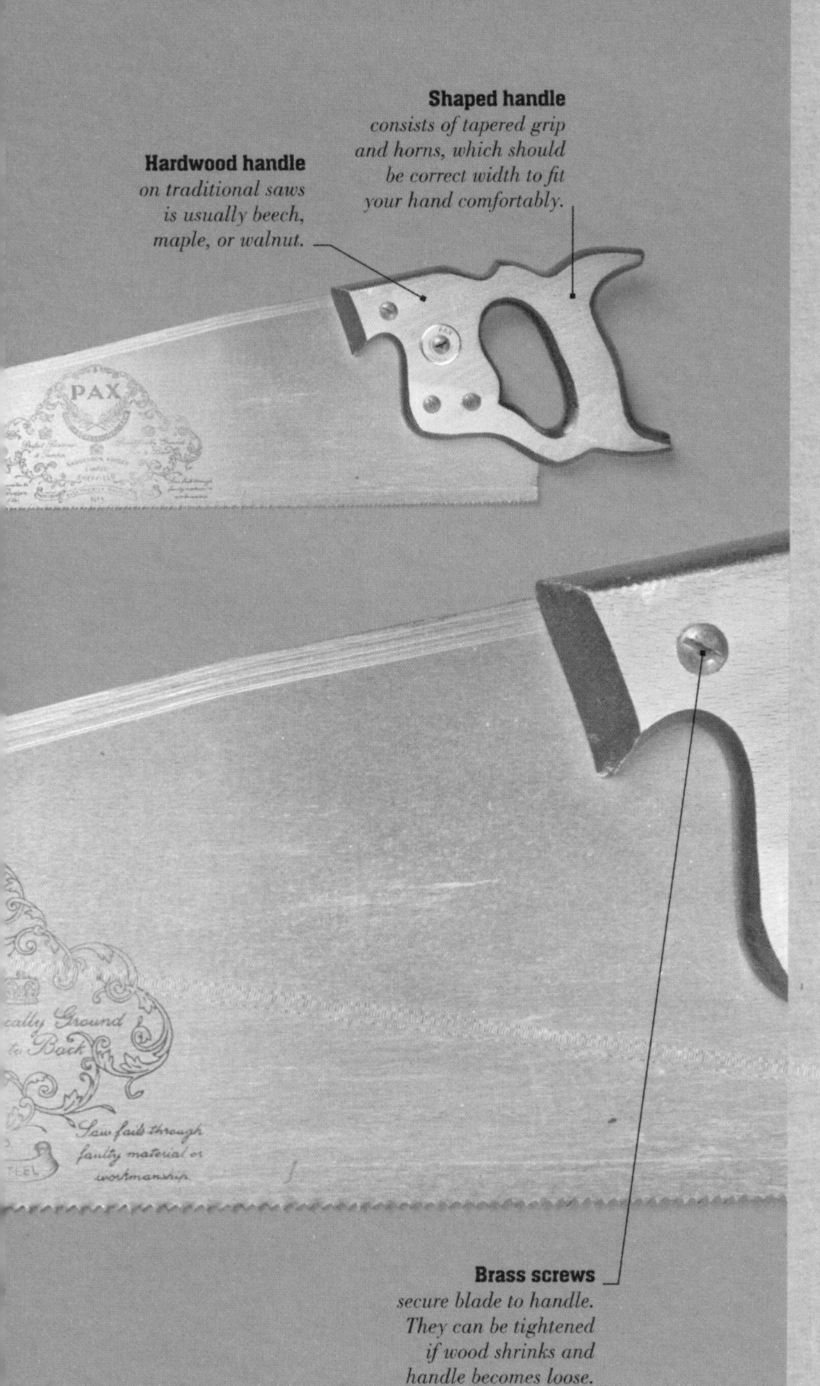

"YOU CANNOT HOPE TO SAW ACCURATELY IF THE TEETH ARE BLUNT"

USING A HANDSAW

Hold the saw with your index finger pointing down the handle. Stand so that your eye is above the line of cut, to judge if the blade is square. Use the entire blade when sawing, rather than just those teeth near the centre.

The Process

Before you start

Mark the wood Always saw to a guide line, whether cutting across or with the grain. Timber should be marked on both the face side and the face edge.

Secure the wood Cramp timber to the top of a workbench before sawing. Place long boards across one or two sawhorses.

1 Line up

Position the teeth of the saw next to the pencil line on the rear edge of the timber. Place the thumb of your other hand next to the blade; this helps to adjust the exact start of your cut.

2 Make a kerf

Pull the saw back gently a few times to make a shallow groove, or kerf. Use your thumb to move the blade sideways if necessary.

3 Saw timber

Saw down through the timber with a push stroke, pulling the tool back, then repeat the process. Follow the pencil lines, keeping the blade square. Hold timber on the waste side of the line as you gently complete the cut. This prevents splitting as the saw exits the wood.

After you finish

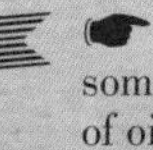

Clean up If the saw is unlikely to be used for some time, wipe the blade lightly with a thin coating of oil to prevent rust.

Store it safely Slide the saw back into its protective plastic sleeve, both to protect the tool and to prevent injury to you or others.

"I spend a lot of TIME doing CARPENTRY. Sometimes there is NOTHING that gives me the CONTENTMENT that SAWING A PIECE OF WOOD does."

ABBAS KIAROSTAMI

Choosing an
Axe

Axes come in many shapes and sizes, each one designed and suited to a specific task. A multipurpose axe or hatchet will be all most people need for chopping firewood; however, specialist tasks can be catered for. There are several different axe types, so first identify the result your want to achieve, then choose the right axe to fit the job.

“The AXE is an ANCIENT TOOL that has evolved into many different FORMS, becoming one of the most VERSATILE”

Side axe

Multipurpose axe

Splitting axe

Adze

☛ **What it is** Ancient cutting tool with a long, scooped head that "shaves" wood.

☛ **Use it for** Shaping or finishing wood. The short handle allows for fine finishing.

☛ **How to use** Take small swings with the grain to start the cut, then larger ones to finish off. This avoids gouging.

☛ **Look for** A very sharp blade to shape timber, create curves.

Side Axe

☛ **What it is** Sharp axe with one flat edge to the head, for accurate shaping and cutting.

☛ **Use it for** Fine crafts, such as hedge-laying or shaping timber.

☛ **How to use** Hold wood with one hand, higher than axe-hand. Chop with flat side of blade facing the wood surface.

☛ **Look for** Either a left- or right-hand blade shape – both are available.

Hatchet

☛ **What it is** A lightweight, short-handled axe, ideal for domestic and camping use.

☛ **Use it for** General cutting, such as chopping kindling and splitting small logs.

☛ **How to use** Grip handle with one or both hands. Swing down from shoulder.

☛ **Look for** A hard-wearing blade that keeps its edge, and a sheath to protect it.

Multipurpose Axe

☛ **What it is** A basic axe with a handle varying from curved to straight, traditional wood to carbon fibre, medium to long.

☛ **Use it for** Splitting and chopping. The head's moderate angle makes it less good for felling and shaping.

☛ **How to use** Bring axe over shoulder, sliding top hand close to head. Swing axe down, sliding top hand down the handle.

☛ **Look for** Lighter or heavier heads, shorter or longer handles. Choose those that best suit your build and needs.

Splitting Axe

☛ **What it is** An axe with a wedge-shaped head for splitting apart wood fibres.

☛ **Use it for** Splitting logs to burn in an open fireplace or woodburner.

☛ **How to use** Grip handle with both hands, swing axe over your shoulder to strike log. Split wide logs from outside in.

☛ **Look for** A longer handle makes splitting logs easier for taller people.

CONTINUED ☛

Billhook

Staffordshire billhook

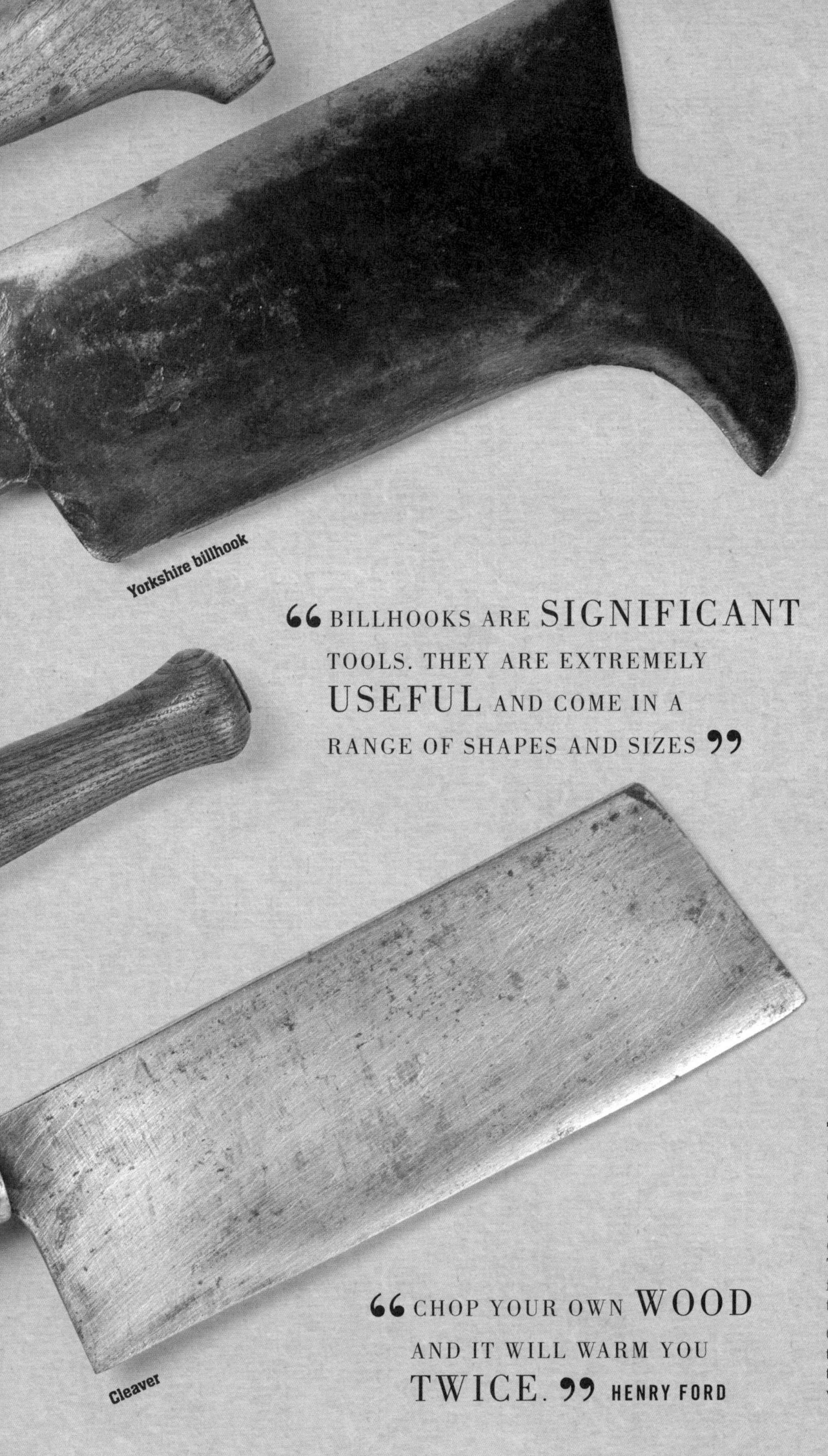
Yorkshire billhook

Cleaver

“BILLHOOKS ARE SIGNIFICANT TOOLS. THEY ARE EXTREMELY USEFUL AND COME IN A RANGE OF SHAPES AND SIZES”

“CHOP YOUR OWN WOOD AND IT WILL WARM YOU TWICE.” HENRY FORD

Billhook

☛ **What it is** A short-handled tool with a deep, flat, hooked blade.

☛ **Use it for** A wide variety of chopping tasks, for woody material 2–10cm (1–4in) in diameter.

☛ **How to use** Swing at material, away from the body. Use gloves if holding stems.

☛ **Look for** Old, well-maintained forged steel with a good, weighty head.

Yorkshire Billhook

☛ **What it is** A long-handled tool with a head that has a hooked blade on one side and a flat edge on the other. Total tool length of 90cm (35in).

☛ **Use it for** Cutting thicker material in a single swing, particularly when hedge-laying. Requires a strong arm!

☛ **How to use** Use one or two hands, swinging blade to the base of stems. Keep free hand high if holding stems.

☛ **Look for** Smooth fitting of the handle to metalwork, ensuring there are no snags or rough edges.

Staffordshire Billhook

☛ **What it is** Similar to the standard billhook, but with an additional flat cutting blade on the back side.

☛ **Use it for** An even wider range of chopping tasks. The flat blade is good for sharpening stakes and kindling.

☛ **How to use** When sharpening stakes, cut vertically, holding the stake at the right angle. Cut onto a chopping block.

☛ **Look for** A head mounted well into the handle, without any play or movement.

Cleaver

☛ **What it is** Similar to a billhook but with a single flat cutting edge.

☛ **Use it for** Cutting points on small stakes and splitting kindling.

☛ **How to use** Swing single-handedly. The main concern is not hitting your free hand, so always keep it higher than your cutting hand.

☛ **Look for** Enough weight in the head to cut smoothly, and no rust pitting in the metalwork. The edge should be very sharp.

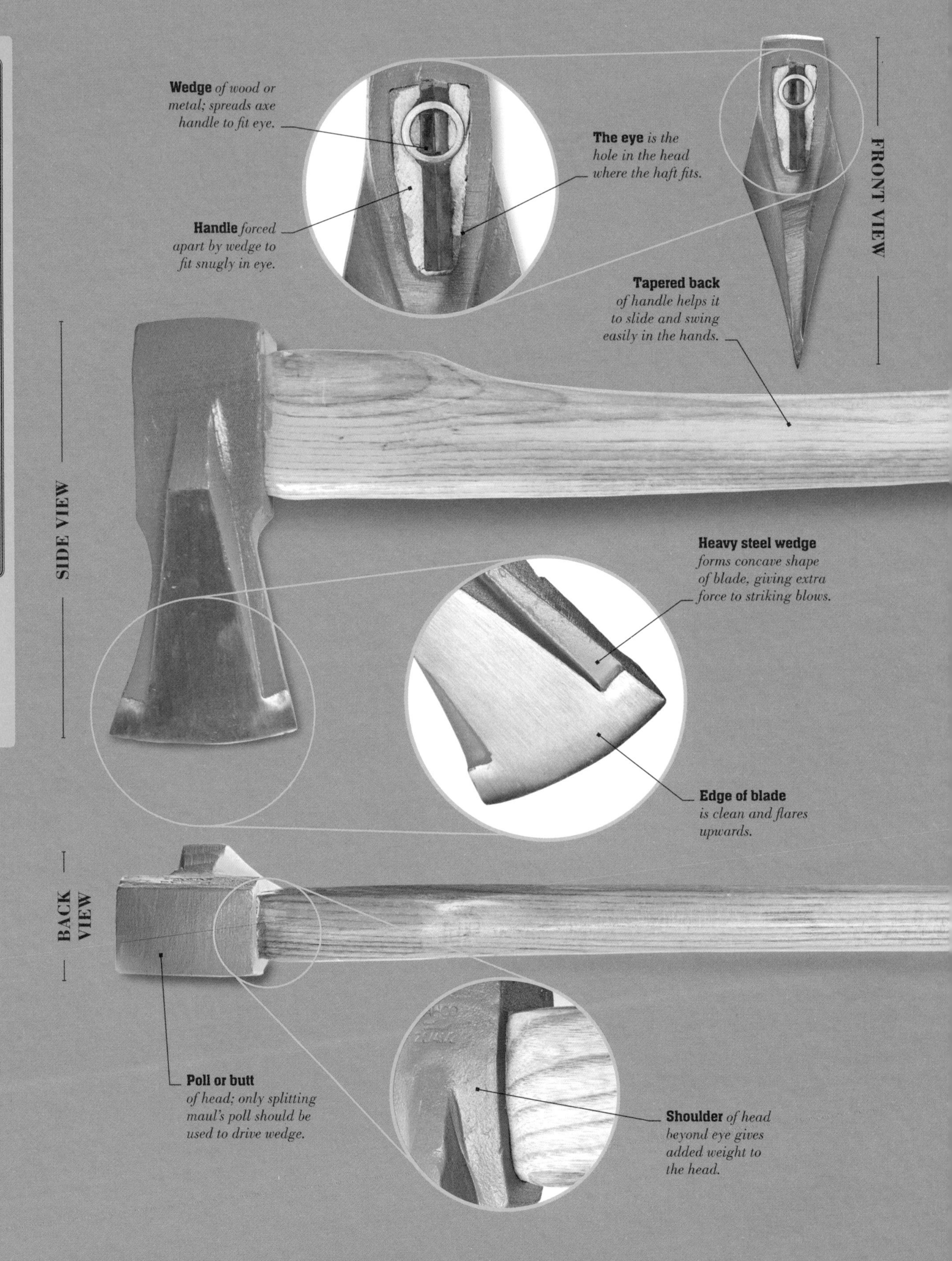
Wedge *of wood or metal; spreads axe handle to fit eye.*
The eye *is the hole in the head where the haft fits.*
Handle *forced apart by wedge to fit snugly in eye.*
FRONT VIEW
Tapered back *of handle helps it to slide and swing easily in the hands.*
SIDE VIEW
Heavy steel wedge *forms concave shape of blade, giving extra force to striking blows.*
Edge of blade *is clean and flares upwards.*
BACK VIEW
Poll or butt *of head; only splitting maul's poll should be used to drive wedge.*
Shoulder *of head beyond eye gives added weight to the head.*

STRUCTURE OF A

SPLITTING AXE

A splitting axe or maul is specially designed and forged to achieve the maximum result with the least physical effort. Its wide, wedge-shaped head, with its thin, sharp blade, forces woodgrain apart – unlike a felling axe, which cuts across the grain. Both the shape and weight of a splitting axe head combine to produce a heavy and powerful strike, splitting the toughest of logs with ease. The blade is also less likely to catch in wood than that of a standard axe.

Wide haft end *or swell knob prevents axe from slipping out of hands.*

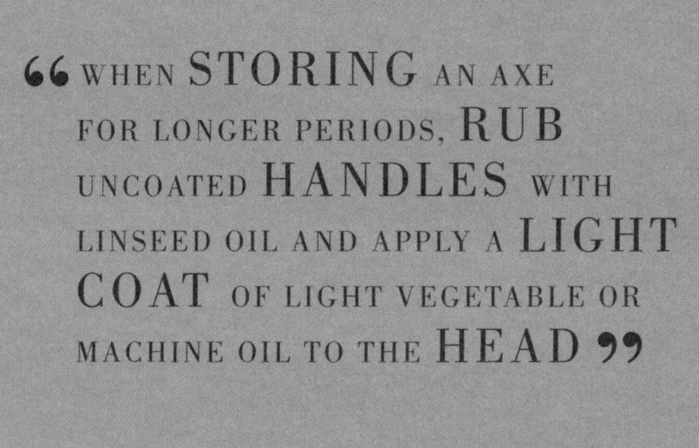

“ WHEN **STORING** AN AXE FOR LONGER PERIODS, **RUB** UNCOATED **HANDLES** WITH LINSEED OIL AND APPLY A **LIGHT COAT** OF LIGHT VEGETABLE OR MACHINE OIL TO THE **HEAD** ”

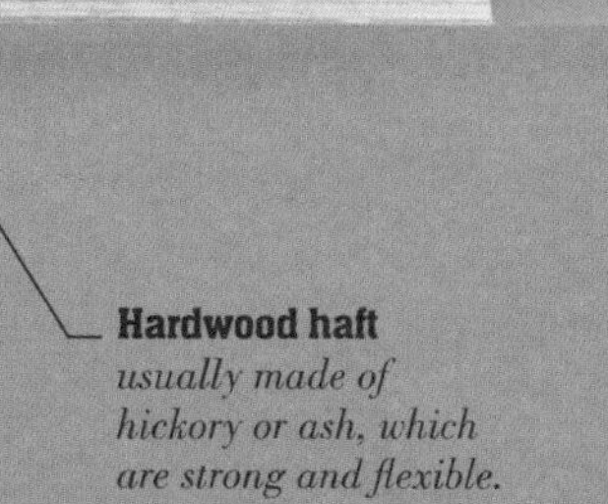

Hardwood haft *usually made of hickory or ash, which are strong and flexible.*

FOCUS ON...

AXE HEAD SHAPES

Axes vary greatly in size, head shape, handle shape, and use. Most people are familiar with the multi-purpose axe for chopping and splitting; however, there are also special felling, splitting, and craft axes. The size, shape, and angle of these different axe heads make them suitable for specific tasks.

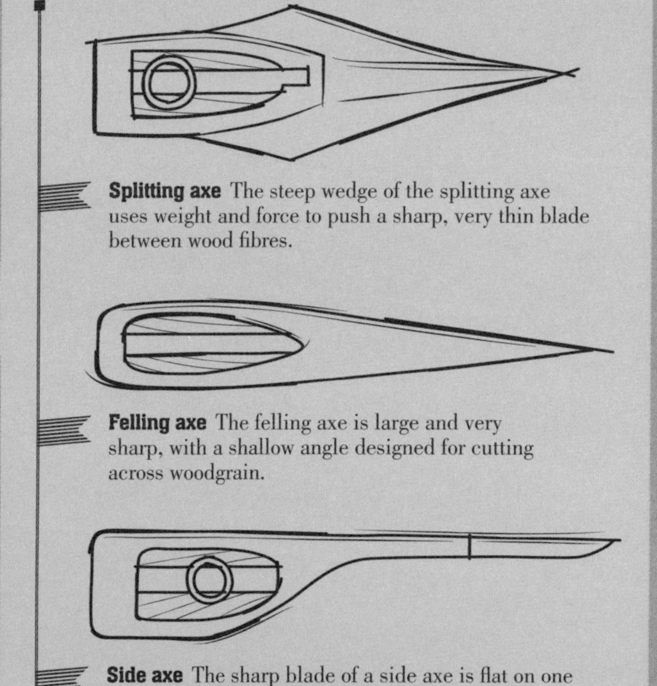

Splitting axe The steep wedge of the splitting axe uses weight and force to push a sharp, very thin blade between wood fibres.

Felling axe The felling axe is large and very sharp, with a shallow angle designed for cutting across woodgrain.

Side axe The sharp blade of a side axe is flat on one side, bevelled on the other. It is used by craftsmen for accurate, close cutting and shaping of wood.

Using a
Splitting Axe

A splitting axe is the most efficient type of axe to use for cutting firewood to size, and its design makes the task easy. It's technique, rather than brute force, that counts here. The heavy, wedged head forces wood fibres apart, working with both gravity and the timber's own grain to make the cut.

The Process

Before you start

- **Check your surroundings** Ensure that your work area is clear and open, without overhead obstructions or trip hazards. Make sure there is plenty of room for you to swing the axe.
- **Inspect the axe** Check that the axe head is fixed firmly on the handle, without any play.
- **Dress appropriately** Wear stout, protective footwear, long heavy trousers, and safety glasses.
- **Arrange your space** Plan out the work area, so that the material to be split is within reach. Split close to where your wood will be stacked.

1 Set up
Choose a large log to use as a chopping block. This elevates the work area which avoids back strain and makes for easier splitting. Centre the log to be split on the chopping block, ideally top up (in the same direction it grew), as it will split more cleanly. Avoid aiming for knots.

2 Line up
Check your stance by touching the log with the axe to set your body position and distance from the log. You will need to end up here so make sure the swing is comfortable – eventually this will become second nature. Raise the axe and settle it on your shoulder.

FOCUS ON...

THE HEAD

The splitting axe has a wider, wedge-shaped head than traditional axes. It is also quite heavy, and this weight, when combined with a strong swing, drives a splitting force with considerable Newton mass. Because wood fibres run parallel to each other, they separate easily when driven apart, more so when cut in the direction in which they grew. Some wood splits more readily when green (fresh) than when seasoned.

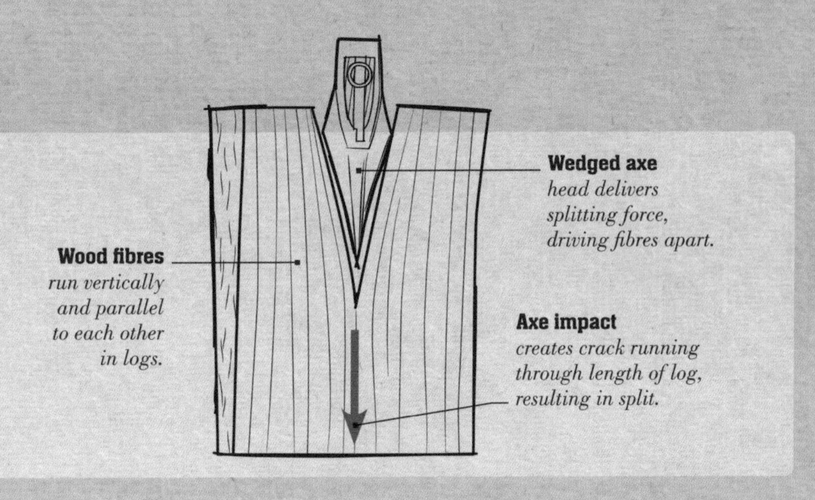

Resting position *settles the axe between swings and allows you to aim before swinging.*

3 Take the swing

You want to achieve a confident, powerful swing, as this makes the task much easier. Grip the haft with two hands, one higher than the other; the upper hand must start near the shoulder. Watching the log always, swing the axe forwards and downwards in an arc to strike the log. Let the weight of the axe do the work. Slide your upper hand down to meet the other near the haft's end. Let the weight of the axe do the work.

4 Make the split

If the wood you're splitting has a long, parallel grain, it should split easily. You are aiming to drive through the log, not stop at the top. If the blade gets stuck, try levering it out first, or bang the axe and log upside down on the base log.

After you finish

- **Examine the axe** Check the axe handle and head for damage, such as chinks or handle splinters.
- **Clean up** Wipe the blade clean of any debris. Store the axe in a lockable workshop or garage, especially if children are nearby.
- **Stack the logs** Stack your split logs off the ground in a log store or similarly dry space with good air circulation.

“PEOPLE LOVE CHOPPING WOOD. IN THIS ACTIVITY ONE IMMEDIATELY SEES RESULTS.”

ALBERT EINSTEIN

Choosing a
Knife

A knife is an essential part of anyone's toolkit, although choosing the right one to fit your needs can often be confusing due to the wide variety available. If you simply want to make straight cuts now and again, then a fixed-blade knife will do the job just as effectively as fancier models with many different blades. Multitools, however, are ideal to have with you during outdoor pursuits such as camping or hiking.

OLFA JAPAN

Craft knife

“A SHARP KNIFE IS NOTHING WITHOUT A SHARP EYE”

LEATHERMAN WAVE

Multitool

“THE BEST KNIFE IS THE ONE YOU HAVE WITH YOU WHEN YOU NEED IT”

Swiss Army knife

Retractable folding utility knife

Pruning knife

Craft Knife

What it is A slim, light-duty tool with a sliding blade. Blade tips snap off when blunt to expose sharper cutting edge.

Use it for Crafts and model-making. Cuts balsa wood, cardboard, and thin sheet plastics, and can trim wallpaper.

How to use Slide the blade out to reveal the first break-line, lock the handle, and snap off tip with pliers.

Look for A die-cast metal body, which is more substantial and long-lasting than cheaper plastic versions.

Swiss Army Knife

What it is A combination knife tool that includes two blades, a corkscrew, a can opener, a screwdriver, and more.

Use it for Travel, camping, fishing, and other outdoor pursuits. It's also handy as an emergency household toolkit.

How to use Select the desired blade or specific gadget and unfold carefully. Each blade should lock into position.

Look for Essential functions. You may never use some of the gadgets on the most elaborate models.

Multitool

What it is Gadget that combines a knife with tools such as pliers, slotted and Pozi screwdrivers, and a serrated blade.

Use it for General maintenance and DIY work, camping and outdoor trips. It's also handy as an emergency toolkit.

How to use Press to unlock or slide to open. If your model has pliers, you may need to push them out first to use the tool.

Look for The tool's primary function. It may be pliers, rather than a knife blade, so carefully check options available.

Pruning Knife

What it is A curved, folding blade with a shaped hardwood, plastic, or metal handle designed for general garden use.

Use it for Grafting plants, removing shoots, stems, pruning. Also cuts string and plastic plant ties and compost bags.

How to use Unfold the blade until it locks into its open position. Carefully wipe clean after use, then close.

Look for A stainless-steel blade to prevent corrosion. Ideally, it should be compact enough to slip into a pocket.

Retractable Folding Utility Knife

What it is The blade retracts into a sturdy metal body. Tool can be folded to fit in a pocket. Blades store in the handle.

Use it for General cutting, including plasterboard, roofing felt, vinyl flooring, carpet, cardboard. Also scoring cuts.

How to use Unfold the knife so it locks in open position. Slide the button to reveal the blade. Retract blade after use.

Look for Rubber grips for a more comfortable hold when in use. Blades in the handle should be easy to access.

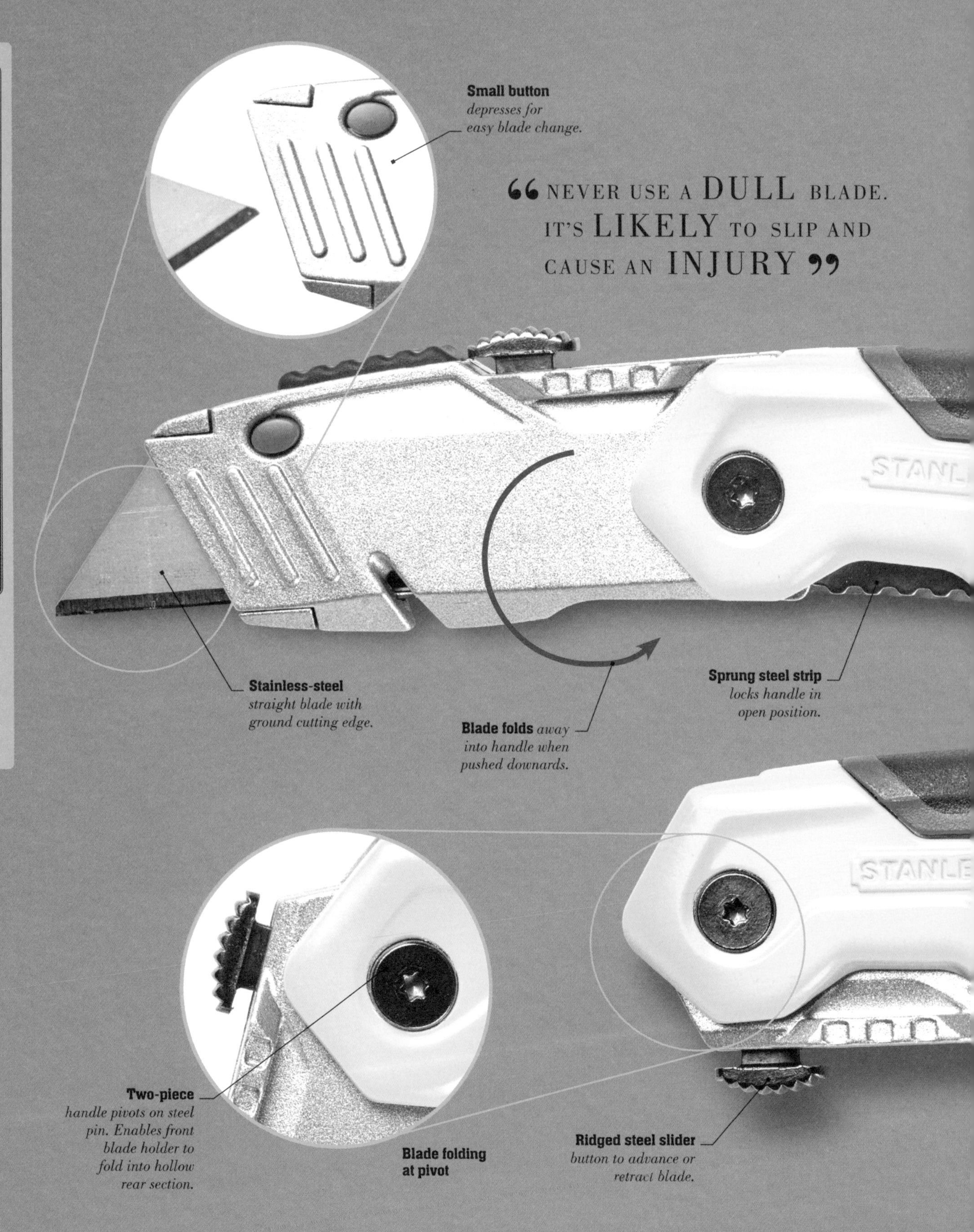

“NEVER USE A DULL BLADE. IT'S LIKELY TO SLIP AND CAUSE AN INJURY”

STRUCTURE OF A
UTILITY KNIFE

As its name suggests a utility knife can be used for everything from cutting rope to scraping hides and craft projects. A folding model is relatively safe because its blade retracts, which protects both it and its user when it is not needed. In addition, the die-cast metal body folds in half, so it can be carried easily and safely in a pocket or toolbox.

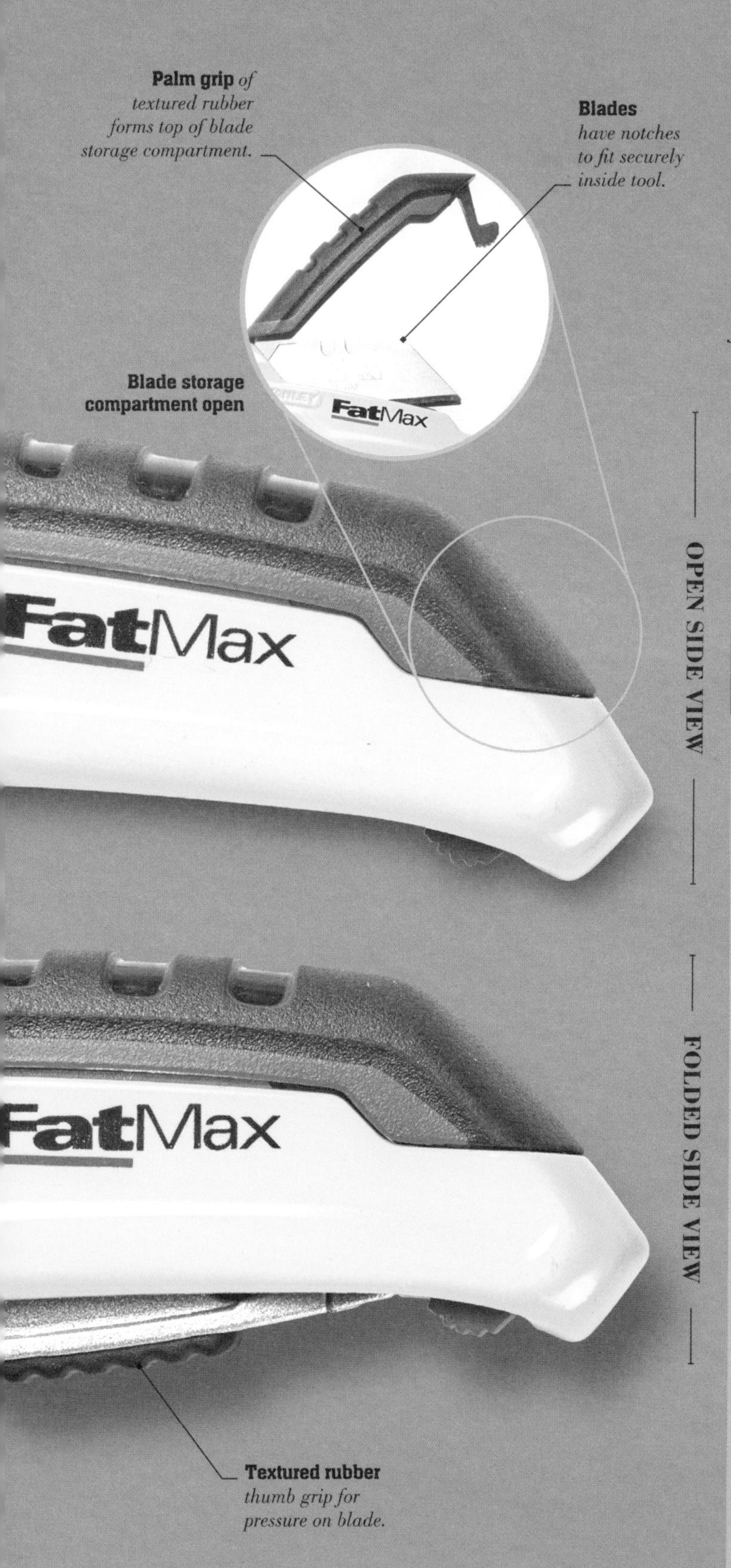

FOCUS ON...

BLADE TYPES

Many blades for utility knives are made of carbon steel, and are available in a standard format that consists of notches on the upper edge to fit securely inside the front of the tool. Longer-lasting bi-metal blades have a spring-steel backing, which provides some flexibility when cutting and are virtually unbreakable in use. All blades should be discarded when blunt.

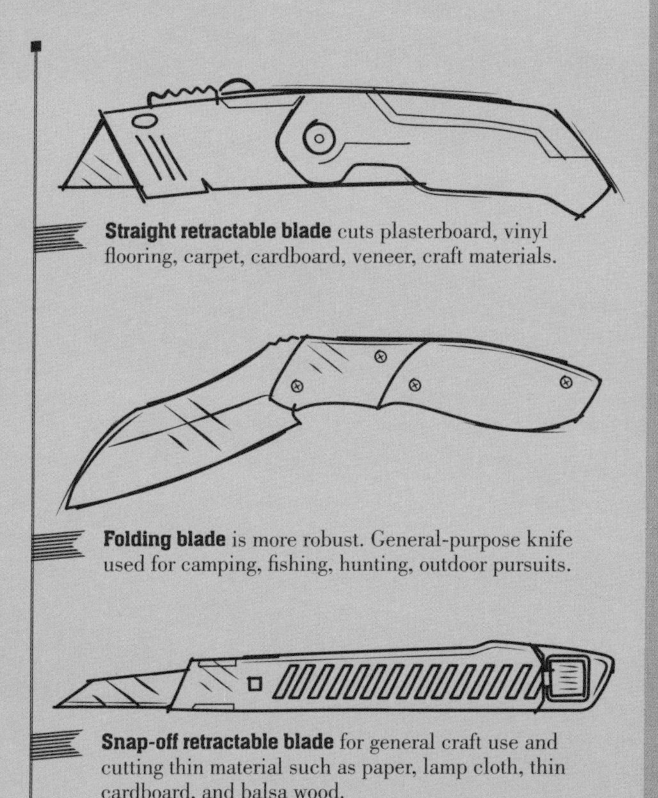

Straight retractable blade cuts plasterboard, vinyl flooring, carpet, cardboard, veneer, craft materials.

Folding blade is more robust. General-purpose knife used for camping, fishing, hunting, outdoor pursuits.

Snap-off retractable blade for general craft use and cutting thin material such as paper, lamp cloth, thin cardboard, and balsa wood.

USING A

UTILITY KNIFE

Because of its light weight and portability, this common household tool may seem like the simplest to use, but a utility knife should always be treated with respect. It packs a very sharp blade within its small body, so always keep your fingers away from the line of cutting – especially when holding down a straightedge when trimming items to size.

The Process

Before you start

- **Stay sharp** Make sure the knife blade is sharp before you start. A blunt blade is more dangerous than a sharp one when trying to cut objects.
- **Select the right blade** Check that you have the correct type and length of blade fitted for the required task.
- **Check your surroundings** Make sure your work area is tidy. This will prevent you from slipping or tripping over items and injuring yourself.

Angle the blade *away from your fingers and body when cutting.*

1 Position the rope

Measure out the rope, if necessary, and mark where it needs to be cut. Choose a stable surface to work on, such as a desk, table, or workbench, but always use a self-healing cutting mat to protect work surfaces when cutting.

2 Release the blade

Unfold the knife and slide the blade outwards to its full capacity (depending on thickness of rope). Check that the blade is locked rigidly in place. Note where your cut-stroke will be and keep your body and hands away from this line of cutting. Hold down the rope with one hand, making sure that your fingers are not too close to where you plan to cut.

“KEEP A SAFETY BLADE DISPENSER IN YOUR TOOLBOX. THAT WAY THERE'S NO EXCUSE TO USE A BLUNT BLADE”

FOCUS ON...

The Cut

A knife's cutting edge is created by grinding a fine bevel on both faces of the blade. As you use the knife, pressure is exerted on the blade, concentrating it on the small surface area of the edge, allowing the bevel to sever fibres or molecules, forcing them apart. The sharpest knives are those with the thinnest edges. When you sharpen a blade, you make it thinner, so the thinner the steel, the sharper the edge – think razor blade.

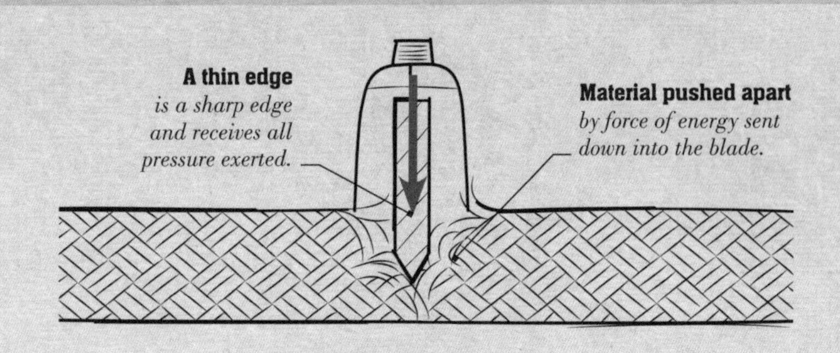

3 Cut correctly

Gripping the knife in your other hand, draw the blade carefully across the rope strands with even pressure. Pull the blade, rather than using a sawing motion, and let the pressure through the blade do the work. Do this for as many times as is necessary.

4 Finish up

Once you complete the cut, retract the blade fully into the tool handle in order to avoid injury. Always store utility and other knives in a lockable cupboard or a secure toolbox if there are likely to be children around.

After you finish

- **Bind the edges** To prevent fraying, melt the cut ends of polypropylene rope by heating them with a match flame.
- **Dispose of blades safely** If you changed blades, wrap the blunt one in masking tape to prevent injury during disposal.
- **Inspect the knife** Make sure the knife blade is fully retracted before you put it away. Wipe down the handle to remove any debris.

Choosing a

Scythe or Sickle

Scythes, sickles, and weed hooks vary enormously. Many are designed for specific tasks, but all are useful and hard to surpass with machinery. The old ones are often the best, forged well and ergonomically designed with durable handles. Hooked blades are still popular with many modern craftsmen.

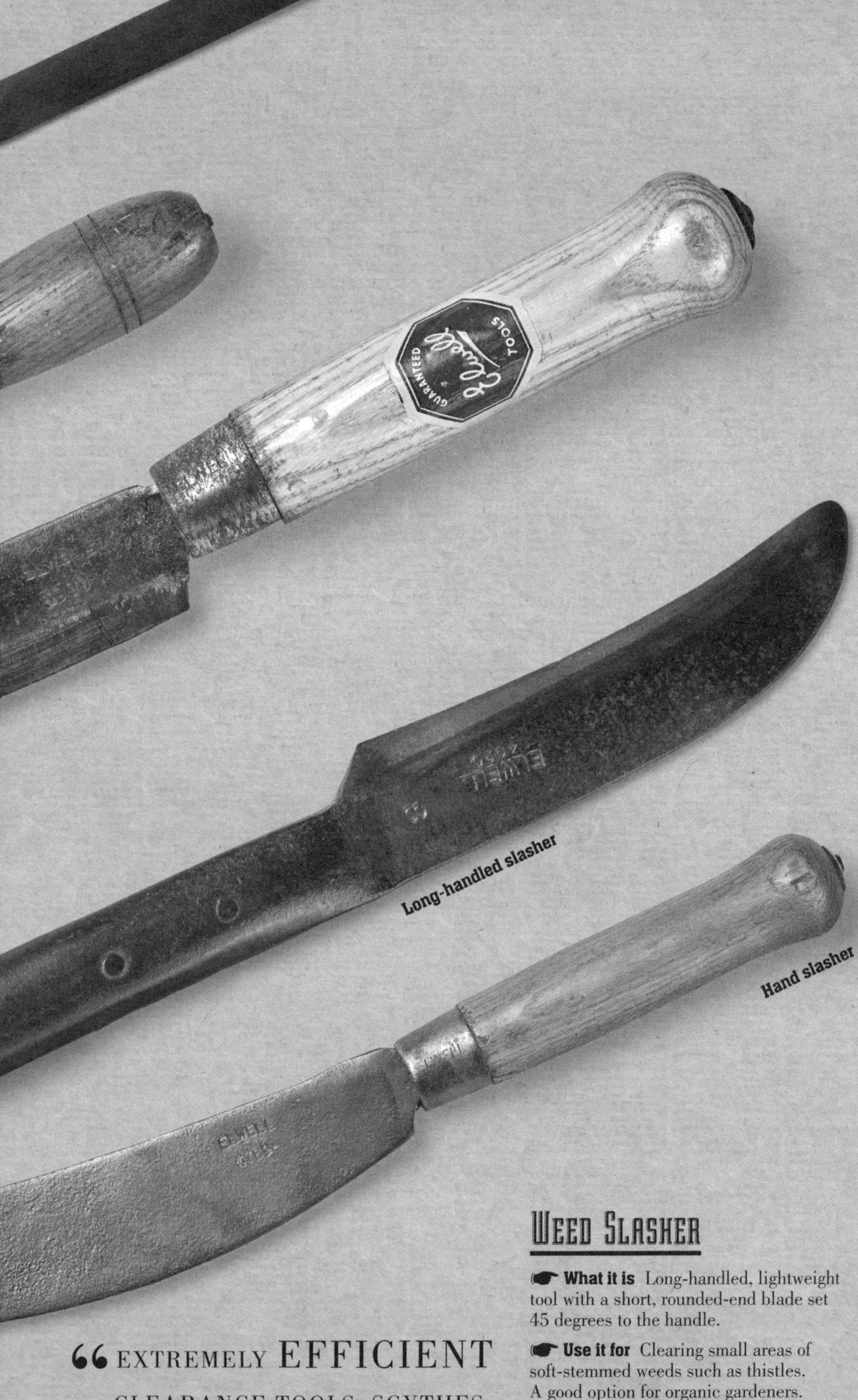

Sickle

What it is Cutting tool with a short handle for one hand and curved blade.

Use it for Clearance of plant growth, or harvesting at close proximity to the user.

How to use Wearing gloves, swing tool with one hand to slash plants. Keep free hand clear, and use a stick to hold back plants for cutting.

Look for A high-quality forged-steel blade and a solid, well-fixed handle. Test the feel for comfort and weight.

Beet Knife

What it is Short-handled tool originally for harvesting sugar beets. Long, flat, razor-sharp blade with short, angled spike at the end.

Use it for General cutting/harvesting, pointing small stakes, chopping kindling.

How to use Spike beet or other item with prong on end, swing up to collect, then use blade to cut off top. Or use like hatchet to chop kindling, stakes.

Look for A blade that's in good condition, without too much pitting or "dents" and "chinks" in the edge.

Long-handled Slasher

What it is A hand tool with a long, strong handle attached to a tough, straight, or slightly curved blade.

Use it for Chopping and slashing course plant material at arm's length, particularly thorns and saplings.

How to use Hold firmly in two hands; swing powerfully at the base of plants.

Look for Weighty, forged-steel head and well-maintained handle if older.

Hand Slasher

What it is Short-handled version of the weed slasher, long enough for two hands.

Use it for Getting a bit closer to material than the weed slasher, and for denser, woodier plants than a sickle.

How to use Use a strong, controlled slashing action, with two hands always on the handle for safety.

Look for A style, shape, and feel that suit you.

Weed Slasher

What it is Long-handled, lightweight tool with a short, rounded-end blade set 45 degrees to the handle.

Use it for Clearing small areas of soft-stemmed weeds such as thistles. A good option for organic gardeners.

How to use Swing using one or two hands to cut weeds at ground level.

Look for Weighty, forged-steel head and well-maintained handle if older.

“ EXTREMELY EFFICIENT CLEARANCE TOOLS, SCYTHES AND SICKLES ARE QUIET AND SUSTAINABLE ”

CONTINUED

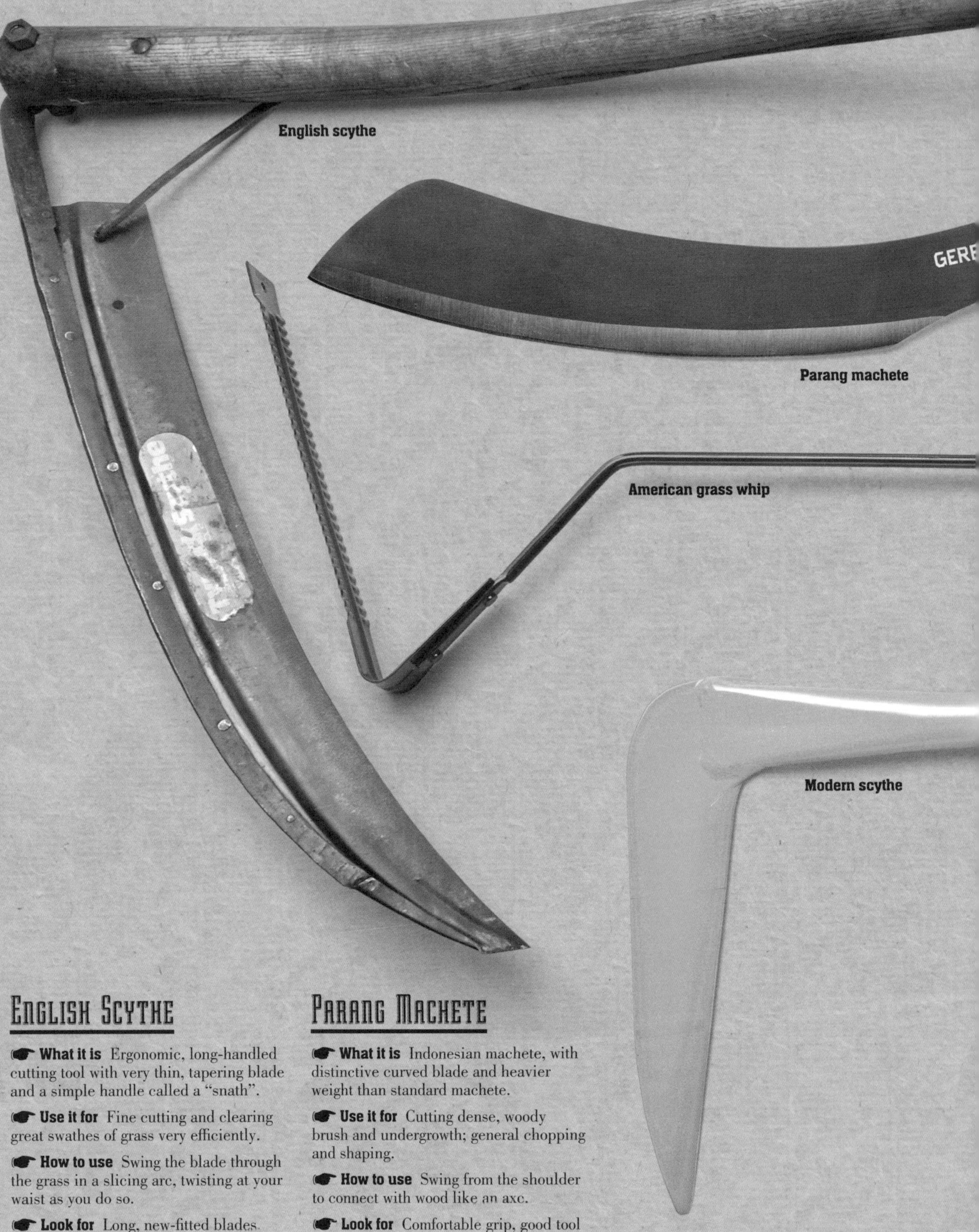

ENGLISH SCYTHE

What it is Ergonomic, long-handled cutting tool with very thin, tapering blade and a simple handle called a "snath".

Use it for Fine cutting and clearing great swathes of grass very efficiently.

How to use Swing the blade through the grass in a slicing arc, twisting at your waist as you do so.

Look for Long, new-fitted blades. Stable, well-maintained handles.

PARANG MACHETE

What it is Indonesian machete, with distinctive curved blade and heavier weight than standard machete.

Use it for Cutting dense, woody brush and undergrowth; general chopping and shaping.

How to use Swing from the shoulder to connect with wood like an axe.

Look for Comfortable grip, good tool balance, and a sharp, heavy steel blade.

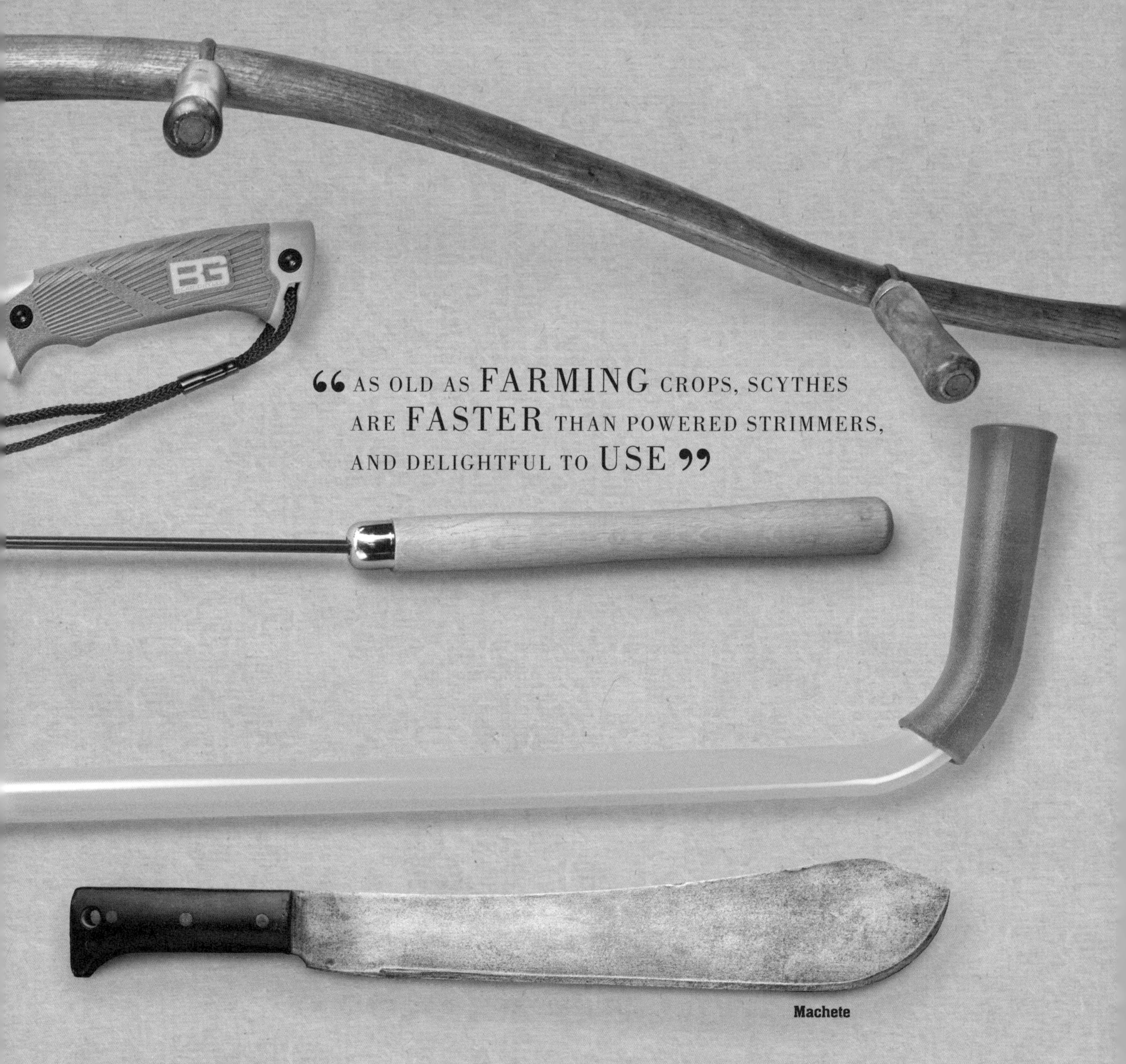

American Grass Whip

☛ **What it is** Long, lightweight cutting tool with a hardwood handle and serrated double-edged blade.

☛ **Use it for** Clearing tall grass from ditches and other overgrown areas.

☛ **How to use** Hold handle with one hand, keep swing clear of body and legs. Swing to cut both ways.

☛ **Look for** Comfortable grip, good tool balance, and a tempered steel blade.

Modern Scythe

☛ **What it is** Shorter handled tool, with a similar shape to a traditional scythe, often with metal handle.

☛ **Use it for** Slashing soft weeds and long grass over small areas.

☛ **How to use** Hold with one hand, swing blade at ground level, well clear of your body.

☛ **Look for** Correct handle length for user height, good weight, and balance.

Machete

☛ **What it is** A long, knife-shaped blade with a short handle.

☛ **Use it for** Clearing and cutting brushy material like thorns and saplings.

☛ **How to use** Hold in one hand and swing downwards from the shoulder, flicking wrist up or down, depending on angle and height of stems to be cut.

☛ **Look for** Correct handle length for user height, good weight, and balance.

STRUCTURE OF AN ENGLISH SCYTHE

Regardless of the type or model, the scythe is still a strong performer when it comes to mowing long grass, even when compared with modern strimmers. Both speedy and silent, it is made of a long, shaped handle, or "snath", with adjustable hand grips that allow the handling of it to be tailored to suit the user. A long, curved blade is mounted at the end.

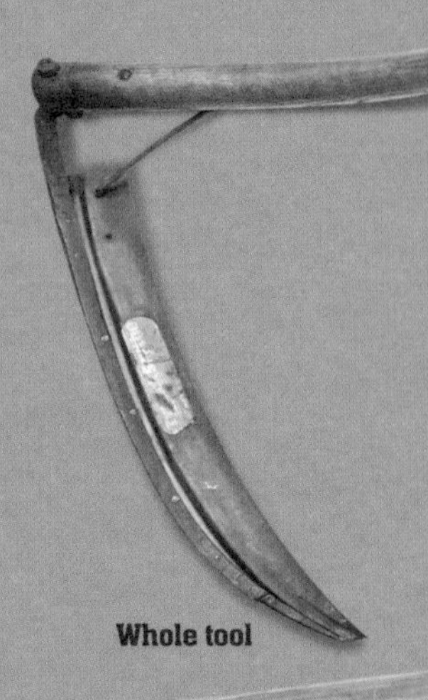

Whole tool

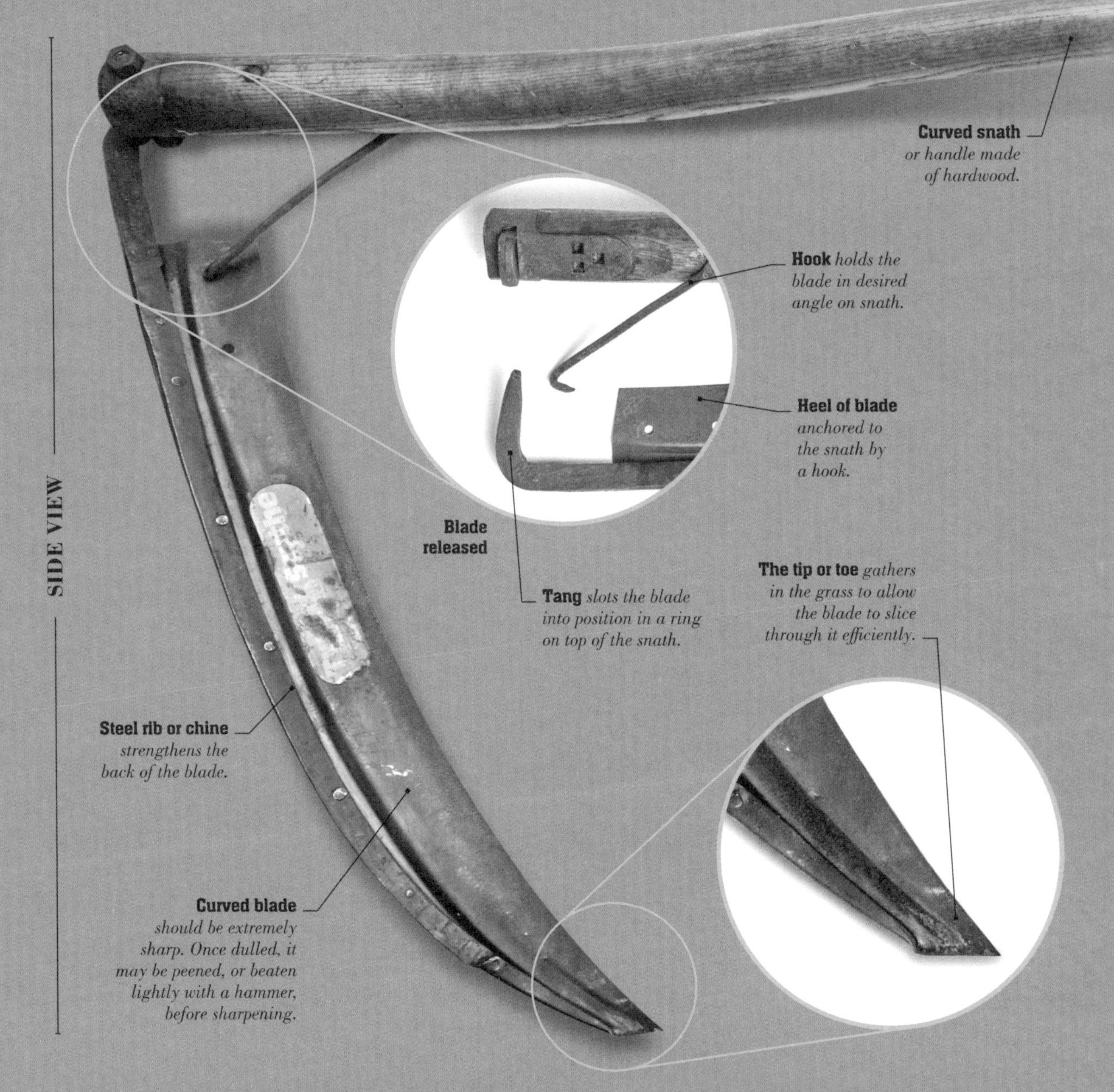

SIDE VIEW

Curved snath *or handle made of hardwood.*

Hook *holds the blade in desired angle on snath.*

Heel of blade *anchored to the snath by a hook.*

Blade released

Tang *slots the blade into position in a ring on top of the snath.*

The tip or toe *gathers in the grass to allow the blade to slice through it efficiently.*

Steel rib or chine *strengthens the back of the blade.*

Curved blade *should be extremely sharp. Once dulled, it may be peened, or beaten lightly with a hammer, before sharpening.*

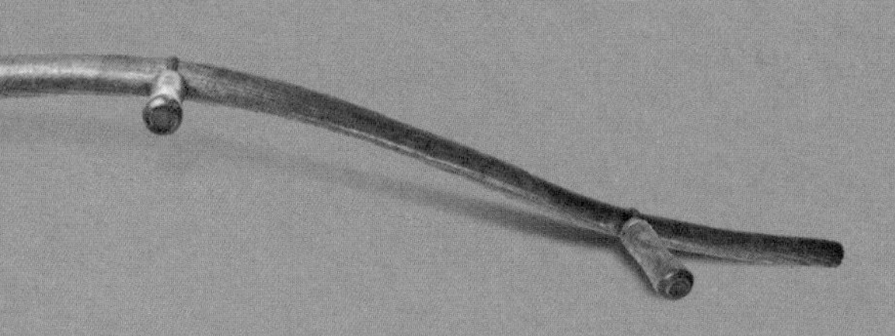

“A SCYTHE IS AN EXTREMELY EFFICIENT WAY TO CUT LONG GRASS, EVEN IN TODAY'S MECHANIZED WORLD”

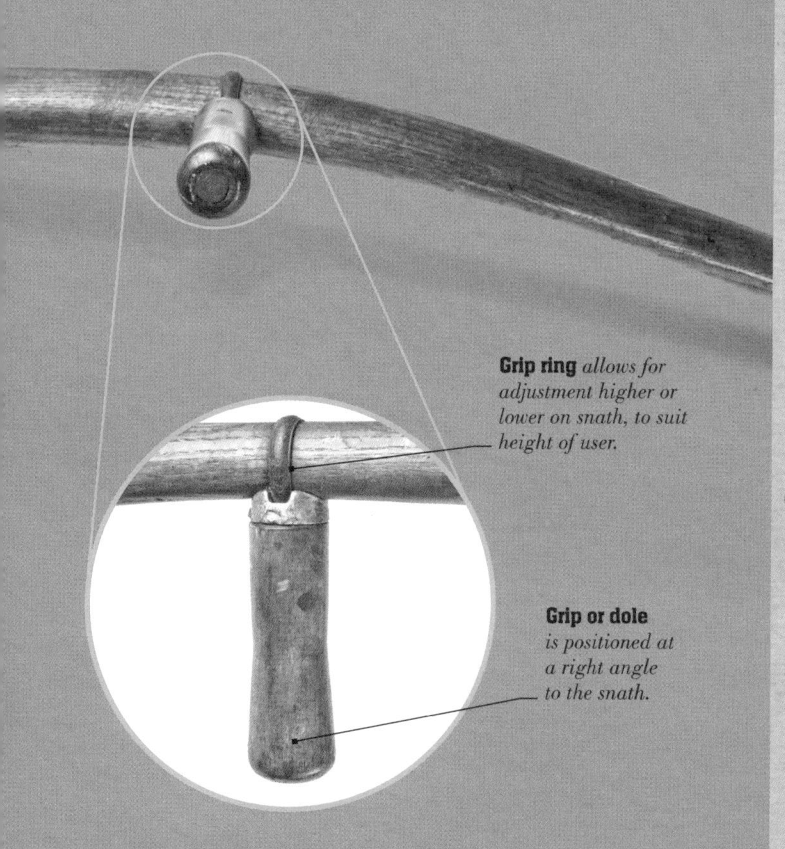

Grip ring *allows for adjustment higher or lower on snath, to suit height of user.*

Grip or dole *is positioned at a right angle to the snath.*

FOCUS ON...

Cutting Motion

The curved design of the scythe has been perfected over centuries, and it places the hand grips in the optimal position while at the same time presenting the blade at just the right angle. Instead of slashing horizontally, a scythe blade cuts in a sweeping arc, beginning as the operator swings it from his or her right side. As the blade moves left, it is swung under grass and long weeds, slicing cleanly through the stems, and depositing them to the operator's left.

USING AN

English Scythe

Every scythe must be adjusted to the user for best results, and for comfortable use. The key to success is getting your body rotations just right, performing a smooth arc, and a gentle shuffle forwards. Maintenance is essential, including frequent sharpening of the blade.

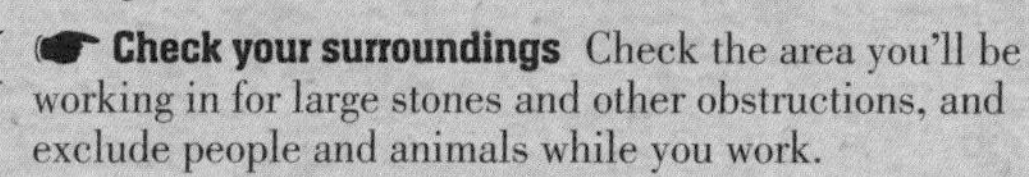

The Process

Before you start

- **Check your surroundings** Check the area you'll be working in for large stones and other obstructions, and exclude people and animals while you work.
- **Adjust the scythe** Make sure that the blade is sharp, and the handles have been adjusted to suit your height and grip.

1 Start slowly

The first cuts are the hardest. Hold the blade low and swing backwards to the right, rotating your body at the hips. Lightly swing the scythe and practise this action until it feels comfortable. Begin to shuffle slowly forwards with each swing. Ensure that the blade arcs in front of you, cutting in narrow bands.

2 Widen your arcs

As you progress, try rocking a little on your feet to make the arcs wider. Don't rush, and keep the blade tilted slightly upwards to avoid hitting the ground. Stop and sharpen it often, especially when working on rough grass. With the right action, you'll soon enjoy the job.

After you finish

- **Clean your tool** Clean the blade and inspect it for damage. Wipe the handle clean of any debris and check it over for cracks or splits.
- **Sharpen up** Before storing the scythe, sharpen the blade and wipe it over lightly with vegetable oil to prevent rust. Replace the blade's protective cover.

"THE BEST KNIFE
IS THE ONE YOU HAVE
WHEN YOU NEED IT."

ANONYMOUS

Choosing

Shears

Lawn shears

There are many different shears for many different tasks, but the same principle applies to them all: good quality endures and gives great results, while poor-quality tools frustrate and break. Hand shears often yield better results than mechanized versions because they achieve a finer finish. And forged steel can always be sharpened, while strong handles last a lifetime.

> "PROFESSIONAL GARDENERS USE JAPANESE SHEARS TO OBTAIN A PERFECT FINISH"

Grass shears

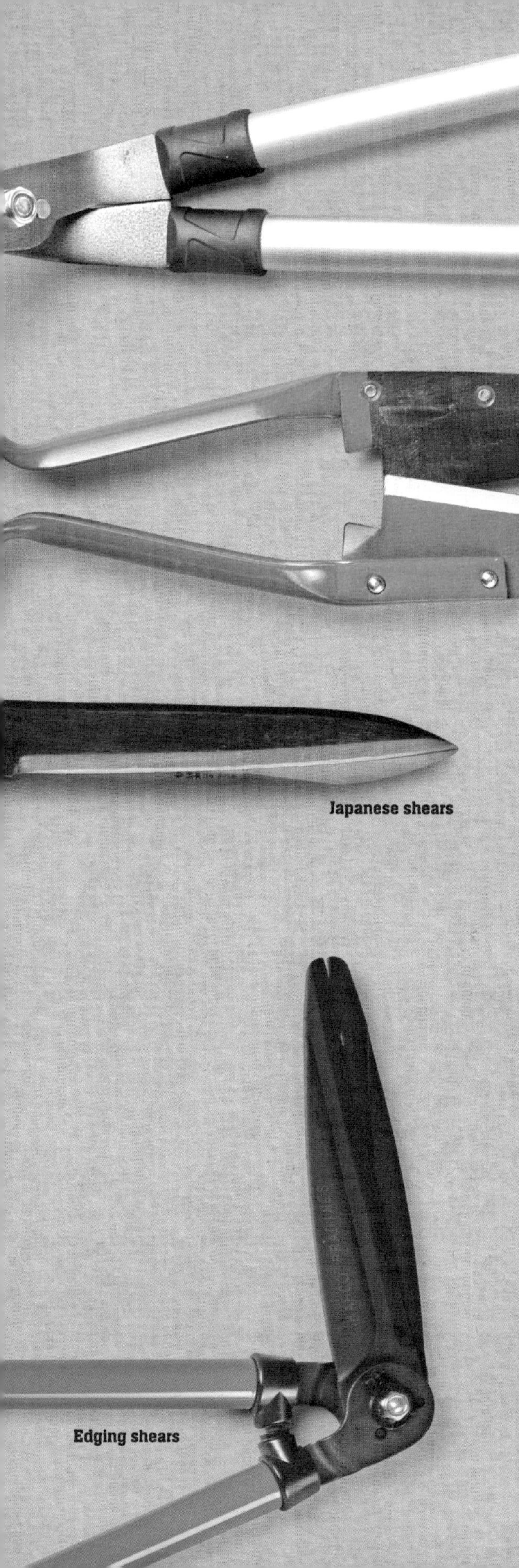

Topiary shears

Japanese shears

Edging shears

Japanese Shears

☛ **What it is** Shears made of the highest quality Japanese steel, with long, wooden handles. Simple, sharp, and efficient.

☛ **Use it for** Everything from detailed topiary trimming and cloud pruning to hedge cutting and heavier garden pruning and shaping.

☛ **How to use** Grasp handles in both hands and cut with scissor action. Keep them clean and sharp.

☛ **Look for** Correct blade length for task: choose short purely for fine topiary, longer for general use.

Lawn Shears

☛ **What it is** Very long-handled shears, with flattened blades set at 45 degrees to the handle, creating scissor action.

☛ **Use it for** Trimming lawn edges not easily accessed by a mower, such as under overhanging plants.

☛ **How to use** Position blades facing away from you. Use the scissor action to cut grass to lawn level.

☛ **Look for** Steel blades for sharpening and an adjustable pivot action. A comfortable handle length for your height.

Grass Shears

☛ **What it is** A tough, heavier version of scissors, but with one moving blade and one fixed.

☛ **Use it for** Trimming lawn edges in corners, cutting herbaceous plants, general tidying.

☛ **How to use** Use with one hand as you would scissors, but keep your free hand clear.

☛ **Look for** A smooth, serviceable pivot action as well as a size and weight that suit your hand.

Topiary Shears

☛ **What it is** Very small, sharp hand shears, ranging from simple "scissors" to more complex designs.

☛ **Use it for** Shaping topiary plants such as box or yew, as well as cutting back herbaceous plants.

☛ **How to use** Hold in one hand and cut slowly and carefully with scissor action to shape material. Keep free hand clear.

☛ **Look for** Steel blades that can be sharpened, and the right size handles that feel comfortable in your hands.

Edging Shears

☛ **What it is** Long-handled shears with blades set at 90 degrees, facing one direction, and touching the ground.

☛ **Use it for** Trimming lawn edges for a very tidy finish.

☛ **How to use** Keep shears as upright as possible, move only the handle attached to the upper blade, cut left to right. Watch your toes!

☛ **Look for** Best-quality material and construction and the correct handle length for your height – particularly important if you're very tall or very short.

Structure of

Topiary Shears

Hand topiary shears are simple tools for the close and precise pruning and finishing of topiary shapes, as well as for other delicate clearing and maintenance tasks. The simplest models are made from a single piece of high-quality sprung steel, and can be very sharp and durable. A wide mouth allows for efficient trimming.

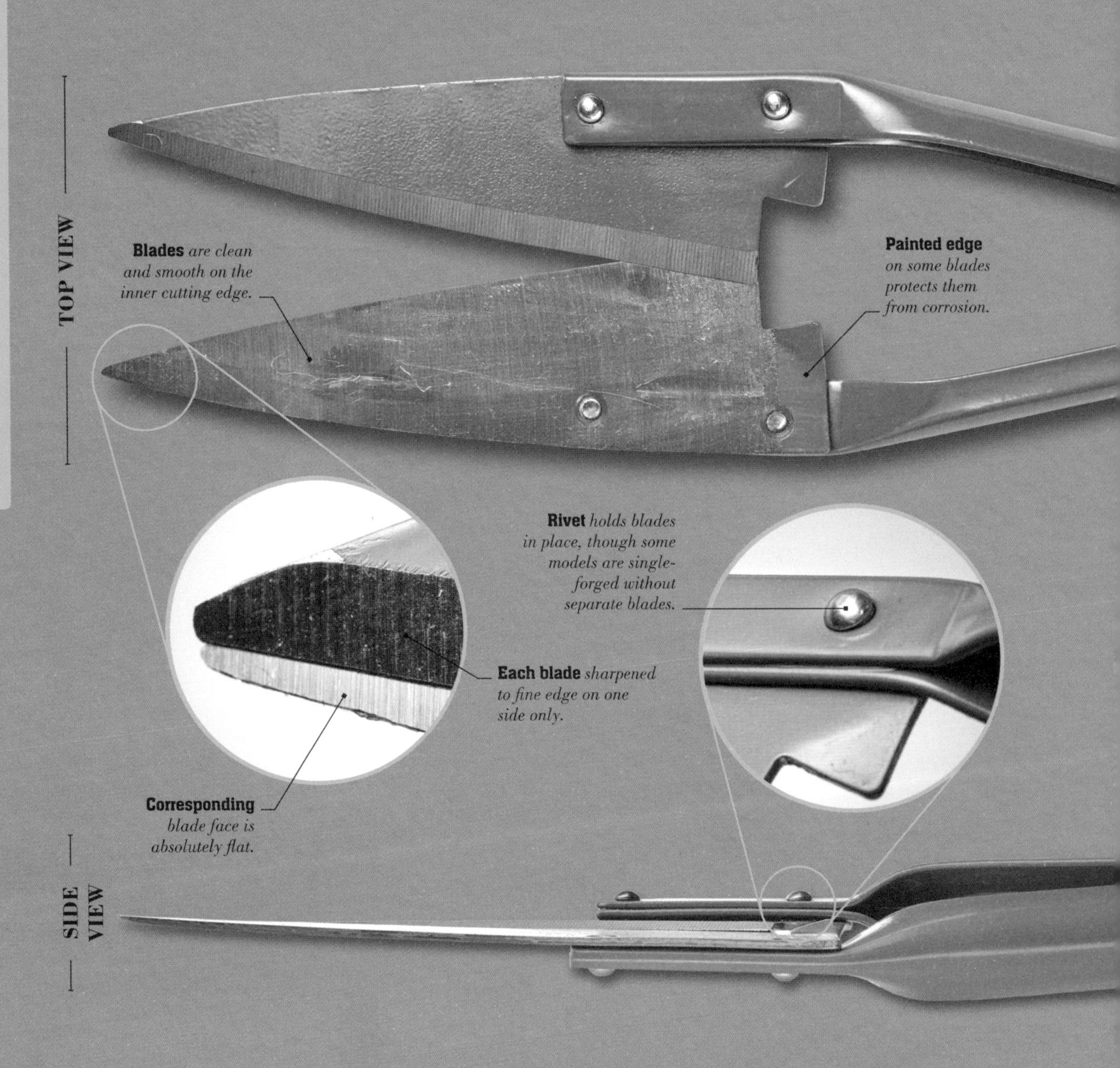

FOCUS ON...

The Springs

Although topiary shears are a simple tool in construction terms, there are a range of complex forces at play in their operation. The spring loop within the handle not only provides the scissor action up and down, but a second spring force draws the overlapping flat blade edges together. This ensures that the cutting edge stays tight from base to tip, enabling it to deliver a steady, clean cut with minimal effort.

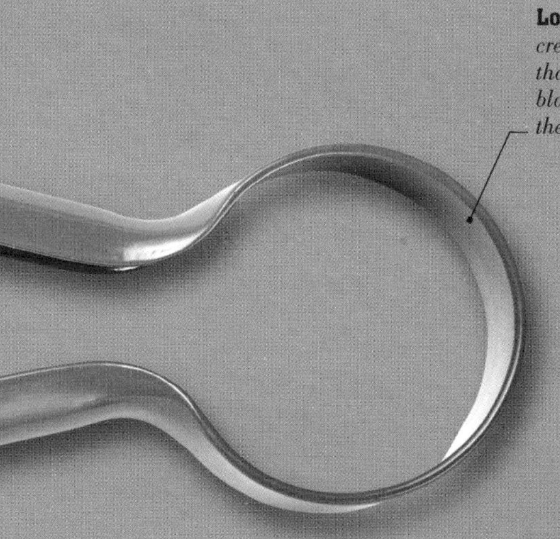

Looped handle *creates the spring that operates the blades, pushing them tightly together.*

“TOPIARY SHEARS ARE GREAT FOR FINE FINISHING, AND FOR REGULAR MAINTENANCE THROUGHOUT THE YEAR”

Handle *is comprised of single piece of shaped steel.*

USING

Topiary Shears

Topiary shears are best used on fine foliage such as box and yew, and are designed for routine tidying as well as fine detail work over small areas. They are also useful for trimming ornamental grasses, lavender, and soft herbaceous plants in borders.

The Process

Before you start

Inspect the blades Carefully check that the blades are sharp by cutting a piece of paper. Sharpen if needed. Look for damage. Check the scissor action is smooth.

Protect your hands These shears are very sharp, so wear gloves with a good grip and be conscious of where your free hand is at all times.

1 Plan your approach

Plants used in topiary often cut better on a cool, dewy morning, when the moisture content makes the foliage soft and more pliable. Begin by contemplating the desired effect, and have a target shape or form in your mind before making the first cut.

2 Make the shape

Work confidently, taking off less than you initially planned overall. Snip at the soft growth methodically, gradually creating the required shapes. Remember to step back and check results often, making corrections as you work. Remove cut debris as you go.

After you finish

Clean up Clean any debris off your shears and check for damage. Sharpen them if required, and carefully apply a layer of vegetable oil for protection.

Store the shears Wrap the open blades in a thick cloth for safety, allowing the spring to sit in the resting position when stored.

Choosing a

Pruner or Lopper

Lopping and pruning tools vary greatly, but all are made for cutting plant material. Some, such as loppers, are designed for thicker branches while secateurs are made for finer pruning and trimming; others are used for specialist tasks. For general pruning, one or two tools will do the trick.

Bypass secateurs

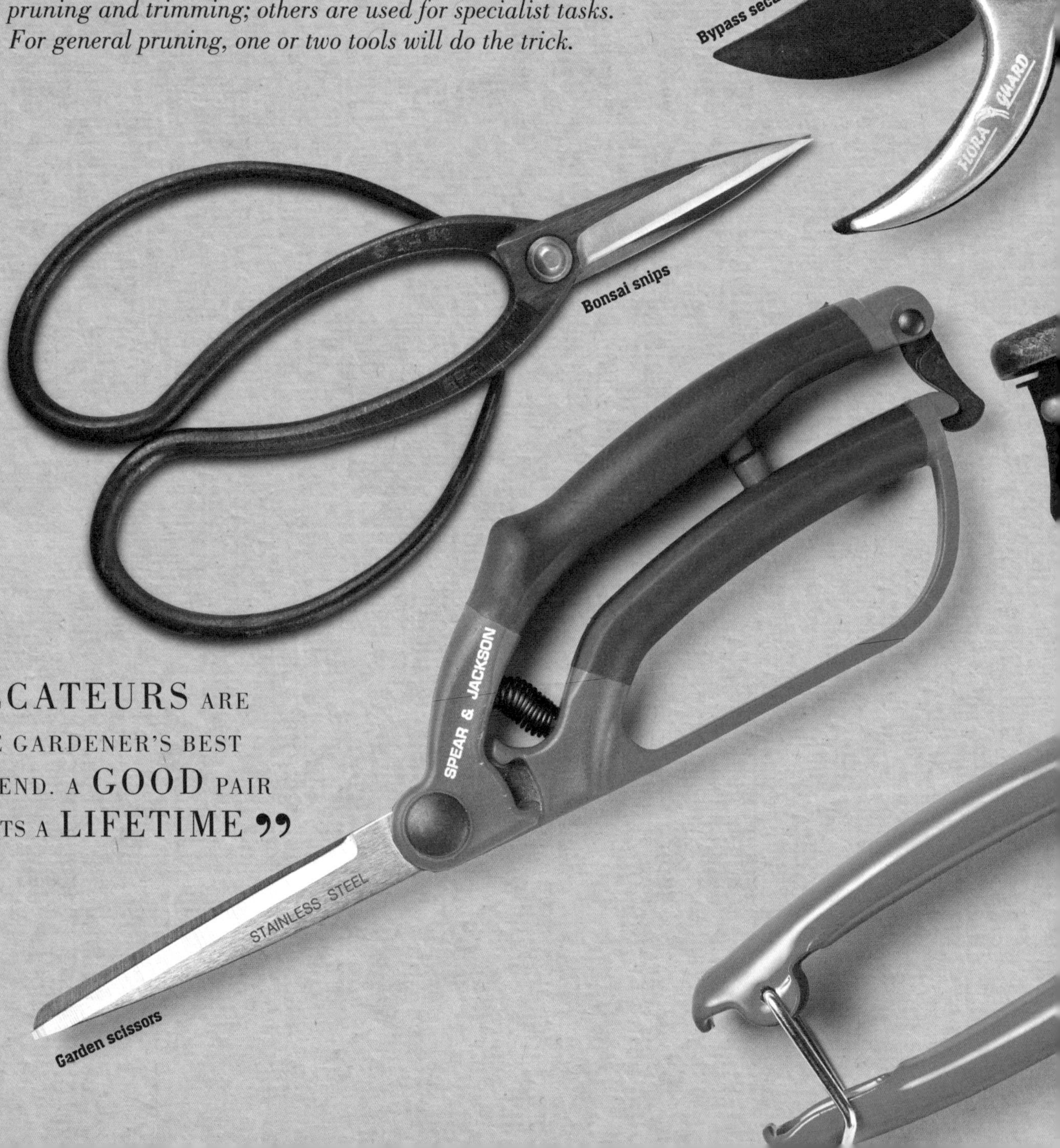

Bonsai snips

"SECATEURS ARE THE GARDENER'S BEST FRIEND. A GOOD PAIR LASTS A LIFETIME"

Garden scissors

> **"THE PURPOSE OF PRUNING IS TO IMPROVE THE QUALITY OF THE ROSES, NOT TO HURT THE BUSH."**
>
> FLORENCE LITTAUER

Bonsai Snips

What it is Like scissors with large handle loops; unsprung with sharp blades.

Use it for Specialist bonsai pruning or fine trimming and general use.

How to use Use like scissors. The large loops in the handle provide fine control of the scissor action.

Look for A fine blade with large handles and a simple action.

Garden Scissors

What it is Standard scissors that have been strengthened for garden use, often with a serrated edge.

Use it for Cutting string, plastics, and fleeces. Useful for harvesting cut flowers and for dead-heading.

How to use As you would general scissors, but don't overwork them in place of secateurs.

Look for Stainless-steel body coated with plastic to protect from the elements. Strong pivot action and large handles.

Flower Snips

What it is A fine hand tool with a scissor action and pointed cutting teeth.

Use it for Cutting flowers for floristry, dead-heading garden plants, and fine pruning of bonsai plants.

How to use Snips are very sharp and work much like scissors. Follow secateur guidelines when pruning.

Look for A clean and simple action, forged to a high standard. Japanese models are some of the best.

Bypass Secateurs

What it is Professional secateurs with a curved cutting blade and a hooked anvil that passes the blade to make the cut.

Use it for General pruning in the garden, propagation, and many cutting and trimming tasks.

How to use The bypass action produces a fine, clean cut. Open blades fully and cut, but don't rock if too thick.

Look for Best quality. Metal body, adjustable action, replaceable blades.

Anvil Secateurs

What it is Common secateurs with one sharp cutting blade that cuts plant material against a flat anvil.

Use it for General pruning and cutting of woody material in the garden. A cost-effective tool.

How to use With blades open wide, use the scissor action to cut through plant material. Do not twist.

Look for A metal body rather than plastic and a blade that touches the anvil completely for a clean cut.

CONTINUED

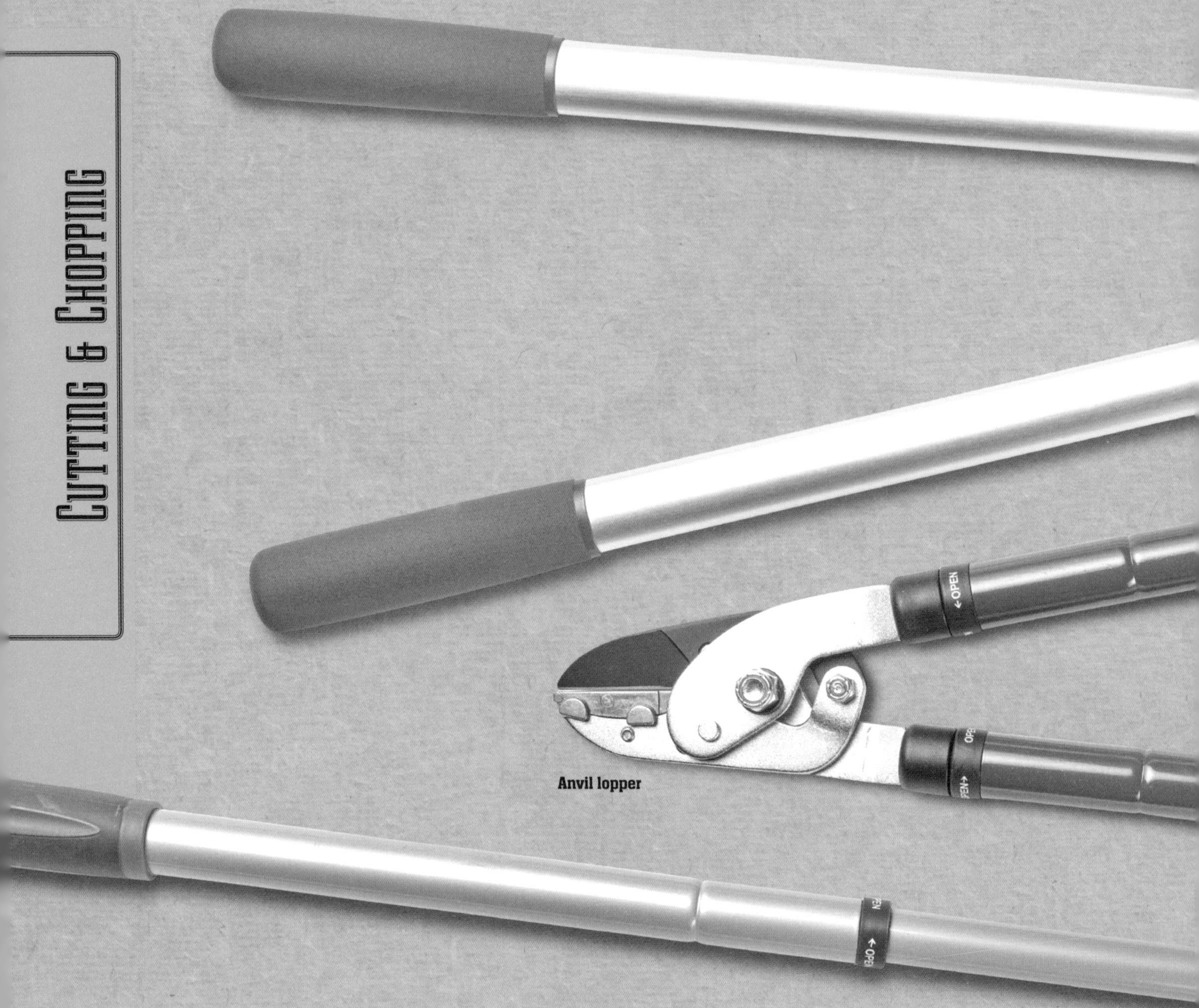

Anvil lopper

“Because of the hard, WOODY nature of the material they deal with, loppers can be among the first tools to LOSE THEIR EDGE. Keep them SHARP for optimal performance”

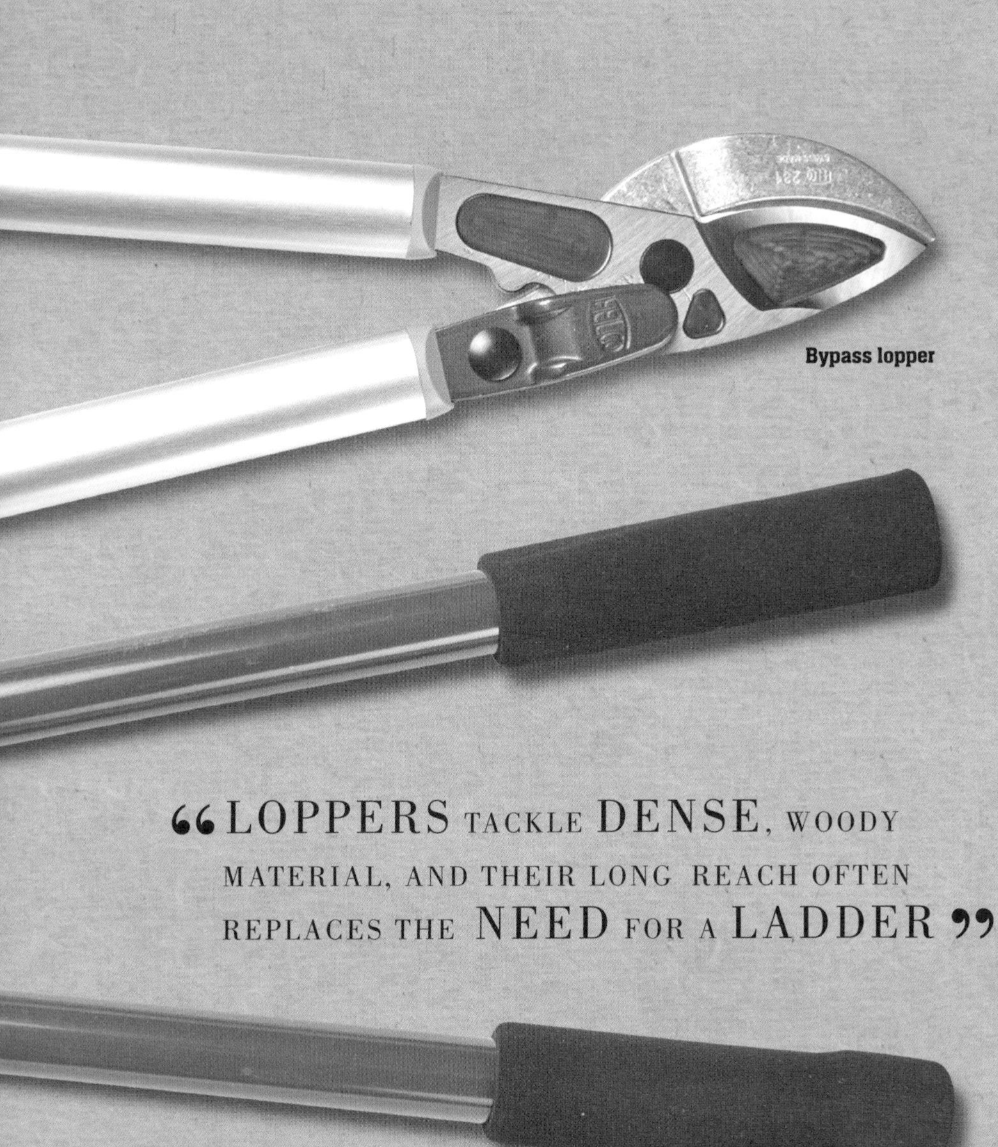

Bypass lopper

Bypass Lopper

☛ **What it is** A long-handled lopper with a bypass cutting head; comes in a range of sizes.

☛ **Use it for** Good, clean cuts when pruning thick, woody material.

☛ **How to use** Push stems deep into jaws, compress the handles.

☛ **Look for** Forged steel heads and blades, mounted or forged into the long handles.

Anvil Lopper

☛ **What it is** The workhorse lopper. Like anvil secateurs, this tool has one sharp blade that crushes plants onto an anvil.

☛ **Use it for** Tough, grubby, hard-clearance jobs. Robust, so good for cutting roots, hedge stems, and suckers.

☛ **How to use** Open handles wide, push the material as far back into the teeth as possible. Do not twist side-to-side!

☛ **Look for** A simple mechanism. Over time this results in less wear and less play in the tool.

> “LOPPERS TACKLE DENSE, WOODY MATERIAL, AND THEIR LONG REACH OFTEN REPLACES THE NEED FOR A LADDER”

Long-handled Lopper

☛ **What it is** A lopper/pruner with a very long shaft and a variety of cutting heads and operating handles.

☛ **Use it for** High pruning of small branches in trees, particularly fruit trees.

☛ **How to use** Reach high to the desired branch, ensure loppers are placed well, then make the cut. Don't try to cut anything too thick or the tool blades may get stuck.

☛ **Look for** A simple mechanism throughout and a strong cutting head. Lightweight build for ease of use.

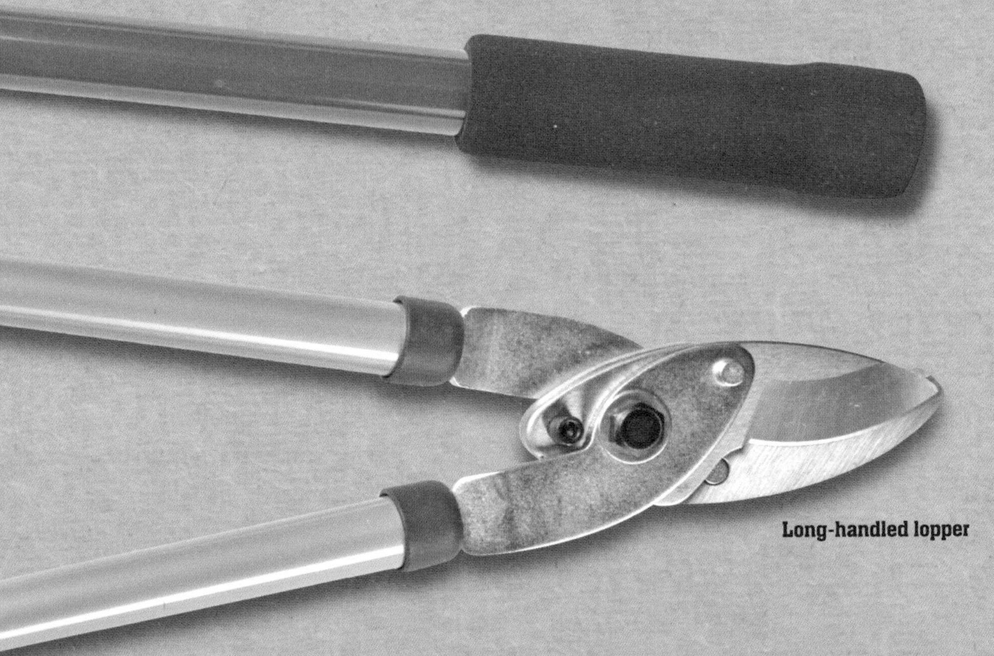

Long-handled lopper

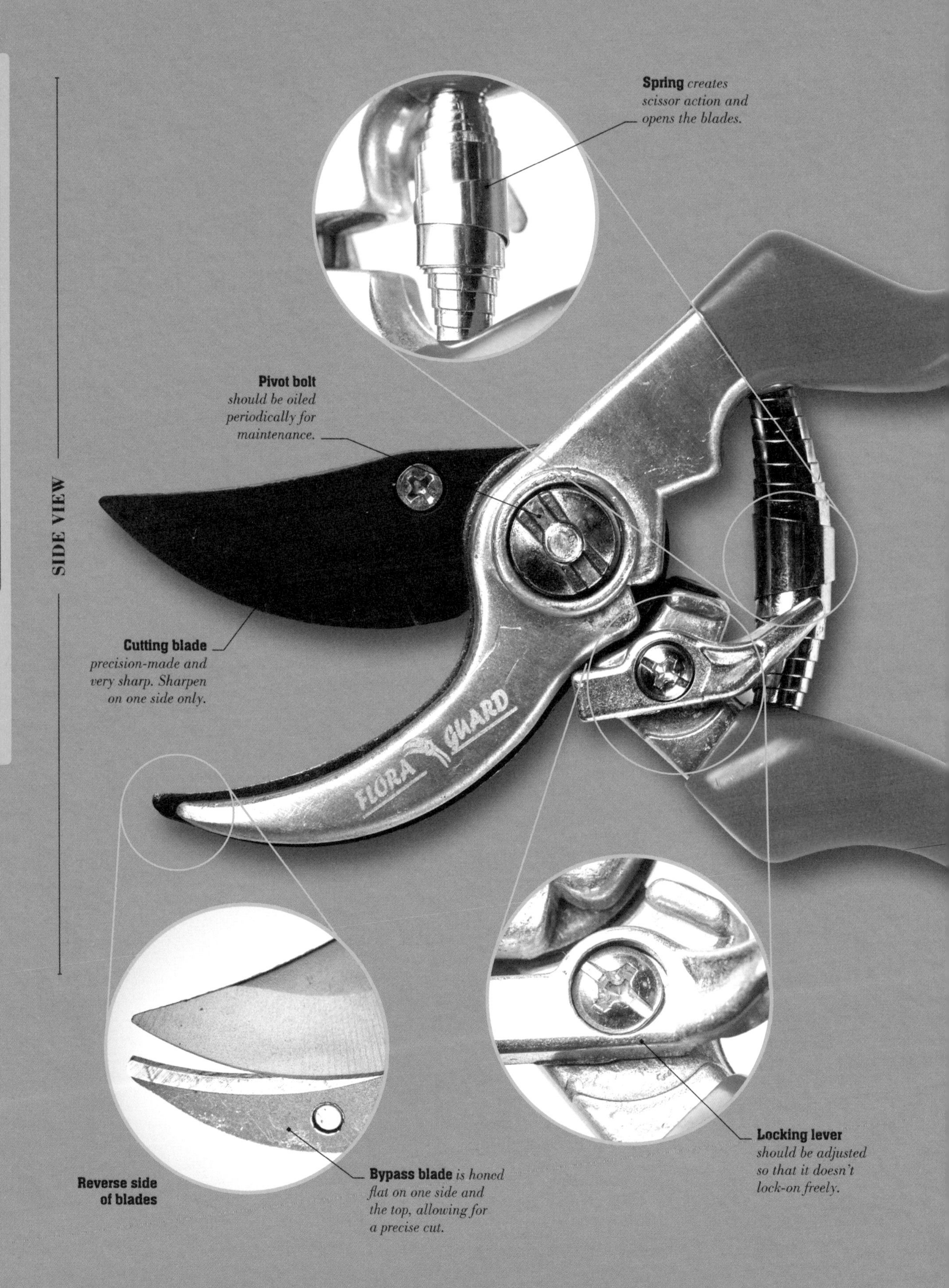
SIDE VIEW
Spring *creates scissor action and opens the blades.*
Pivot bolt *should be oiled periodically for maintenance.*
Cutting blade *precision-made and very sharp. Sharpen on one side only.*
FLORA GUARD
Locking lever *should be adjusted so that it doesn't lock-on freely.*
Reverse side of blades
Bypass blade *is honed flat on one side and the top, allowing for a precise cut.*

STRUCTURE OF Bypass Secateurs

Quite simply the best and most versatile pruning tool for gardeners, bypass secateurs have a sharp, curved blade that passes accurately by a curved anvil section. Ergonomic handles that easily lock and open effortlessly make for easy use.

Bright handle *colours prevent loss when in the garden.*

“PRODUCING CLEAN CUTS, BYPASS SECATEURS ARE THE IDEAL CHOICE FOR ACCURATE AND FINE PRUNING OF TREES AND SHRUBS”

Handles *are ergonomically shaped and often covered with vinyl or plastic for softer grip.*

FOCUS ON…

Secateur Shapes

Secateurs come with a range of heads and handles. Try them out before you buy, either in a shop or garden centre, or ask your gardening friends if you can try theirs. Some are designed for tight work, some for repeated tasks, some for durability over accuracy.

Bypass Bypass heads are the most accurate secateur type. They are also very strong. Some models come with adjustable grip widths.

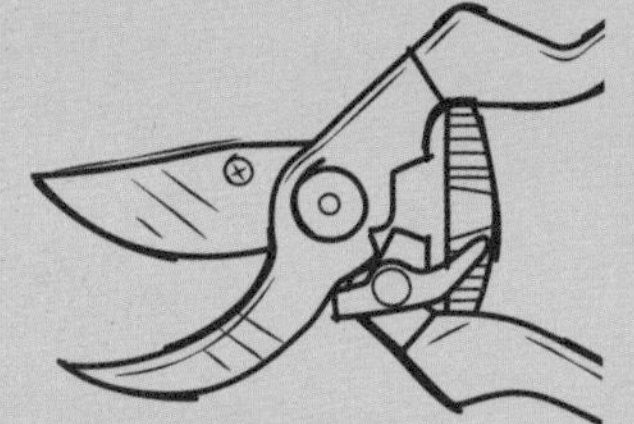

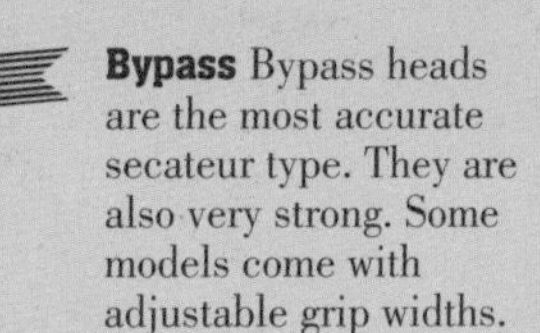

Anvil The anvil head is tough and durable, but less accurate. Its cutting blade presses onto a metal anvil like a knife onto a chopping board.

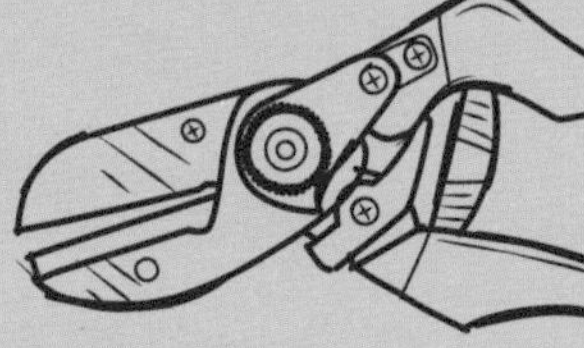

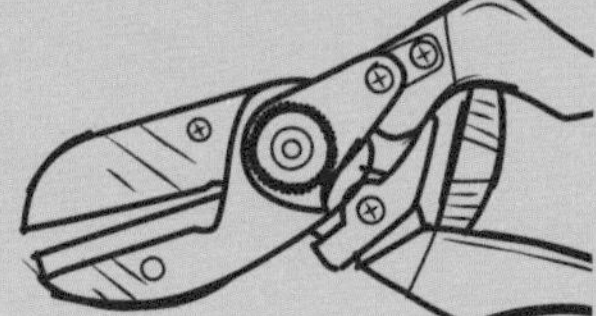

Flower snips Snips are useful pruners to have for delicate stems. The sharp, narrow blade is perfect for fine pruning or dead-heading flowers.

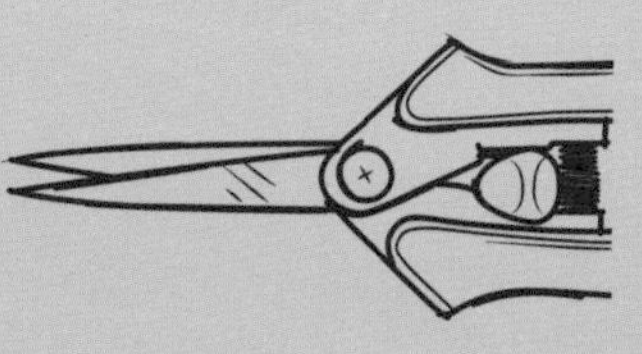

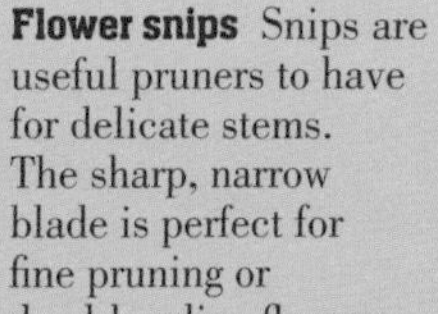

USING

Bypass Secateurs

Bypass secateurs are the gardener's best option for the fine and accurate pruning of anything woody up to 25mm (1in) thick. The simple but strong design means that they are able to slice cleanly through plant material with ease. Although secateurs are usually right-handed, left-handed options are available, as well as a range of handle sizes.

The Process

Before you start

Check the blade Make sure that the blade is clean and sharp; the task will go more smoothly if it is, and a clean blade won't spread disease.

Try the action Ensure the action of the blades is smooth, and that the lock is not loose. Adjust as required.

Get a carrier A holster is very useful for keeping secateurs to hand when working in the garden; if not, use a bucket, tool-belt, or apron.

1 Assess the plant

Before you begin pruning, research the plant if necessary to assess its particular pruning needs – you need to know how much to take off and when as well as where. Decide where the first cut should be located. Aim to prune back to dormant buds or close to roots.

2 Get in position

When pruning to a dormant bud position, the blade should be positioned above and at a slight angle close to the bud, with the bypass blade on the other side of the stem to avoid snagging the bud. Position the stem to the rear of the jaws so maximum leverage can be achieved. This will also give the tightest and cleanest-possible cut.

FOCUS ON...

Bypass Action

Good secateurs reflect the quality of engineering and materials used, and should work like a pair of scissors, with one blade passing the other to make a slicing cut. Plant material sits in the groove between the cutting blade and curved bypass blade. The extremely sharp cutting blade is then drawn down through the material, gliding past the bypass blade and shearing the plant material in the process.

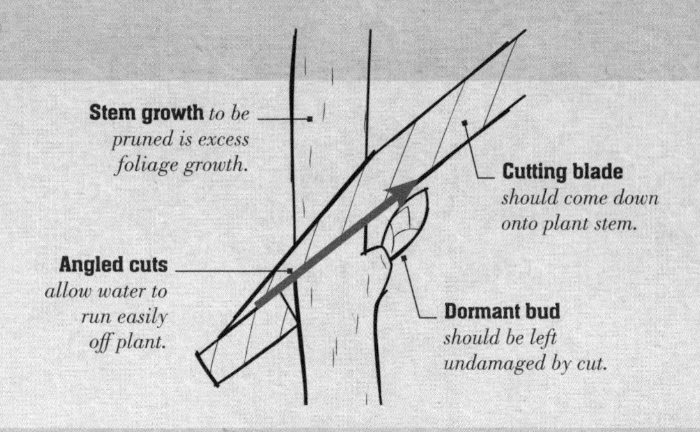

3 Make the cut

With a firm and steady hand, make the cut, which will be smooth and clean if the blades are sharp and well adjusted. Do not be tempted to rotate the handles or twist sideways if the plant stem is too tough. If you do, you will damage both the plant and the secateurs.

4 Close and lock

Once you have made the cut, squeeze the handles again to close the secateurs. Most models have a locking mechanism to keep the blades closed safely. Simply press the locking mechanism with your thumb so that it turns and holds the blades together. Keep blades closed whenever they are not in use.

“BYPASS SECATEURS ARE THE BEST CHOICE FOR PRUNING LIVING PLANT TISSUE, DUE TO THEIR CLEAN, SHARP CUTTING ACTION”

After you finish

- **Clean up** Wipe down both blades and disinfect them if necessary to ensure cleanliness after use. Great results can be achieved by washing secateurs in a dishwasher. Wipe them over with a drop of vegetable oil afterwards before locking them closed.
- **Stow them away** Store the secateurs in a safe place, but one that is easy to access as you go out into the garden.

MAINTAIN TOOLS FOR

CUTTING & CHOPPING

A blunt edge cannot make a clean cut and makes the task difficult and even dangerous. Maintenance of these tools is essential and can be satisfying.

SHARPENING BLADES

Most tools have a simple edge, which is easily sharpened. Little and often is best, combined with correct use.

1 Check blade
Blades with a single edge, without teeth or serrations can easily be kept sharp, and must be inspected for nicks or warping. Note if the blade is has a bevel on one side or both.

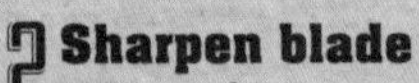

Check edge *of knife blade is sharp before use.*

2 Sharpen blade
Using a sharpening stone, fine file, diamond sharpener, or similar, carefully run the stone or file along the blade. Follow the angle of the edge, moving the stone from the outer edge into the blade.

3 Finish edge
Several strokes with a sharpener after each use are usually enough. More filing may be needed after time to improve the angle. In this case, remove burrs (raised ridges) in the edge with a final stroke of the file on both sides.

TREAT WITH CARE

Saws are vital for many tasks, and the sharper the teeth edges, the cleaner and more accurately they cut. Sharpening individual teeth is time-consuming, but careful use can reduce the need for this.

Cut clean
Saw teeth are blunted by dirt, grit, stones, and sheet materials. Avoid cutting dirty material like roots, laying saws on the ground, or accidentally cutting into soil. Wash any dirt off immediately.

Cut sharp
Care is needed with a sharp saw, but struggling with a blunt saw can cause accidents. It also makes for poor cuts in woodwork, and in the garden using a blunt pruning saw reduces plant health.

TOOLS	INSPECTION
SAWS	▪ Wipe down handle and blade to remove any debris after use
AXES	▪ Check handle for damage after use, including splits or fractures ▪ Check for play in axe head, causing it to wobble slightly or even slip off ▪ Check cutting edge for dents, nicks, and blunting
KNIVES	▪ Check that blade is sharp, as blunt knives are not safe. When using, if a blade is not working as you expect it to, it's not sharp enough ▪ Look for damage to blade
SCYTHES & SICKLES	▪ Scythes and sickles are best maintained while you work, and therefore the extra tasks are infrequent ▪ Check blades for damage, including little dents or missing chunks
SHEARS	▪ Shears rely on scissor action to work well, so check if mechanism needs adjusting ▪ Check blades for sharpness
LOPPERS & PRUNERS	▪ Check mechanism is working smoothly and adjust if necessary ▪ Inspect blades for sharp edge, and for damage to edge caused, for example, by catching wire or stones

	Cleaning/Oiling	Sharpening	Joint Care	Storage
	▪ Wash dirt from saw teeth with water and dry ▪ Use fine wire wool to remove light rusting ▪ Unless lacquered, wooden handles and metal blades should be oiled regularly, especially if wood is dry	▪ Usually only teeth on traditional handsaws can be resharpened – hacksaws and hardpoint saws cannot be resharpened. ▪ File edges of each tooth to create sharp point – keep angle equal for each tooth	▪ Loose handle on traditional saw can be tightened if fitted with screws	▪ Keep storage area clean and dry to prevent rust ▪ If you can't hang your saw, keep it in a toolbox or drawer – make sure the blade can't be damaged by anything that touches it ▪ Put a cork or rubber mat at bottom of drawer so metal doesn't rub ▪ Use blade covers when not in use
	▪ Axes rarely need much cleaning, but they can be lightly washed or brushed off ▪ In damp conditions, smear with vegetable oil to keep rust off surface	▪ Keep shaping or felling axes sharp by using a flat file or whetstone	▪ Play in head/handle is usually caused by wooden handle drying out too much: first, with the head uppermost, tap handle on floor – then soak head/handle in water to swell head	▪ Store sharp axes with axe-head sheath ▪ Store in dry, airy shed but not in heated or exposed store – ambient humidity keeps wood hydrated ▪ Store where head will not fall and cause injury
	▪ Keep knives clean as you go, and work clean to prevent blunting of edge ▪ Use fine wire wool to remove light rust or stuck-on dirt ▪ Smear with vegetable oil to protect blade	▪ For best results, sharpen with a whetstone or diamond stone ▪ Sharpen little and often to keep blade keen ▪ Kitchen sharpeners are often a good alternative	▪ Folding knives need to open smoothly, and not fold too easily – keep mechanism oiled but clean ▪ Ensure locking mechanisms function well to avoid accidental closure	▪ Ensure folding knives are clean and always closed when not in use ▪ Use a sheath to protect rigid knives ▪ A knife, when out working, is best stored in your pocket, and always there when needed ▪ Store knives safely when not in use in a dry place and with edge enclosed
	▪ While working, blade can be cleaned by running a fistful of grass down blunt edge to remove build-up ▪ Alternatively, wipe with a wet rag, thick leather gloves, or a stiff brush	▪ Sharpen hooked blades with whetstone with curved or rounded edge – do this little and often while working, to maintain razor-sharp cutting edge	▪ Check scythes are adjusted for you, the user – any discomfort in use is likely to mean that it is not set correctly	▪ As it is not used frequently, a scythe or hook is best cleaned and then treated with a smear of vegetable oil before storage – use protective sheath if you have one ▪ Scythes are awkward to store, and dangerous if they fall – hang on custom-made hooks in a dry store ▪ Store sickles and other hooks in their own box, or hang on dedicated wall hook
	▪ Wash shears often in clean water to keep cutting edges clean and cutting action smooth ▪ Remove hard dirt with fine wire wool or wire brush	▪ If shears are treated well they should not need sharpening; however, if they become blunt, use a flat file or whetstone on blades	▪ Oil mechanism from time to time ▪ Budget shears often work loose, and play will develop in the "hinge"; tighten and lock-off to adjust	▪ Wipe with vegetable oil, and store safely in protective sheath to keep edge from damage ▪ Large shears like edging shears can be awkward to store and are best kept on a bespoke hook in shed ▪ Store hand shears on shelves or in tool bins, but ensure they are locked shut or put away closed
	▪ Wash with water, and for small hand tools, occasionally in dishwasher ▪ Clean tough dirt with fine wire wool	▪ Only sharpen one edge of tool if that's how it was designed. ▪ Use flat file, whetstone or specially designed sharpener. ▪ Higher-quality blades may be replaceable	▪ Most loppers and pruners allow for subtle adjustment to hinge mechanism – set this so there is no play, but without great friction ▪ Ensure lock mechanism does not keep locking on in use	▪ Using a holster for secateurs is convenient and safe ▪ Store tools locked closed for safety and to protect edges ▪ Hang up loppers or store face-down in tool bucket

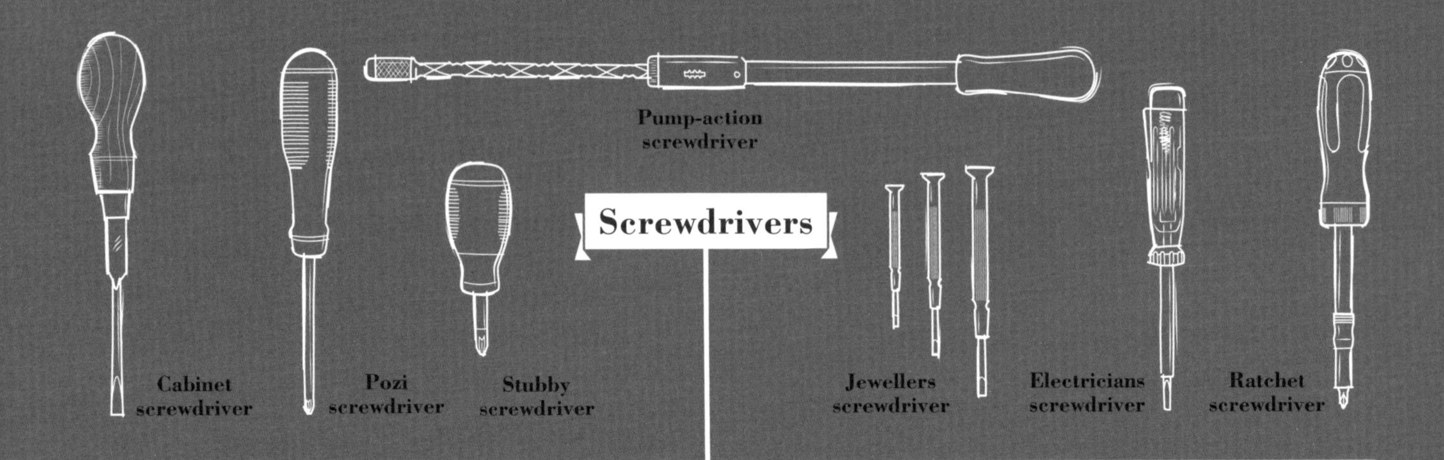

THE TOOLS *for* FIXING & FASTENING

Each toolkit needs a set of screwdrivers and every workshop needs a vice or clamp to secure work. From putting up shelves to changing tyres, fixing and fastening tools are everyday essentials.

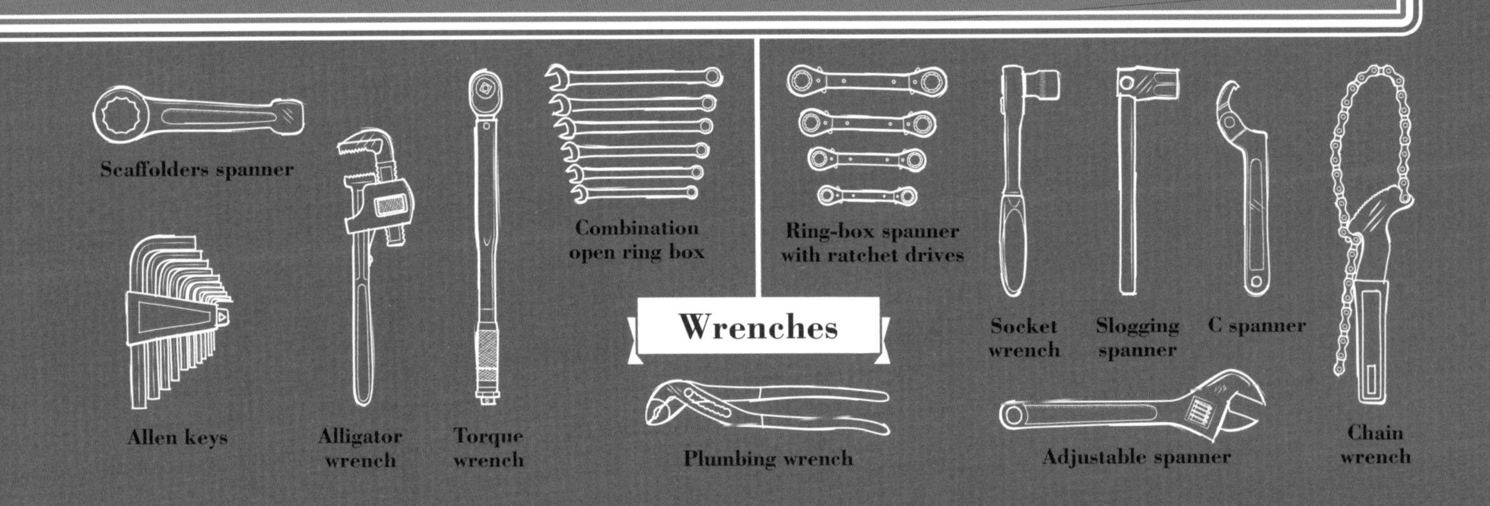

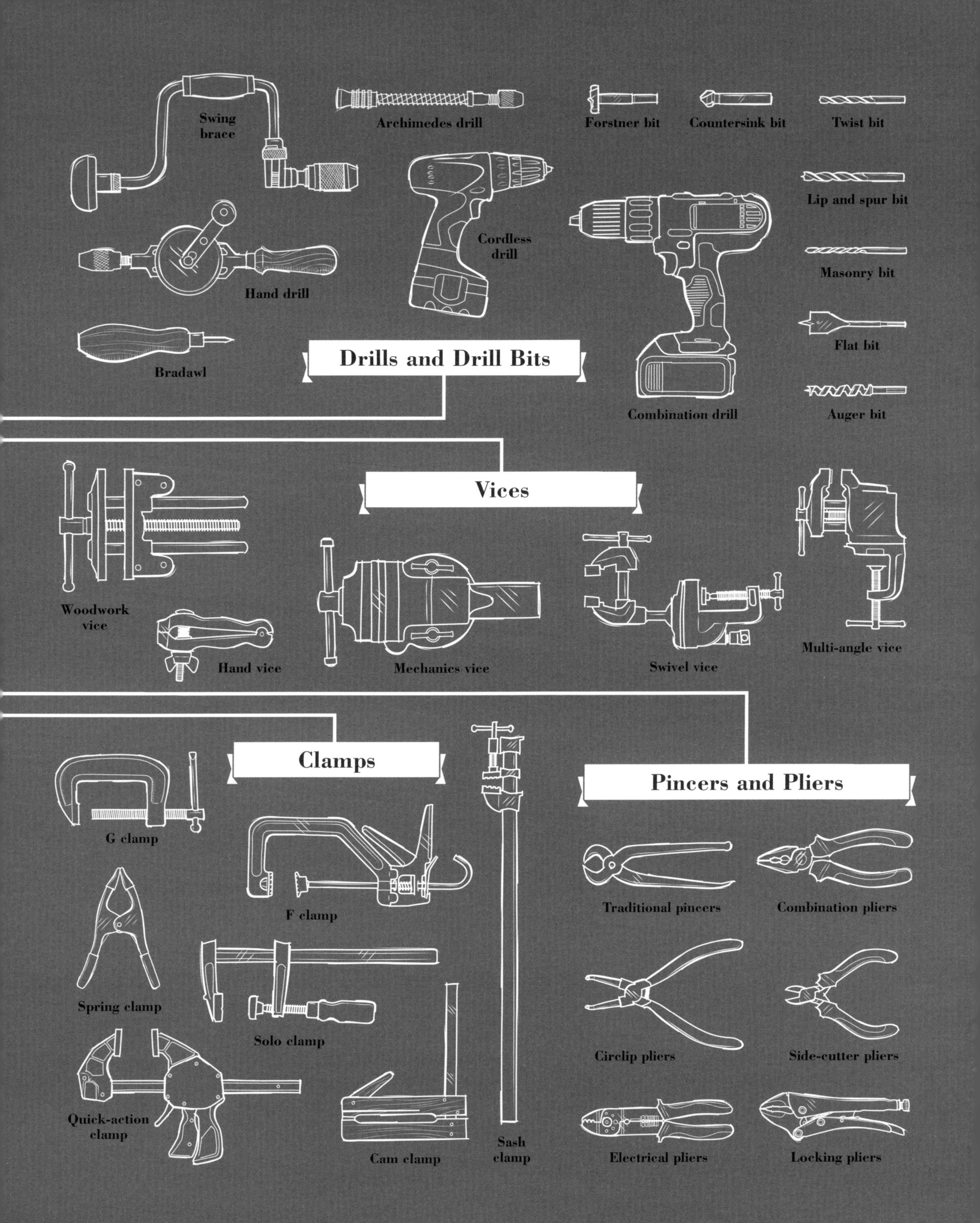

Swing brace
Archimedes drill
Forstner bit
Countersink bit
Twist bit
Hand drill
Cordless drill
Lip and spur bit
Masonry bit
Flat bit
Bradawl
Drills and Drill Bits
Combination drill
Auger bit
Vices
Woodwork vice
Hand vice
Mechanics vice
Swivel vice
Multi-angle vice
Clamps
Pincers and Pliers
G clamp
F clamp
Traditional pincers
Combination pliers
Spring clamp
Solo clamp
Circlip pliers
Side-cutter pliers
Quick-action clamp
Cam clamp
Sash clamp
Electrical pliers
Locking pliers

HISTORY OF FIXING & FASTENING

c.40,000 YA — BOW DRILL

The bow drill was developed in the Paleolithic period, when a slack bowstring was wrapped around a straight stick. By moving the bow back and forth quickly, the string rotated the stick, creating enough friction to start a fire when dry grass was added at the base.

c.3000–1900 BCE — EARLY PINCERS

Smelting was invented in the Bronze Age, when many new or improved tools were created, including the first tweezers, an early form of pincers.

Bronze replaced wooden sticks.

Wide ends used for grasping objects.

Early tweezers

7000 BCE

Small bow drills were being used in the Indus Valley (modern western Pakistan) – for dental procedures.

c. 3000 BCE — EARLY PLIERS

Sticks were probably the earliest form of plier-like holders, but during the Bronze Age, bars of bronze replaced wooden tongs as an early form of pliers, possibly developed as a response to handling hot objects such as coals.

c.1000 BCE — BASIC AUGER

An early form of auger for enlarging holes emerged during the Iron Age. It consisted of a pipe split vertically, which was joined to a crossbar that could be turned with two hands. The end was either a sharpened half-circle or a spoon shape with sharpened edges.

735 BCE–500 CE — PUMP DRILL

The Romans refined the pump drill, which consisted of a bow-like crosspiece that slid up and down a spindle. Cords attached to the crosspiece were wrapped around the spindle. Pushing down on the crosspiece rotated the spindle, and the weight of a flywheel kept it spinning. As the cords reversed direction, the crosspiece lifted as the drill slowed.

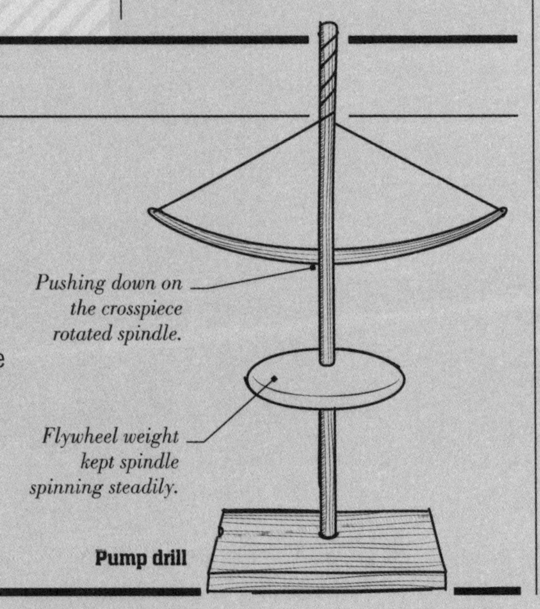

Pushing down on the crosspiece rotated spindle.

Flywheel weight kept spindle spinning steadily.

Pump drill

> "ALL I HAD WAS A DRILL, AN ELECTRIC DRILL. THAT WAS THE ONLY MACHINE I HAD."
>
> JAMES DYSON
> BRITISH INVENTOR AND INDUSTRIAL DESIGNER

c.500–1500 EARLY VICE

Craftsmen in medieval times often strapped their workpieces to trestles or small benches to secure them. The strap was tightened by a craftsman placing his foot into a loop under the table.

1400s NUTS AND SCREWS

Metal nuts and screws were developed around this time. Square and hexagonal nuts and bolt heads were turned with special box wrenches that were designed to fit snugly onto the heads.

26 SCREW DRIVES

The number of different ways a screw or fastener can be turned, ranging from simple slots to Pozidrivs to the five-pointed pentalobe security screw used in personal computers.

1500s SCREW VICE

Metalworkers began using small screw vices to secure their workpieces. The vice, which was tightened with a nut and bolt, consisted of a hinge, with one jaw fastened to a bench while the other was pulled up to hold the workpiece.

1500s SOCKET WRENCH

An early form of socket wrench with a T-shaped handle was developed in the 16th century, but each example fitted only a specific size of nut or bolt. These socket drivers were often used to wind early clocks.

1800s MONKEY WRENCH

A refinement of the sliding-jaw wrench, the monkey wrench was developed in the 19th century. Instead of using a wedge, the adjustable jaw was held in position by a screw. This is the forerunner of the modern crescent wrench, which is a much thinner tool.

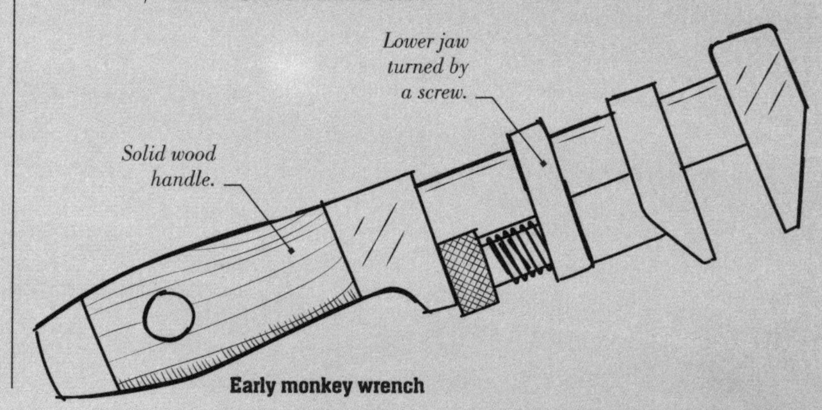

Early monkey wrench

1805 GEARED DRILL

The first geared hand drill was an improvement on bow and pump drills because its unidirectional bit could be turned much faster by the gears.

1860 IRON CRANKS

The commercial introduction of iron "sweeps" or cranks meant holes up to 2.5cm (1in) across could be drilled. Two-handed augers were still needed for larger holes.

POWER DRILLS THEN & NOW

**1916 Black & Decker drill weight: c.10kg (22lb); cost: £173/ US$230 = £4,055/US$5,391 today.
2017 Cordless drill weight: c.1kg (2lb); cost: c.£75/US$100.**

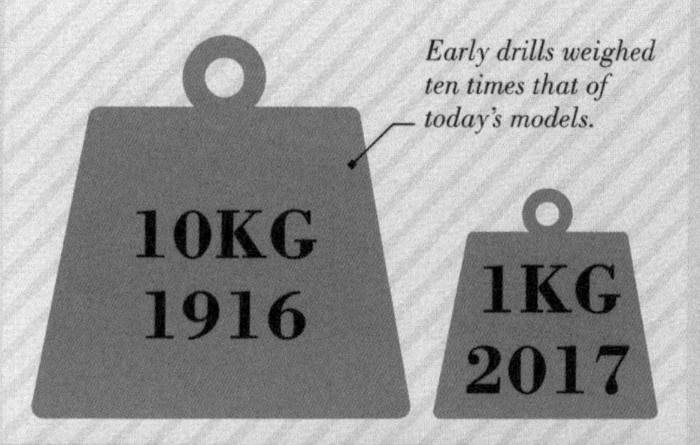

1917 POWER DRILLS

The first electric drill was invented by Australian Arthur James Arnot in 1889, but it was Black & Decker that patented the first portable pistol-grip drill in 1916. The drill had the now-familiar trigger switch of modern cordless drills.

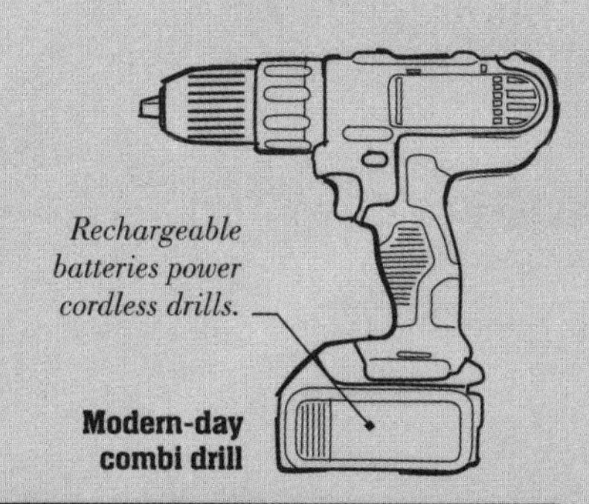

Modern-day combi drill

Choosing a
Screwdriver

With an increasing variety of screw patterns emerging onto the market, it has become more important than ever to match the tip of any driving tool to the screw head. The blade size should also be considered, as you can damage the head of the driver if it doesn't fit properly. It is best to have several different screwdrivers in your toolkit in order to deal efficiently with a variety of day-to-day DIY tasks.

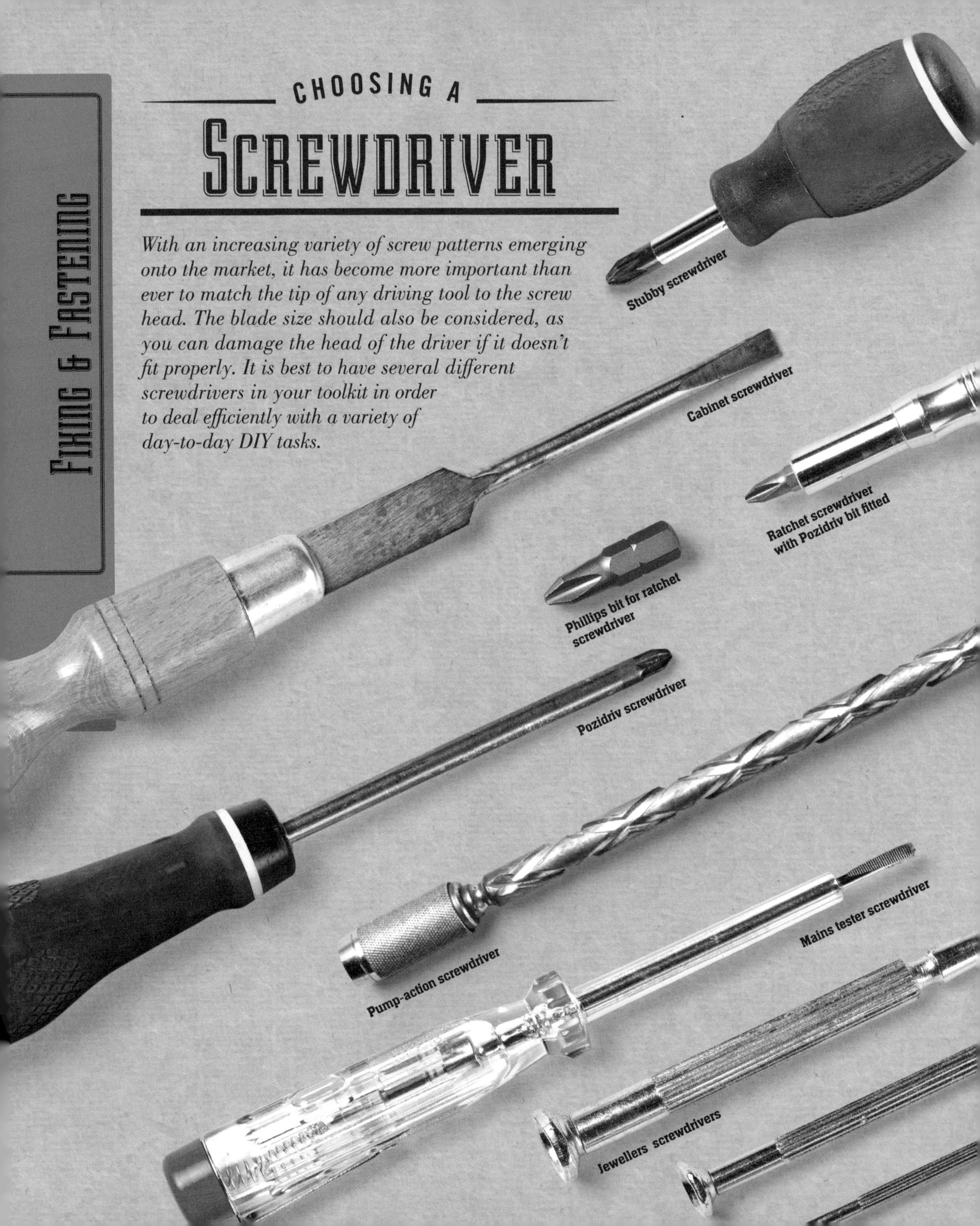

Pump-action Screwdriver

What it is Spiral-action tool designed for speed, with interchangeable bits.

Use it for Rapid insertion of slotted or cross-headed screws.

How to use Insert required bit in end of tool. Select ratchet button and operate handle with up/down action.

Look for Adaptors that enable modern hex bits to be used with this older-style tool.

“CORDLESS DRIVERS ARE POPULAR, BUT YOU’LL ALWAYS NEED A MANUAL SCREWDRIVER”

Cabinet Screwdriver

What it is A traditional tool with slotted or Pozidriv blade; comes in several sizes to match screw heads.

Use it for Inserting or removing screws in furniture; cabinetmaking.

How to use Insert blade tip in screw, turn clockwise or anticlockwise. The handle is designed for increased torque.

Look for An oval section handle will not roll off surfaces. Check slotted-pattern blade tip matches width of screw slot.

Ratchet Screwdriver

What it is Driver with hardwood or plastic handle, slotted or cross-headed blade. Often has interchangeable hex-shank bits, like Pozidriv or Phillips.

Use it for Driving or removing screws with easy change of direction.

How to use Insert blade tip in screw head. Select clockwise or anticlockwise drive and rotate handle.

Look for Hex bit storage in combination models.

Mains Tester Screwdriver

What it is Slim, insulated slotted blade with plastic handle. Detects voltage up to 250 volts via built-in lamp.

Use it for Electrical plugs, sockets, and general maintenance work. Checking if mains electric circuits are live.

How to use Place the tip carefully on electrical item. If there’s alive circuit, the lamp lights up.

Look for Clearly marked voltage rating on the tool.

Stubby Screwdriver

What it is A compact tool with a short slotted or cross-headed blade, often interchangeable. Plastic or rubber handle.

Use it for General maintenance. Working in confined spaces, such as kitchen cupboards.

How to use Match blade tip to screw head; rotate clockwise or anticlockwise.

Look for A textured rubber handle to provide greater grip.

Pozidriv Screwdriver

What it is Cross-headed blade tip for Pozidriv screws, comes in several sizes.

Use it for Driving in or removing Pozidriv screws.

How to use Match blade tip to screw head and rotate.

Look for Avoid mistaking Pozidriv blade with older-pattern Phillips tool, even though the blade tips look similar.

Jewellers Screwdriver

What it is Miniature blade with metal or plastic handle for precision work. The head of handle revolves for greater control.

Use it for Electronics, computers, watches; extremely small screws generally.

How to use Apply pressure to head with forefinger. Grip the shaft with thumb and fingers to rotate.

Look for Tools sold in sets, which are more economical, with a range of tips.

“MATCH THE RIGHT SIZE OF SCREWDRIVER BIT TO THE SCREW. OTHERWISE IT MAY SLIP, DAMAGING THE SURFACE”

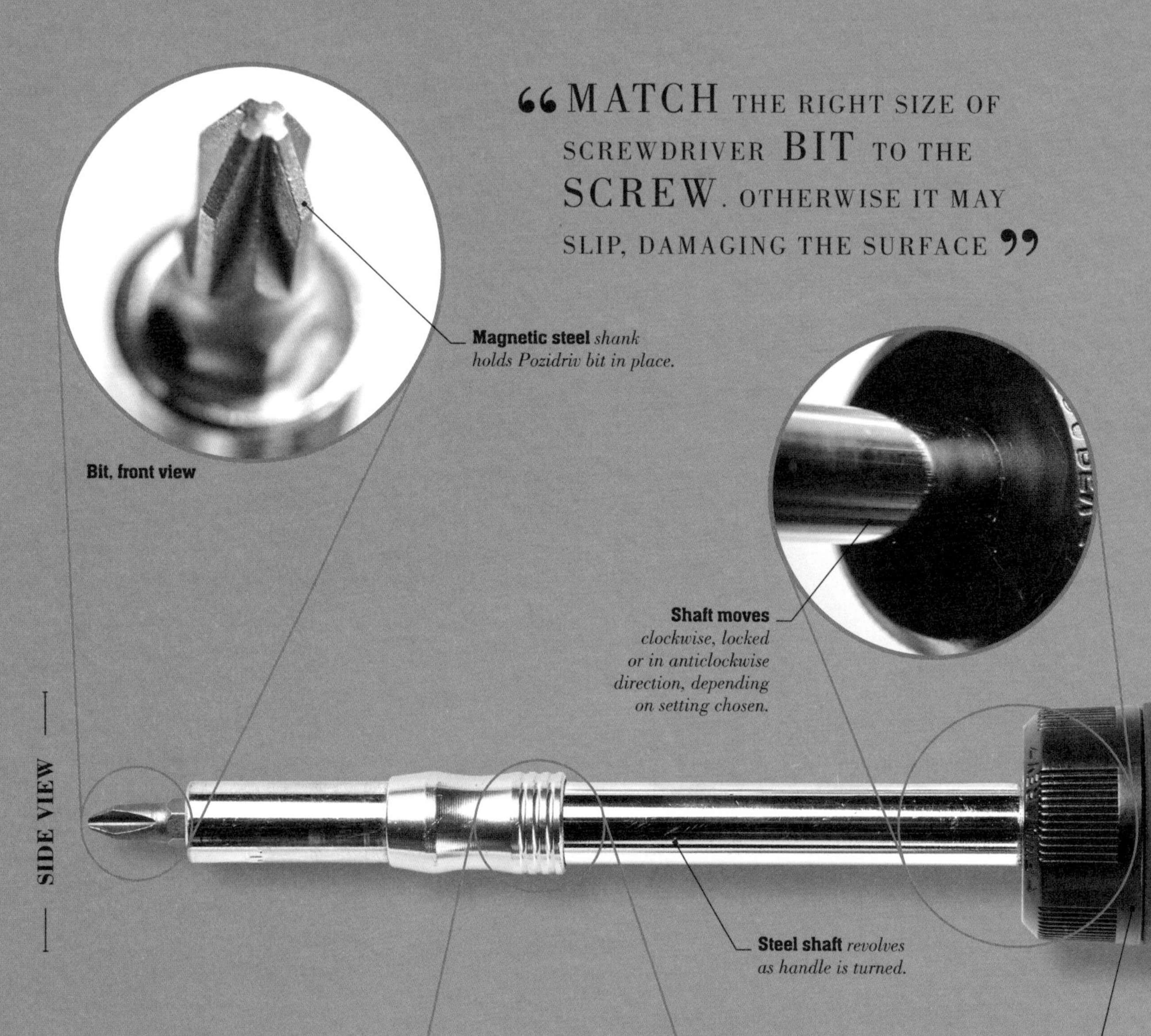

STRUCTURE OF A RATCHET SCREWDRIVER

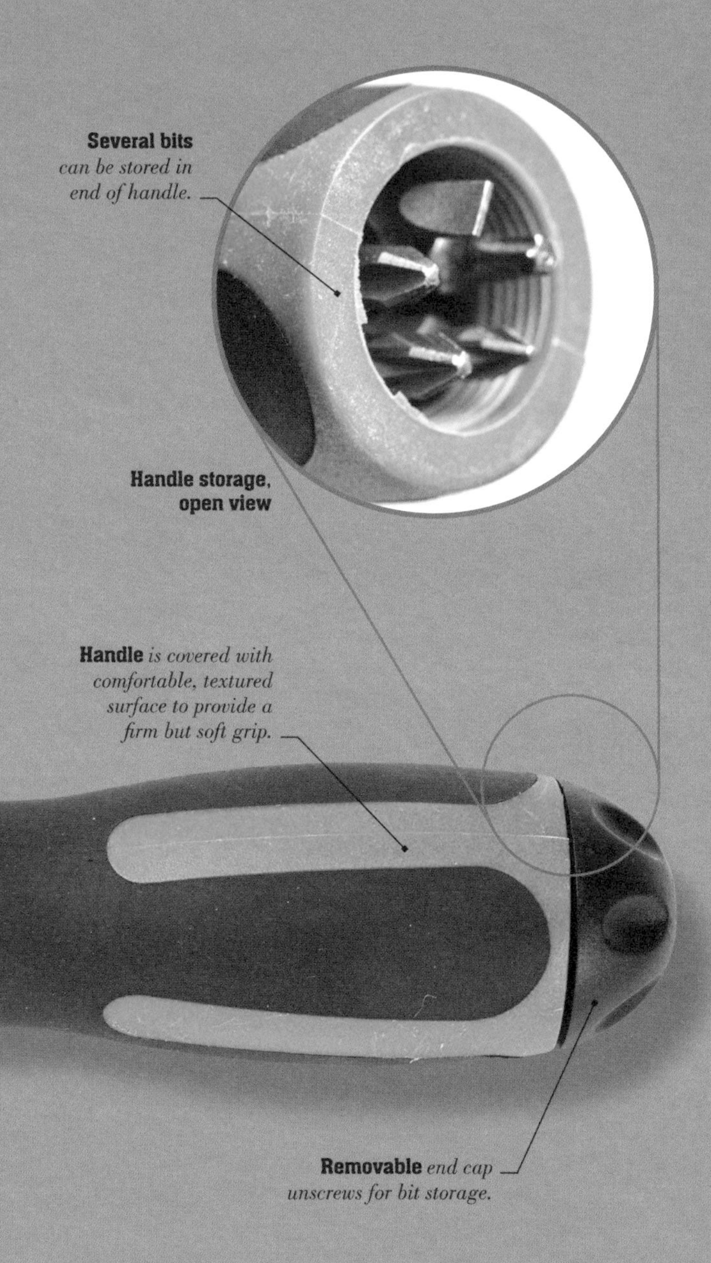

A ratchet screwdriver is faster to use than a traditional cabinet driver, mainly because less wrist movement is necessary. Older tool designs feature hardwood handles with a fixed blade, while popular combination screwdrivers generally provide a standard hex shank holder complemented by interchangeable bits that fit a wide range of screws.

“WITH A GOOD SELECTION OF BITS, AN INEXPENSIVE RATCHET SCREWDRIVER CAN REPLACE A MUCH PRICIER SCREWDRIVER SET”

FOCUS ON...

SCREWS AND DRIVES

Once the only type available, the traditional slotted screw could be problematic as a screwdriver tip could easily slip out of the slot. The introduction of cross-headed screws (Phillips and Pozidriv) meant that it was easier to drive or remove a screw without damaging the head. Today, there is an increasing number of specialist screw patterns to choose from, each requiring a specialist screwdriver bit to fit it correctly.

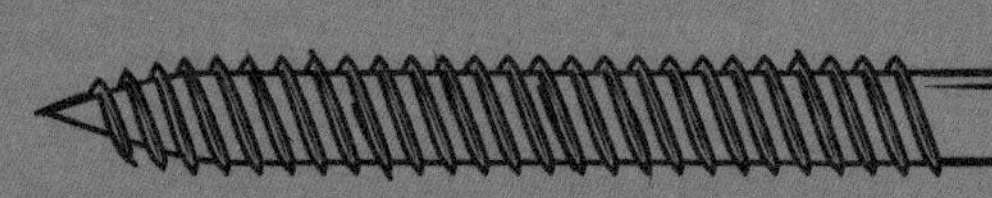

Screw structure Penetrating point, thread, shank, countersunk head so screw is flush with surface.

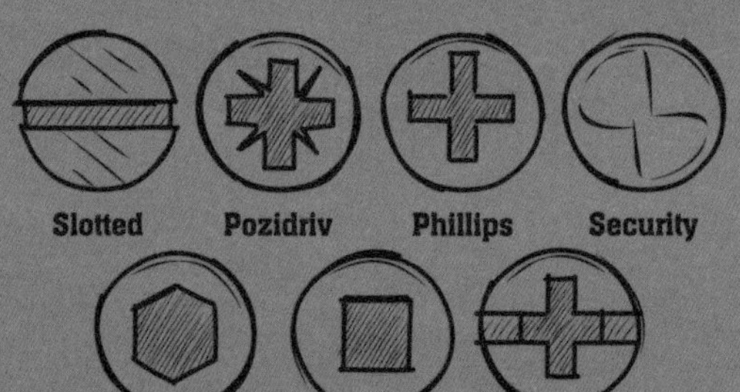

Drives The variety of screws ranges from the basic slotted to those with more specialist drives such as the tamper-proof security, or one-way, screw.

USING A

Ratchet Screwdriver

A combination ratchet screwdriver is particularly versatile, as it can be used for working with most types of screws. At the end of the shaft is a magnetic holder that accepts standard hex-shank bits. The tool usually includes storage for several bits in the handle, which are accessed by unscrewing the end cap – saving storage space and keeping the bits safe.

The Process

Before you start

- **Choose the screws** Check that available screw gauges and lengths are suitable for the task. When fixing timber pieces, screws should be three times the depth of the thinnest piece.
- **Need a pilot hole?** Some screws can be driven without one, although a pilot hole ensures timber will not split.
- **Sink the screw?** Decide if you need a countersink bit to make the screw head flush with the surface.

Magnetic socket *allows fitting of most standard hex-shank bits.*

Pozidriv bit *with a hex shank.*

1 Select bit

Choose a screwdriver socket and bit that corresponds with the screw type you want to use. For example, there are six Pozidriv head sizes (P0 to P5) and six slotted-head sizes, as well as four Phillips head sizes: 0–4 (0 is smallest).

2 Fit the bit

Insert the hex-shank Pozidriv bit into the magnetic holder at the end of the screwdriver shaft. Bits are identified by a number stamped on the side. Always replace bits when they start to show signs of wear, as otherwise they will be unable to turn screws effectively and may even strip them.

FOCUS ON...

The Thread

Each screw has a continuous spiral thread formed along its length. As the screw is driven clockwise, it cuts its own thread in the material. When the shaft of the screw rotates, it moves along its axis relative to the surrounding material. The screw cannot be withdrawn without reversing this rotation (now anti-clockwise). This process prevents any two items that have been screwed together from being forced apart.

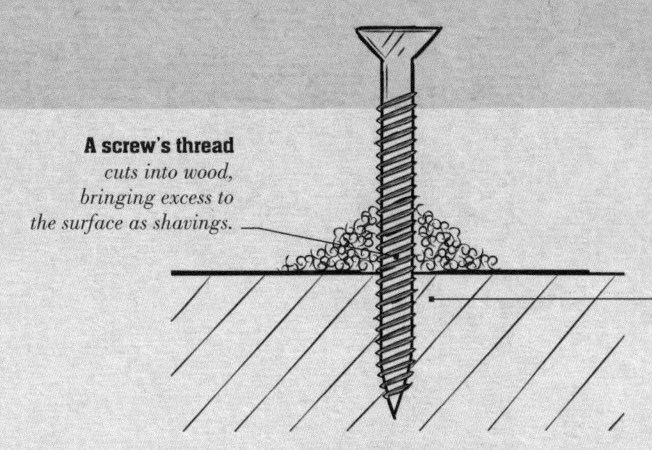

3 Line it all up

If required, countersink the hole before inserting the screw. This enables the head to sit flush with the surrounding timber. Insert the screw tip in the hole and select the clockwise rotation on the screwdriver. Alternatively, select the lock position on the sleeve to use it as you would a traditional (non-ratchet) screwdriver.

4 Turn the screw

Keeping the screwdriver vertical, drive the screw home using a twisting action of the wrist. The greater the wrist movement, the quicker the screw is driven into the wood. If required, the screw can be removed by selecting the driver's anticlockwise rotation and reversing the process.

After you finish

- **Check the surface** When required, check that all screw heads are flush with the timber surface by carefully running your fingers across the surface. Tighten if necessary.
- **Tidy up** Detach the bits and replace them in the handle. Wipe down the screwdriver with a clean, soft cloth before storing.

"ARE THE TOOLS WITHOUT, WHICH THE CARPENTER PUTS FORTH HIS HANDS TO, OR ARE THEY AND ALL THE CARPENTRY WITHIN HIMSELF; AND WOULD HE NOT SMILE AT THE NOTION THAT CHEST OR HOUSE IS MORE THAN HE?"

C A BARTOL

Choosing a
Wrench

As long ago as the 15th century, rudimentary wrenches and spanners were used to turn nuts and other fasteners on cartwheels and armour plates. Modern wrenches and spanners are used to secure every type of rotary fastener, from a tiny hex bolt on a bicycle brake to the nuts that secure huge wind turbines.

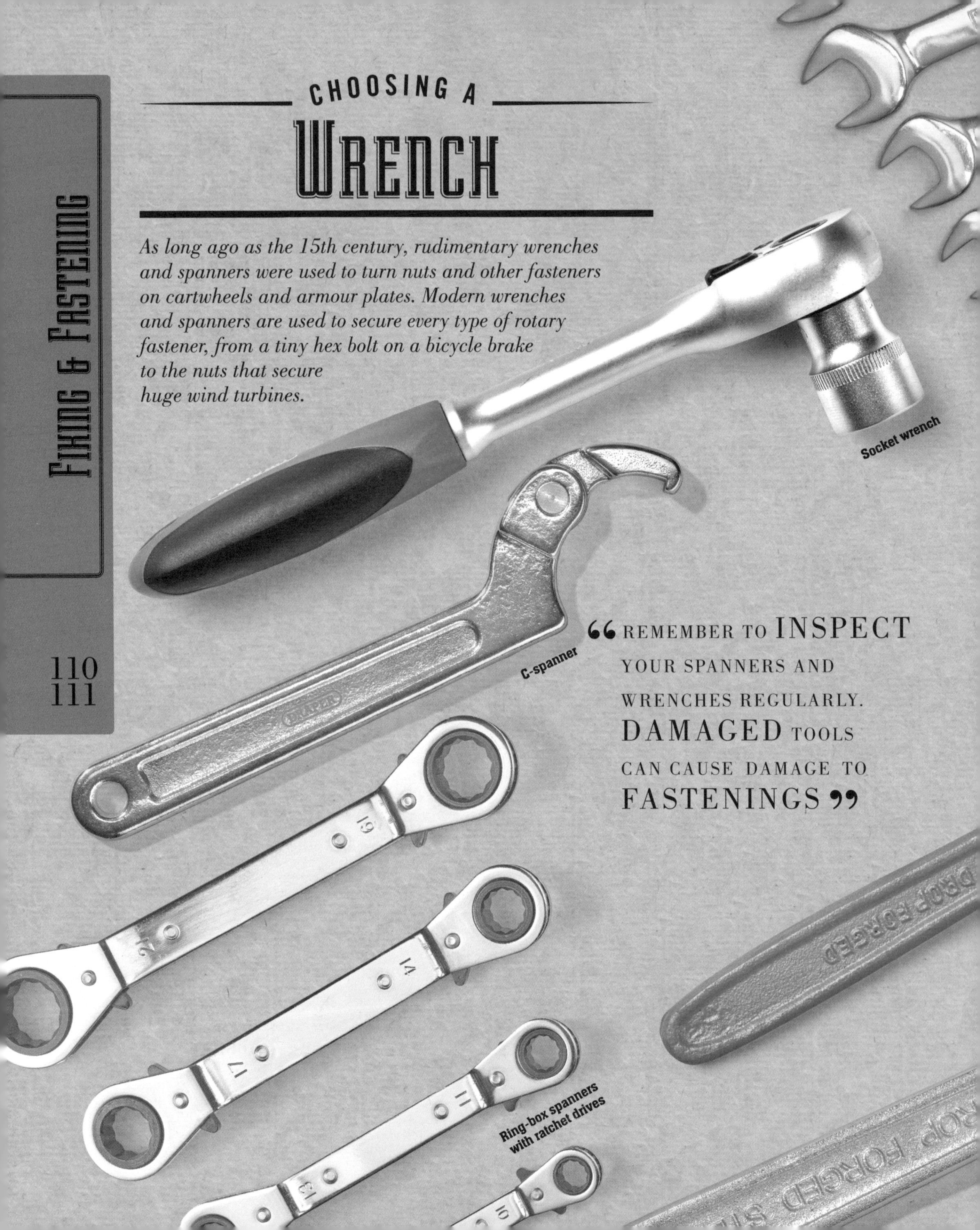

Socket wrench

C-spanner

Ring-box spanners with ratchet drives

“REMEMBER TO INSPECT YOUR SPANNERS AND WRENCHES REGULARLY. DAMAGED TOOLS CAN CAUSE DAMAGE TO FASTENINGS”

C-Spanner

What it is Single- or double-ended spanner with C-shaped ends featuring teeth, pins, or hook.

Use it for Tightening or adjusting larger ring-type fasteners or locking rings.

How to use Ensure all teeth, pins, or hook are engaged and snug on fastener.

Look for Correct-sized tool. A wrong size could damage the fastener.

Ring-Box Spanner with Ratchet Drives

What it is Double-ended ring or box spanner with integral ratchet drives.

Use it for Fasteners in tight spaces where less than a quarter turn is possible.

How to use Fit ring end over fastener. Lever handle back and forth at least one click at a time.

Look for Correct size and a snug fit on fastener head.

Plumbing Wrench

What it is Adjustable wrench with serrated jaws that grip soft pipework.

Use it for Holding or turning soft metal pipes, typically copper or soft iron.

How to use Close moving jaw on pipe. As leverage is applied to the handle, the jaw locks tighter on the pipe.

Look for Ensure pipe is clean and grease-free.

Socket Wrench

What it is Cylindrical socket with 6–12 internal flats that engage with a square ratchet drive on a handle.

Use it for Fastening nuts or bolts.

How to use Select correct socket size and fit to ratchet driver. Turn fastener and returning lever at least one click at a time.

Look for Socket to fit four driver sizes between 6mm and 19mm (0.25–0.75in).

Combination Ring-Box Spanners

What it is Open spanner at one end with ring or box end at the other.

Use it for Rotary fasteners where there is room for a quarter turn.

How to use Place either end on exposed nuts.

Look for A snug fit on the fastener.

Adjustable Spanner

What it is Spanner with jaw adjusted via a screw to fit multiple-sized fasteners.

Use it for Opening or closing basic or non-hexagonal fasteners.

How to use Tighten jaw on fastener, ensuring the movable jaw is forced into, not away from, the direction of rotation.

Look for A tight fit on flat sides of fastener; otherwise there is a risk of rounding off bolt head.

CONTINUED

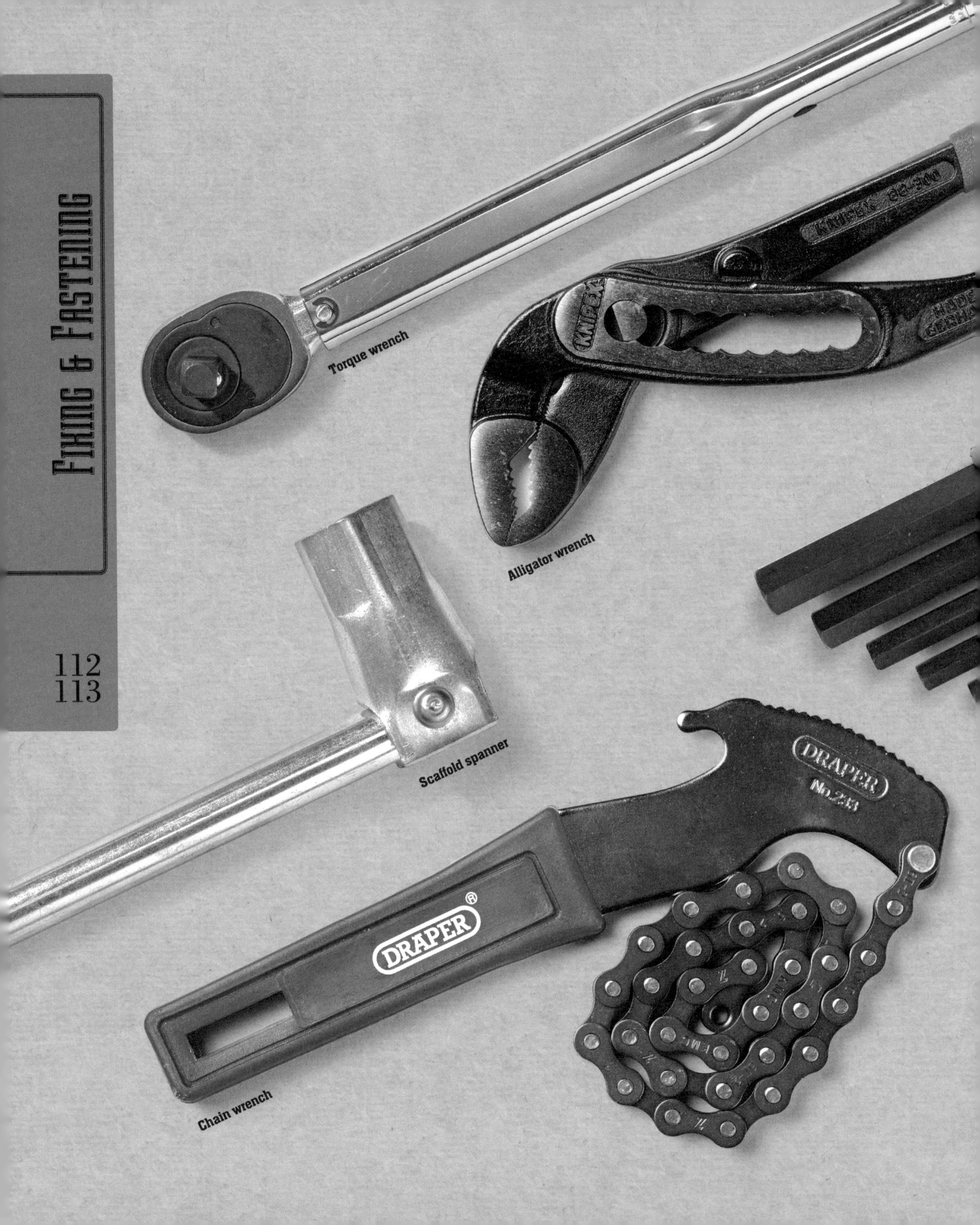

Torque wrench

Alligator wrench

Scaffold spanner

Chain wrench

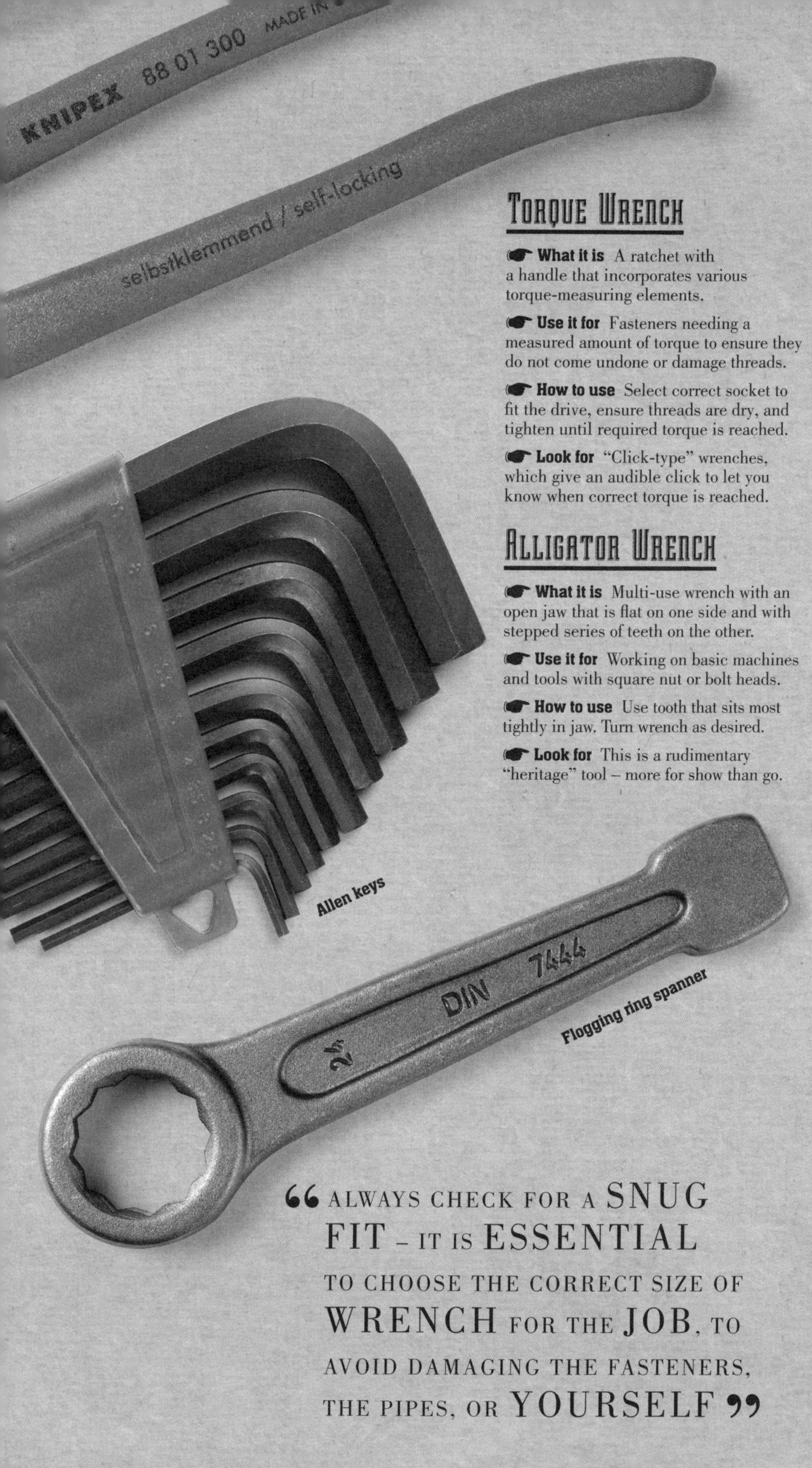

Allen keys

Flogging ring spanner

Torque Wrench

What it is A ratchet with a handle that incorporates various torque-measuring elements.

Use it for Fasteners needing a measured amount of torque to ensure they do not come undone or damage threads.

How to use Select correct socket to fit the drive, ensure threads are dry, and tighten until required torque is reached.

Look for "Click-type" wrenches, which give an audible click to let you know when correct torque is reached.

Alligator Wrench

What it is Multi-use wrench with an open jaw that is flat on one side and with stepped series of teeth on the other.

Use it for Working on basic machines and tools with square nut or bolt heads.

How to use Use tooth that sits most tightly in jaw. Turn wrench as desired.

Look for This is a rudimentary "heritage" tool – more for show than go.

Scaffold Spanner

What it is Fixed-head socket with a fixed or articulated socket at one end.

Use it for Work on many same-sized fasteners, such as scaffolding brackets.

How to use Fit socket over hand-tightened nut. Turn bar a quarter to a half.

Look for An articulated socket head to reach fasteners in inaccessible places.

Chain Wrench

What it is Self-tightening chain or strap with a driver used with a ratchet bar.

Use it for Tight gripping of cylindrical automotive oil filters.

How to use Wrap the chain around the middle of filter, turn driver until tight, then use the ratchet to turn the filter.

Look for Clean oil and dirt from the filter before fitting the chain.

Allen Key

What it is L-shaped bit that fits recessed hexagonal fasteners.

Use it for Usually smaller-sized fasteners on machinery and furniture.

How to use Insert the key fully into the recess.

Look for The Torx type is designed for use with power tools.

Flogging Ring Spanner

What it is Heavy-duty wrench with an open spanner at one end and a block at the other for striking with a hammer.

Use it for Heavy bolts or nuts that require torque figures best achieved by striking the wrench with a sledgehammer.

How to use If fastener has alignment marks, screw down by hand, then strike the wrench until the marks line up.

Look for Strike it as squarely as possible when tightening or releasing.

“ALWAYS CHECK FOR A SNUG FIT – IT IS ESSENTIAL TO CHOOSE THE CORRECT SIZE OF WRENCH FOR THE JOB, TO AVOID DAMAGING THE FASTENERS, THE PIPES, OR YOURSELF”

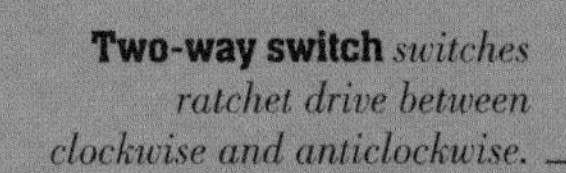

Two-way switch *switches ratchet drive between clockwise and anticlockwise.*

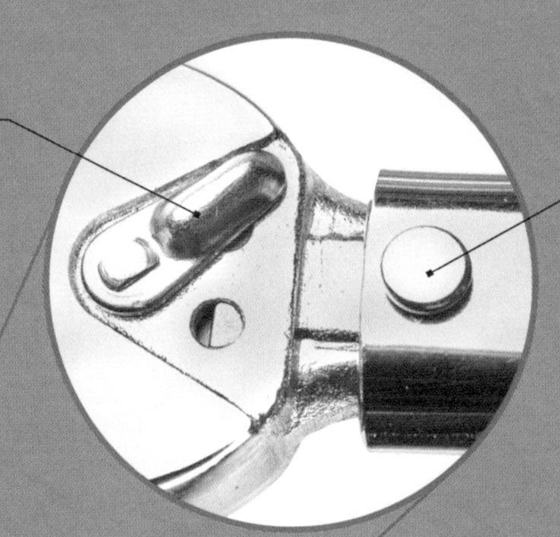

Pivot and detent ball *momentarily releases drive head when torque setting is reached.*

Square drive bit *has 6–19mm (0.25–0.75in) drive sizes for use with appropriate-sized socket sets.*

Flat half of shaft *is not designed to be used as handle.*

TOP VIEW

Hollow shaft *exerts leverage on fastener and can be of varying lengths.*

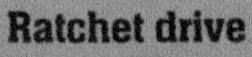

Ratchet drive *under bit allows handle to be pumped without removing the socket from the fastener.*

Ratchet drive, side view

“A TORQUE WRENCH IS ESSENTIAL FOR WORKING ON MACHINES WITH SOFT OR CRUSHABLE MATERIALS”

Structure of a Torque Wrench

Pre-loading a fastener using a wrench that measures torque is the best way to ensure that a bolt is tight enough without damaging threads or components. A torque wrench is essential for working on machines with soft or crushable materials, such as carbon fibre, or on critical components that require high torque settings.

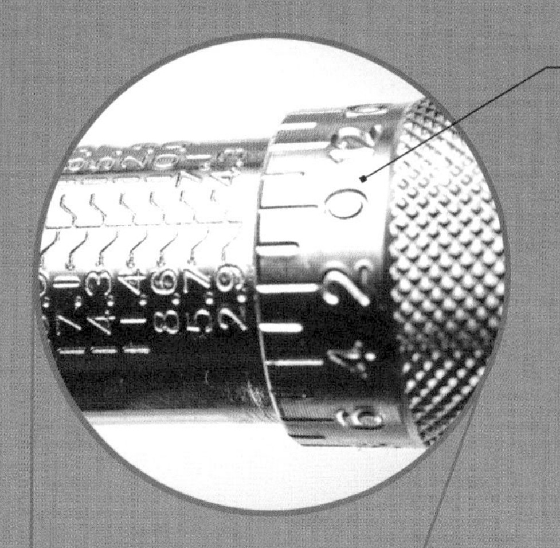

Torque markings *like most instructions will give torque values in both newton metres (N m) and pound-feet (lb-ft).*

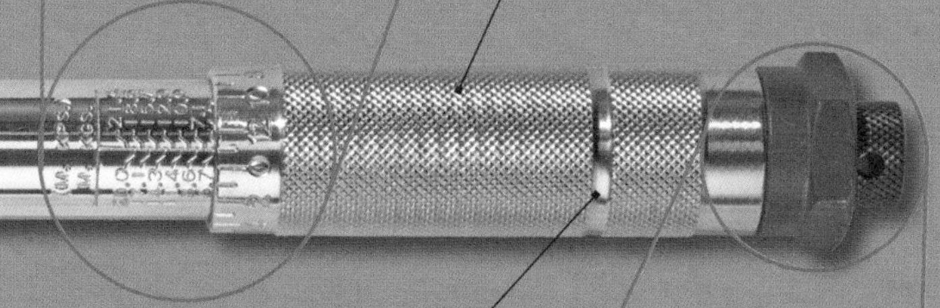

Adjustable handle *screws in and out to align with torque settings marked on shaft.*

Textured grip *must be used to achieve required torque setting.*

Lock-nut *on end of handle locks handle in place at desired torque setting.*

Nut *on end of wrench for dismantling tool for maintenance only.*

FOCUS ON...

Torque Wrench Types

Hand-calibrated, spring-loaded torque wrenches are widely used but there are other types of torque wrench: beam; electronic or digital; and T-handle, no-ratchet type. The latter is really a mini torque wrench designed for a single, very low torque setting of about 5 N m, which makes it popular for clamping together fragile carbon-fibre components.

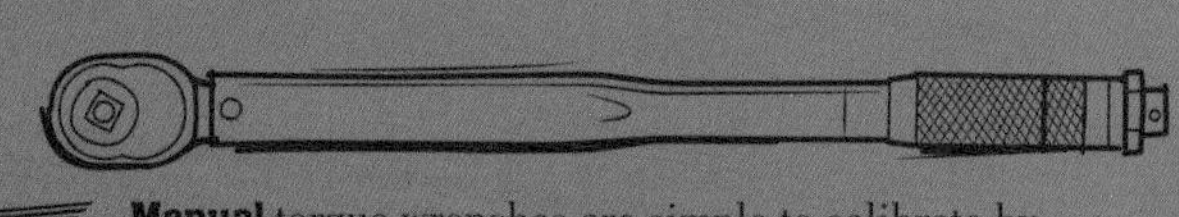

Manual torque wrenches are simple to calibrate by hand and they are spring-loaded for ease for use.

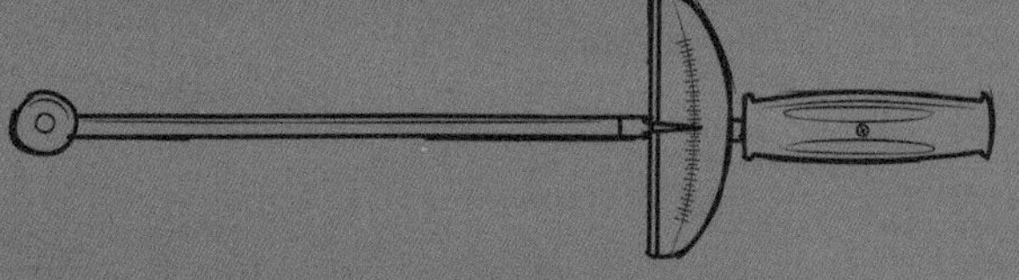

Beam type has a deflecting beam that acts as a handle, and a fixed indicator to show fastener load.

T-handle type is designed for single or very low torque loadings and may have digital readout.

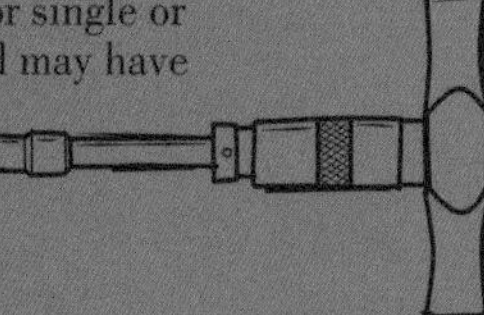

Using a Torque Wrench

Loading a fastener to a specified torque setting is common practice when working with precision components, or when a loose or stripped nut or bolt could result in a dangerous failure. A torque wrench works like a standard ratchet wrench and is similarly easy to use – as long you know the torque setting required for your project.

The Process

Before you start

- **Check loading** Take a careful note of the required torque setting for the fastener involved, either from the workshop manual or as marked on the fastener itself.
- **Choose torque measure** Torque settings may be given in newton metres (N m), pound-feet (lb-ft), or both. Often the figure will have a small amount of leeway either side.
- **Keep it clean** To load the fastener accurately, make sure the threads are clean and free from grease or excess oil.

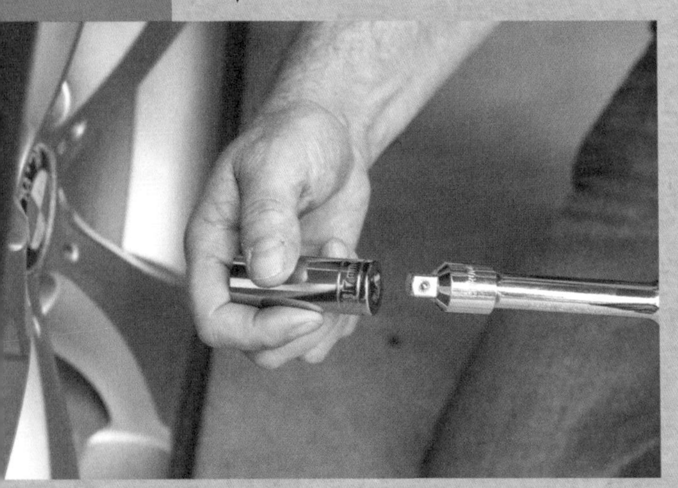

1 Select correct socket

Check size of drive, from 6–19mm (0.25–0.75in), and select correct size socket for the fastener. Automotive bolts or screws may need a hex or Torx bit. Push the socket firmly onto the square drive until it clicks into place.

Make sure the handle *matches the torque scale you require– either N m or lb-ft.*

Lock *the handle again to secure settings.*

2 Dial it in

The wrench should have been stored with the rotating handle screwed outwards. Undo the locking bolt on the end and screw in the handle until the leading edge aligns exactly against the required torque setting. Double-check you are using the correct N m or lb-ft scale on the barrel.

FOCUS ON...

Torque

Torque measures rotational force: how much force on an object causes it to rotate around a pivot point. In this example, the pivot point is the nut or bolt, and the force comes from your hands and arms via the torque wrench. The longer the distance from the force to the pivot point, the greater the torque exerted – which is why it's easier to turn a stuck nut by using the end of a wrench rather than the middle.

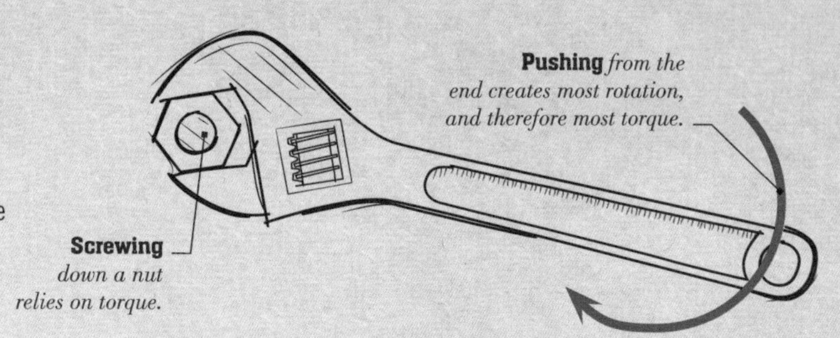

Pushing *from the end creates most rotation, and therefore most torque.*

Screwing *down a nut relies on torque.*

Tighten the bolt *to full torque but no further. Take care not to overtighten as it could damage the threads or bolt.*

3 Load it

Screw the bolt in by hand until snug. Check that the switch on the back of the ratchet is locked in the correct direction. Fit the socket onto the fastener and, holding the adjusting handle, slowly tighten the bolt. The handle may need to be pumped while the ratchet gently ticks on the return stroke.

4 Get the setting right

When the desired torque setting is reached, you should hear a "clunk" sound from the head of the wrench, which will also give a little. Remove the wrench from the fastener.

When you have finished

- **Unlock** Release the socket and return it to its appropriate place in its case or in your toolbox.
- **Undo** Release the locking nut on the end of the handle.
- **Unscrew** Screw out the handle until you can feel the tension release on the spring, to a point where it exerts minimal pressure on the internal spring and mechanism. Always store it this way.

Choosing a

Drill & Drill Bit

Though they are not perhaps as efficient as their powered counterparts, specialist bits are still available for traditional, reliable hand drills – so you can experience their satisfying whirr, or feel the brute force of a swing brace. However, for tasks in which time and torque are key, cordless power drills really do fit the bill.

Hand drill

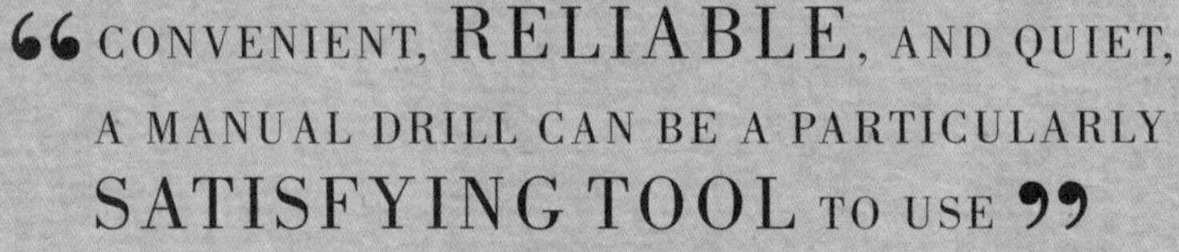
“CONVENIENT, RELIABLE, AND QUIET, A MANUAL DRILL CAN BE A PARTICULARLY SATISFYING TOOL TO USE”

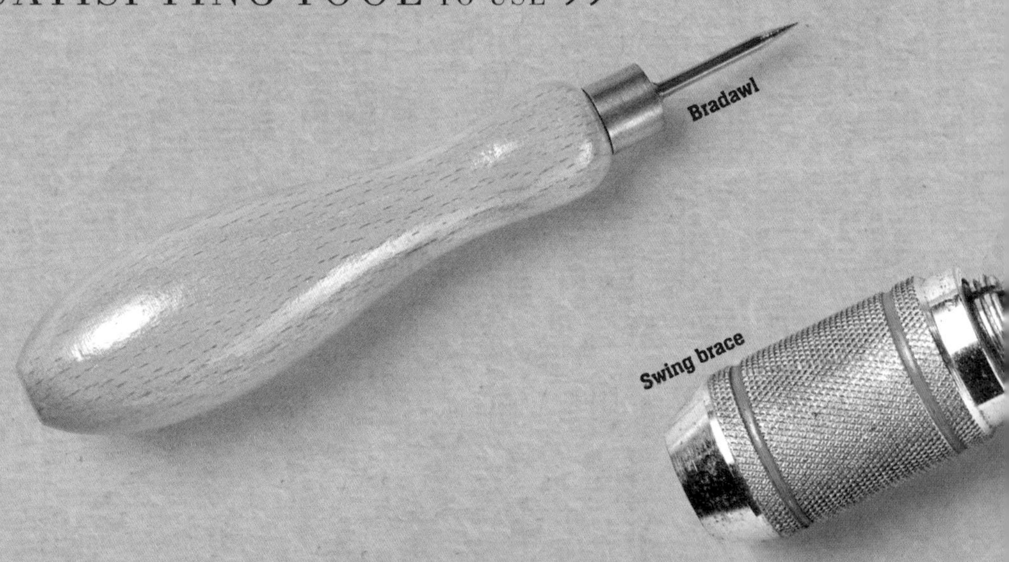
Bradawl

Swing brace

> “THINK ABOUT IT: EVERYTHING WITH A POWER CORD EVENTUALLY WINDS UP IN THE TRASH.”
>
> **JOHN SARGE, TIMBER FRAMER**

Hand Drill

What it is Toothed hand wheel that meshes with smaller gears on steel or aluminium frame to rotate chuck. Has self-centring chuck, hardwood handle.

Use it for Drilling holes in most materials up to 9mm (0.35in) in diameter. Good for small holes in delicate surfaces.

How to use Insert twist bit in jaws and tighten chuck. Turn side handle to rotate chuck clockwise or anticlockwise.

Look for Check all three chuck jaws are working. Side knob is often missing from older tools.

Bradawl

What it is Fixed square or pointed steel blade with hardwood or plastic handle. Forces wood fibres apart.

Use it for Making starter holes for small screws or larger drill bits instead of creating pilot holes.

How to use Position the tip of blade on your pencil mark, holding tool upright. Push down and use a twisting action to make hole.

Look for The tip should be sharp, so use edge of an oilstone, if necessary, to maintain the point.

Swing Brace

What it is U-shaped steel frame with removable chuck. Hardwood or plastic handle. Most have chuck ratchet action.

Use it for Drilling deep, large-diameter holes in dense timber where extra torque is necessary.

How to use Insert bit in V grooves in jaws; tighten sleeve. Grip rear handle and swing centre grip clockwise to turn chuck.

Look for Check jaw pattern (two or three) as some braces only accept bits with square taper shanks.

Archimedes Drill

What it is A small steel or brass tool with a spiral shaft, a spring-loaded sliding collar, and a collet (collar-type) chuck.

Use it for Model-making; other small-scale work. Drilling tiny holes up to 1mm (0.04in) diameter in wood or fragile surfaces.

How to use Insert bit into collet and tighten. Push down end of tool with index finger; slide collar up and down shaft.

Look for Replacement micro drill bits, as these are easily lost or broken.

CONTINUED

Cordless drill

"FAST AND FURIOUS, A CORDLESS TOOL IS A REAL TIME SAVER, SPEEDING UP BOTH THE DRILLING AND THE SCREWDRIVING PROCESSES"

Combi drill

Cordless Combi Drill

☛ **What it is** Battery-powered tool with an added hammer-action function for drilling masonry.

☛ **Use it for** Drilling holes in most materials, including concrete and brickwork. Driving or removing screws.

☛ **How to use** Insert appropriate bit in chuck, select drilling, hammer action, or screwdriving function, choose the appropriate speed.

☛ **Look for** A variable-speed trigger aids drilling process.

Cordless Drill / Driver

☛ **What it is** Two-speed power tool with rechargeable battery. Chuck revolves when motor is activated by on/off trigger.

☛ **Use it for** Drilling holes in timber, metals, plastics, and similar material. Driving or removing screws.

☛ **How to use** Insert appropriate bit, tighten chuck, and select drilling mode and speed. Adjust torque and speed for screws.

☛ **Look for** A fast charger with removable batteries. Budget tools with built-in batteries generally have a slower recharge time.

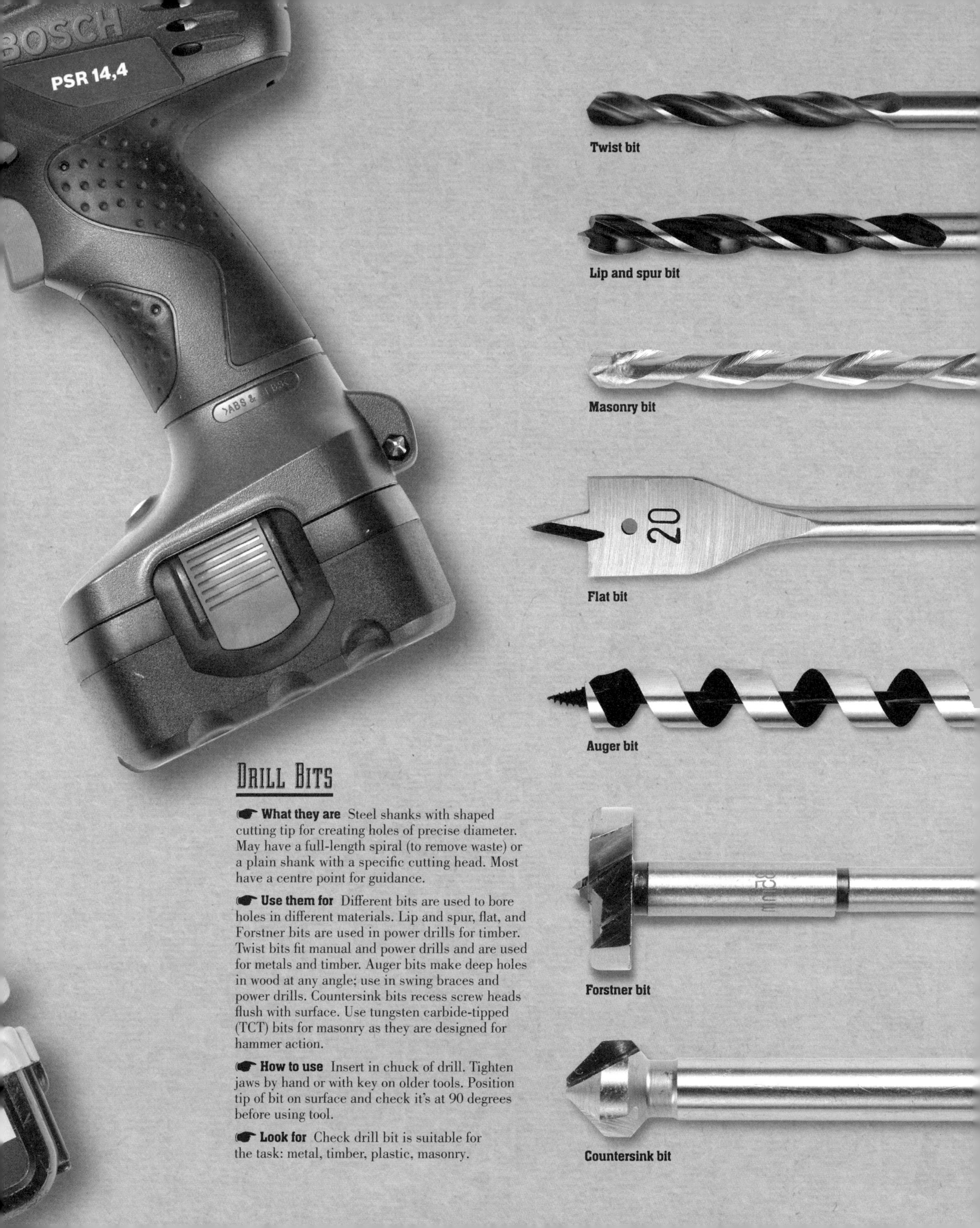

Twist bit

Lip and spur bit

Masonry bit

Flat bit

Auger bit

Forstner bit

Countersink bit

Drill Bits

What they are Steel shanks with shaped cutting tip for creating holes of precise diameter. May have a full-length spiral (to remove waste) or a plain shank with a specific cutting head. Most have a centre point for guidance.

Use them for Different bits are used to bore holes in different materials. Lip and spur, flat, and Forstner bits are used in power drills for timber. Twist bits fit manual and power drills and are used for metals and timber. Auger bits make deep holes in wood at any angle; use in swing braces and power drills. Countersink bits recess screw heads flush with surface. Use tungsten carbide-tipped (TCT) bits for masonry as they are designed for hammer action.

How to use Insert in chuck of drill. Tighten jaws by hand or with key on older tools. Position tip of bit on surface and check it's at 90 degrees before using tool.

Look for Check drill bit is suitable for the task: metal, timber, plastic, masonry.

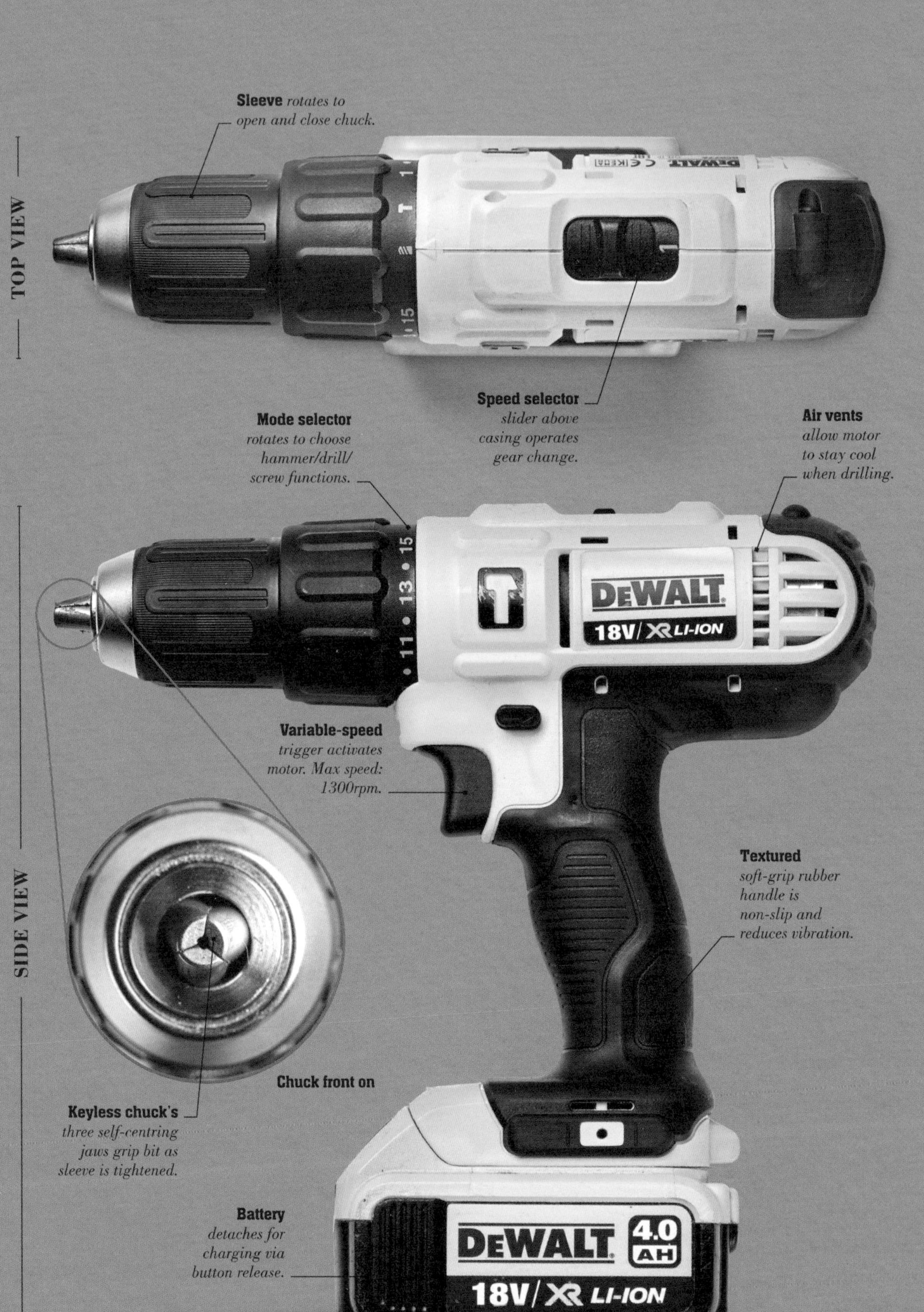
TOP VIEW
SIDE VIEW
Sleeve *rotates to open and close chuck.*
Speed selector *slider above casing operates gear change.*
Mode selector *rotates to choose hammer/drill/ screw functions.*
Air vents *allow motor to stay cool when drilling.*
DEWALT
18V/XR LI-ION
Variable-speed *trigger activates motor. Max speed: 1300rpm.*
Textured *soft-grip rubber handle is non-slip and reduces vibration.*
Chuck front on
Keyless chuck's *three self-centring jaws grip bit as sleeve is tightened.*
Battery *detaches for charging via button release.*
DEWALT
4.0 AH
18V/XR LI-ION

"SAFER AND MORE CONVENIENT THAN A PLUG-IN EQUIVALENT, A CORDLESS COMBI DRILL IS A VERSATILE TOOL"

STRUCTURE OF A COMBI DRILL

The heart of a cordless drill is its brush or brushless motor, encased in a plastic shell. Metal or plastic gears create two or three speed settings, controlled via a trigger. The chuck revolves on a spindle, driven by the motor. Standard rotary action is used for drilling metals, timber, and driving screws; hammer action is for drilling masonry.

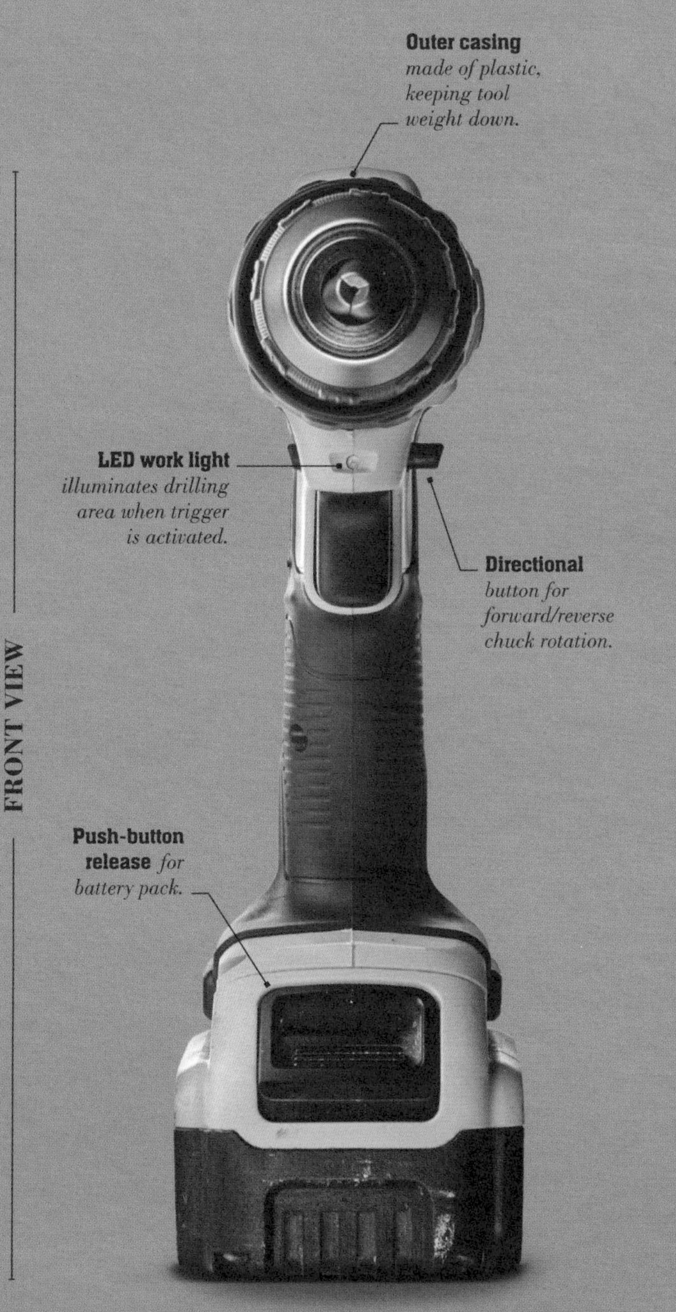

FOCUS ON...

SIZES

Compact cordless tools are ideal for working in confined spaces and are much more manageable for those with smaller hands. However, they're not as powerful as combi drills with hammer action. The battery capacity is rated in amp hours (Ah) and is generally up to 3.0Ah in 10.8-volt tools. Lithium (Li-ion) batteries are more environmentally friendly than older NiCd or NiMH types.

10.8V Combi drill
A lightweight drill and powered screwdriver, offering drilling and driving functions with two variable speeds. Fine for lighter work, but not suitable for heavy masonry or concrete.

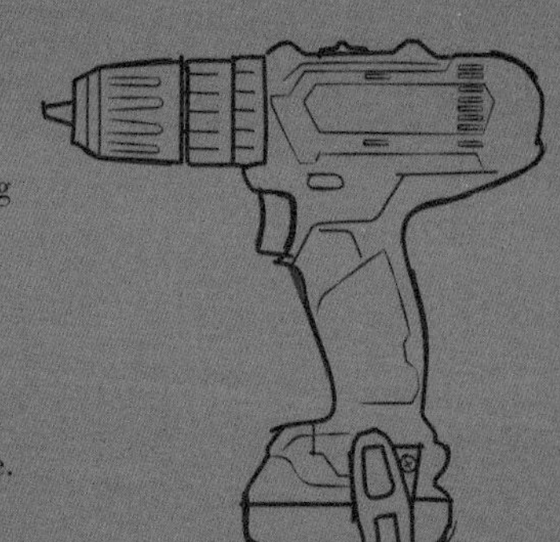

18V Combi drill
Higher-voltage combi drills are larger and heavier, but offer greater torque for driving larger screws into a wider variety of materials. They also take bigger-capacity batteries of up to 5.0Ah or more.

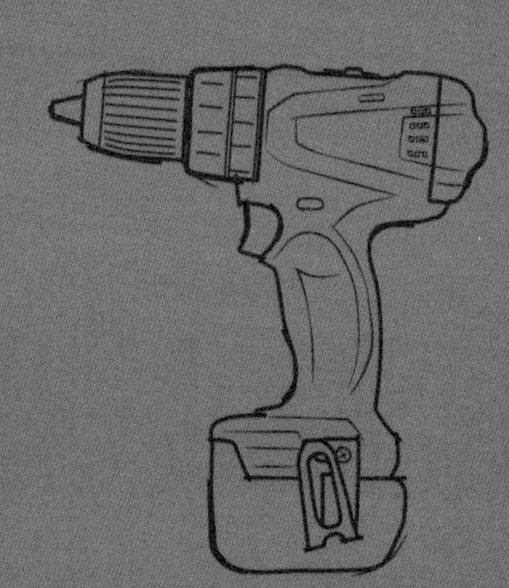

USING A

COMBI DRILL

A combination, or "combi", drill can be used for virtually any drilling task due to its additional hammer function. Once the drill is fitted with a tungsten carbide masonry bit, the combi's percussive hammer action can tackle concrete, while switching to standard rotary action makes the tool suitable for wood, metal, and most other materials. And, because it's battery powered, the tool can also be used safely outdoors.

The Process

Before you start

- **Safety** Always wear eye protection and a dust mask – ear defenders are also a good idea with sustained hammer drilling.
- **Battery** Make sure the battery is charged. A second pack means you can swap when one is flat.
- **Bits** If drilling many holes, save time with a quick-release bit holder to make switching easy. Secure bit holder in chuck, then swap bits as necessary.

1 Mark out a guide

Using a steel centre punch and hammer, mark the metal. This acts as a guide for the drill bit and prevents it from skidding during the drilling process. Make sure the metalwork is secured and resting on a sturdy, flat surface.

2 Insert the bit

After you've selected a twist bit of the correct diameter, insert it into the chuck's jaws and tighten the sleeve securely. If using a quick-release bit holder, make sure the bit's shank is seated correctly in the chuck before drilling.

“ONE MACHINE CAN DO THE WORK OF 50 ORDINARY MEN. NO MACHINE CAN DO THE WORK OF ONE EXTRAORDINARY MAN.”

ELBERT HUBBARD

FOCUS ON...

Torque Settings

Like a car going uphill or pulling a load, you need to change down a gear when using a cordless drill for certain tasks. A lower speed produces higher torque, the rotational force required to turn a screw or drill a large diameter hole. Conversely, when drilling softwood you need high speed and low torque. When selected, a large gear transmits torque to a smaller gear, making it turn faster but with less force.

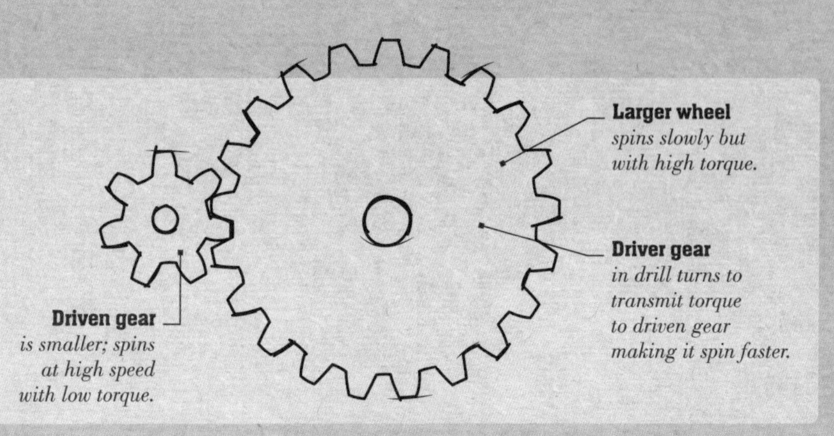

3 Select correct setting

Choose the appropriate drilling function by rotating the torque collar to the recommended setting. Select the right speed for the job: generally, you need a high speed for most twist bits, although larger ones may require a reduction in speed because they need more rotational force – torque – to get them moving (see box, above).

4 Drill the hole

Place the tip of the drill bit on the mark and ensure the drill is held square to the metal. Squeeze the trigger gently to activate the tool, then increase speed to complete the hole.

Swarf *or metal debris.*

After you finish

- **File it off** There will be swarf (metal waste) around the hole, so remove this with a flat file or similar. Wear lightweight work gloves so you don't pick up any metal splinters.
- **Clean it up** Remove all drilling debris from the drill carefully, making sure it does not enter the tool's cooling vents.

Choosing a
Vice

A vice is a basic necessity for anyone interested in or about to start using hand tools. Installed on a bench, it's a sturdy device that is designed for holding items securely, whether you're working with wood, metal, or plastics. A smaller, portable vice can be fitted virtually anywhere and may be the solution if you don't have a dedicated workspace.

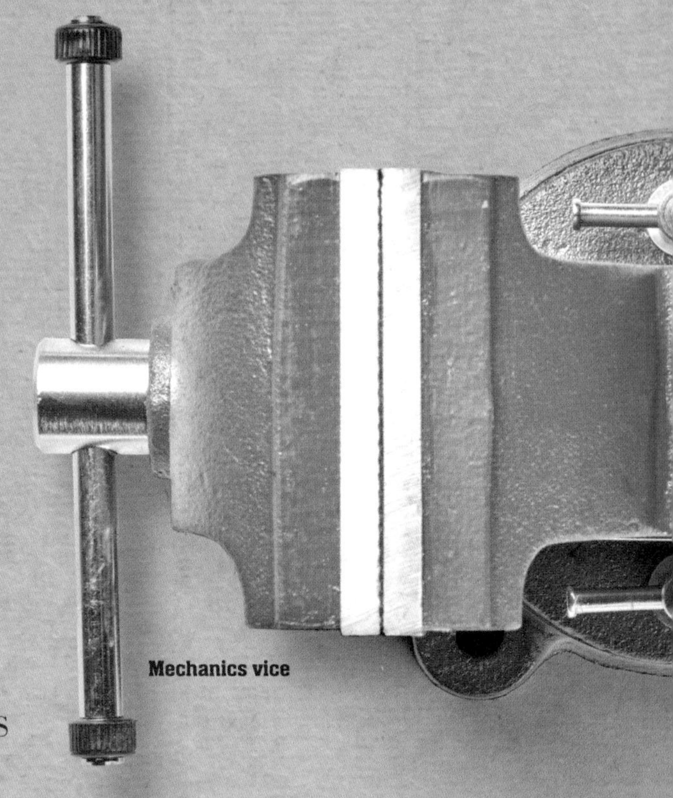
Mechanics vice

“Line jaws with hardwood facings to PREVENT vulnerable surfaces becoming DENTED”

Woodwork vice

Swivel vice

Hand vice

Multi-angle vice

Swivel Vice

What it is Portable tool similar to mechanics vice, with clamp for fixing to table. Jaws rotate through 90 degrees.

Use it for Angling small items when soldering, wiring, sawing, model-making. Useful where work space is tight.

How to use Clamp vice to bench, then release base locking levers. Swivel to desired position and retighten.

Look for Check that the clamp's capacity is sufficient for the thickness of the table it clamps on.

Mechanics Vice

What it is Heavy cast-iron vice screwed to workbench, with serrated jaws, rear anvil. Some models can be rotated.

Use it for Gripping pipes, cylinders, and square workpieces. Its high jaws make it possible to cut rods or bars without fouling saw blade.

How to use Slacken off the jaws, position the workpiece in them and tighten firmly with the tommy bar.

Look for Rubber facings to clip over the jaws to protect vulnerable surfaces when tightening.

Woodwork Vice

What it is A heavy-duty, cast-iron unit screwed to underside of workbench. Big-capacity jaws with large surface area.

Use it for Planing timber or holding workpieces horizontally or vertically. General-purpose work-holding.

How to use Open jaws with the tommy bar. Hold workpiece in place while tightening jaws.

Look for A quick-release mechanism makes jaws faster to open and close.

Multi-angle Vice

What it is Lightweight cast-alloy tool with clamp adjuster for fixing to table. Jaws tilt to any angle, rotate 360 degrees.

Use it for Greater access to smaller items when soldering, wiring, sawing, and model-making.

How to use Clamp to bench, then slacken rear tommy bar. Swivel jaws to desired position and tighten front bar.

Look for Some plastic jaw facings can sometimes work loose easily.

Hand Vice

What it is Narrow tool, with forged-steel jaws hinged at base. Tightened via spring-loaded wing nut.

Use it for Gripping very small items such as jewellery before grinding, filing, drilling. Also mounts in regular vice jaws.

How to use Grasp in hand, position component, and tighten jaws with wing nut on threaded bolt.

Look for Vertical and horizontal V grooves in the jaws for gripping circular items.

Structure of a
Mechanics Vice

At its most basic level, a vice consists of a pair of hardened steel jaws that are adjusted by a lever to grip a workpiece tightly. Heavy-duty versions are made of cast iron, while smaller, lighter, and more portable vices tend to be of cast-alloy construction. A swivelling base is useful, as it makes the tool much more versatile in terms of its work-holding capacity.

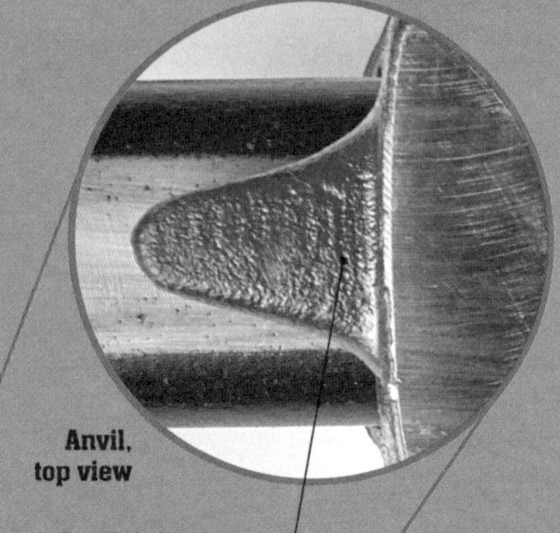

Anvil, top view

Anvil *behind rear jaw for small-scale hammering tasks.*

"BUY THE STURDIEST, HEAVIEST VICE YOU CAN AFFORD, THEN FIT IT TO A SOLID BENCH"

U-shaped channel *slides through opening in body.*

Fixing holes *around base for bolting vice to bench top.*

Locking bolt *allows base to swivel and lock in position.*

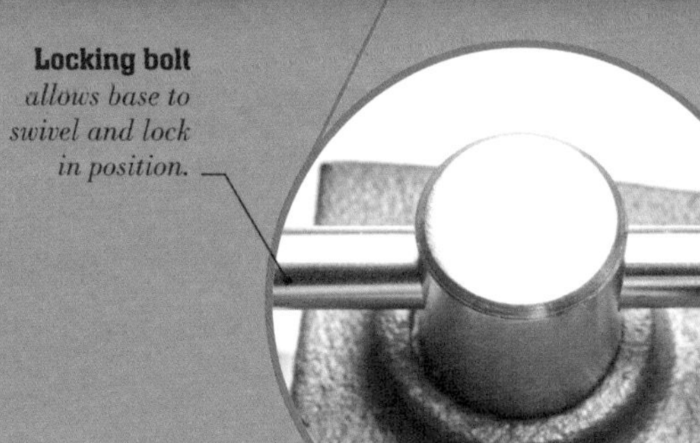

Locking bolt, top view

FOCUS ON...
How It Works

Different vices operate in slightly different ways. On a woodworking vice, dual steel bars either side of a central screw thread guide the jaws, preventing them from racking and keeping their faces parallel. A mechanics vice also uses a screw thread, but in these models a substantial square-sectioned, U-shaped steel channel is used to guide the front jaw. This passes through a square opening on the cast body of the vice to maintain rigidity.

USING A

Mechanics Vice

Position is important when installing a heavy-duty vice permanently on a bench. A right-handed worker will find it most convenient on the left side and vice versa. A portable vice can be easily repositioned.

The Process

Before you start

☛ Prepare the jaws Clip on hard rubber/aluminium vice jaws to protect timber or softer materials, if needed. Magnetic strips are usually included to secure to jaws.

1 Check the screw

Make sure the screw mechanism is operating smoothly and lightly oil the thread if it seems at all sluggish.

2 Adjust the jaws

Adjust the jaws to open slightly greater than the workpiece thickness. Hold the workpiece in position and tighten the front jaw by turning the handle in a clockwise direction.

3 Position the base

Rotate the jaws to the most convenient working position if the vice is equipped with a swivel base. This is done by slackening off two small locking bolts on either side of the base, altering the vice's angle to your desired position, then relocking the bolts.

After you finish

☛ Clean the vice Use a cloth to wipe down the vice, removing any wood shavings, grit, or metal swarf from exposed screw threads.

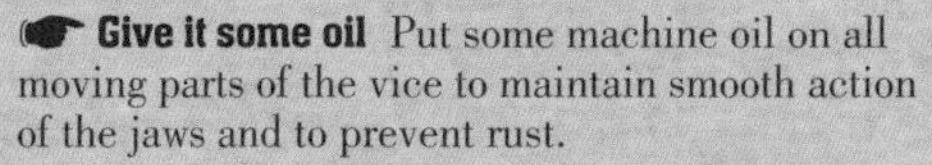

☛ Give it some oil Put some machine oil on all moving parts of the vice to maintain smooth action of the jaws and to prevent rust.

SIDE VIEW

Jaw faces *are cross-hatched to increase grip.*

Screw thread *mechanism hidden beneath channel.*

Steel handle *or tommy bar used to adjust front jaw.*

Handle caps *prevent handle bar from slipping out of hole.*

Choosing a
Clamp

There are probably as many types of clamps available as there are tasks for them to do. Whether you want to glue timber components together, hold metalwork in position for welding or brazing, or simply grip items on the workbench, you'll probably need several of these essential items in your toolkit.

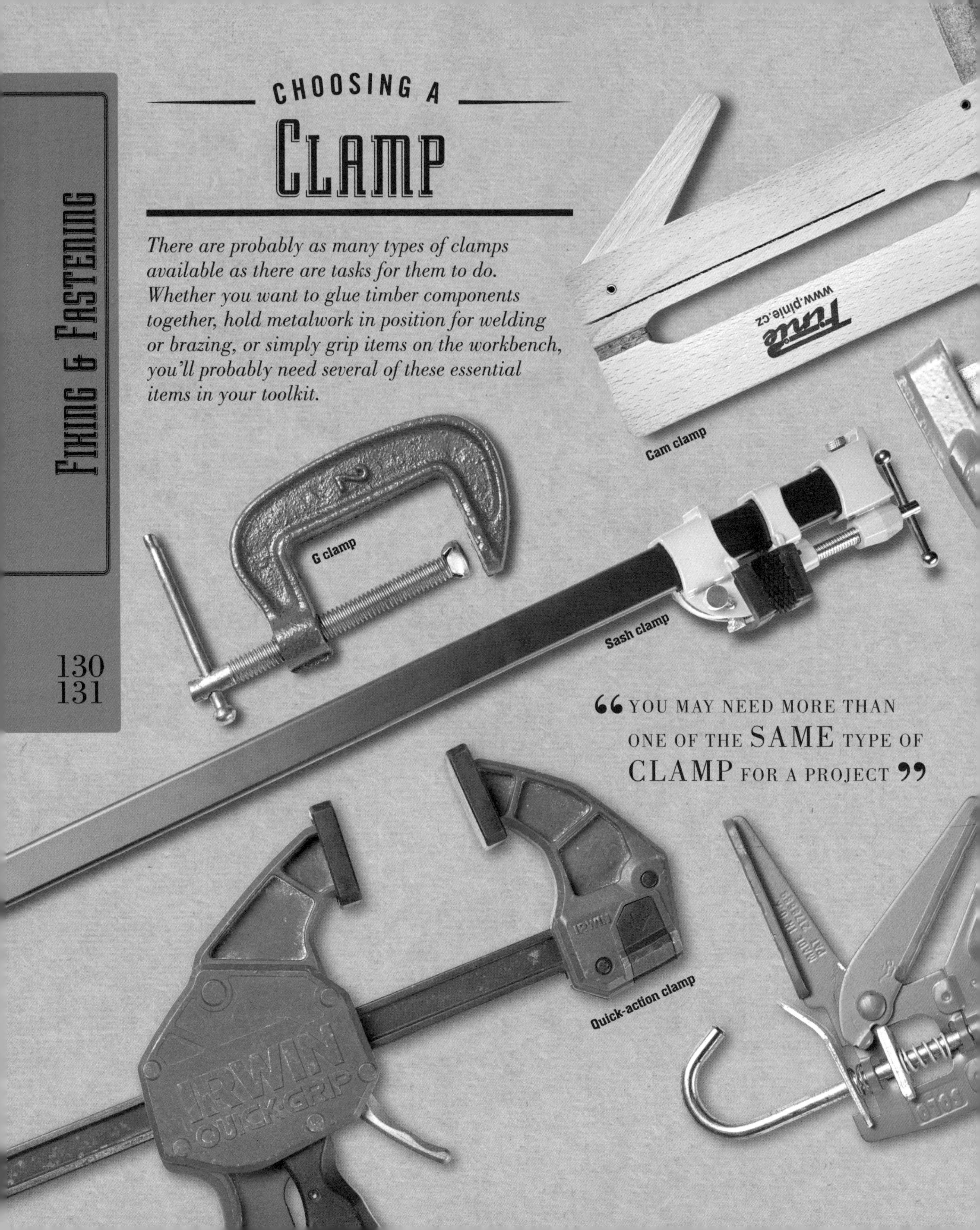

> “YOU MAY NEED MORE THAN ONE OF THE SAME TYPE OF CLAMP FOR A PROJECT”

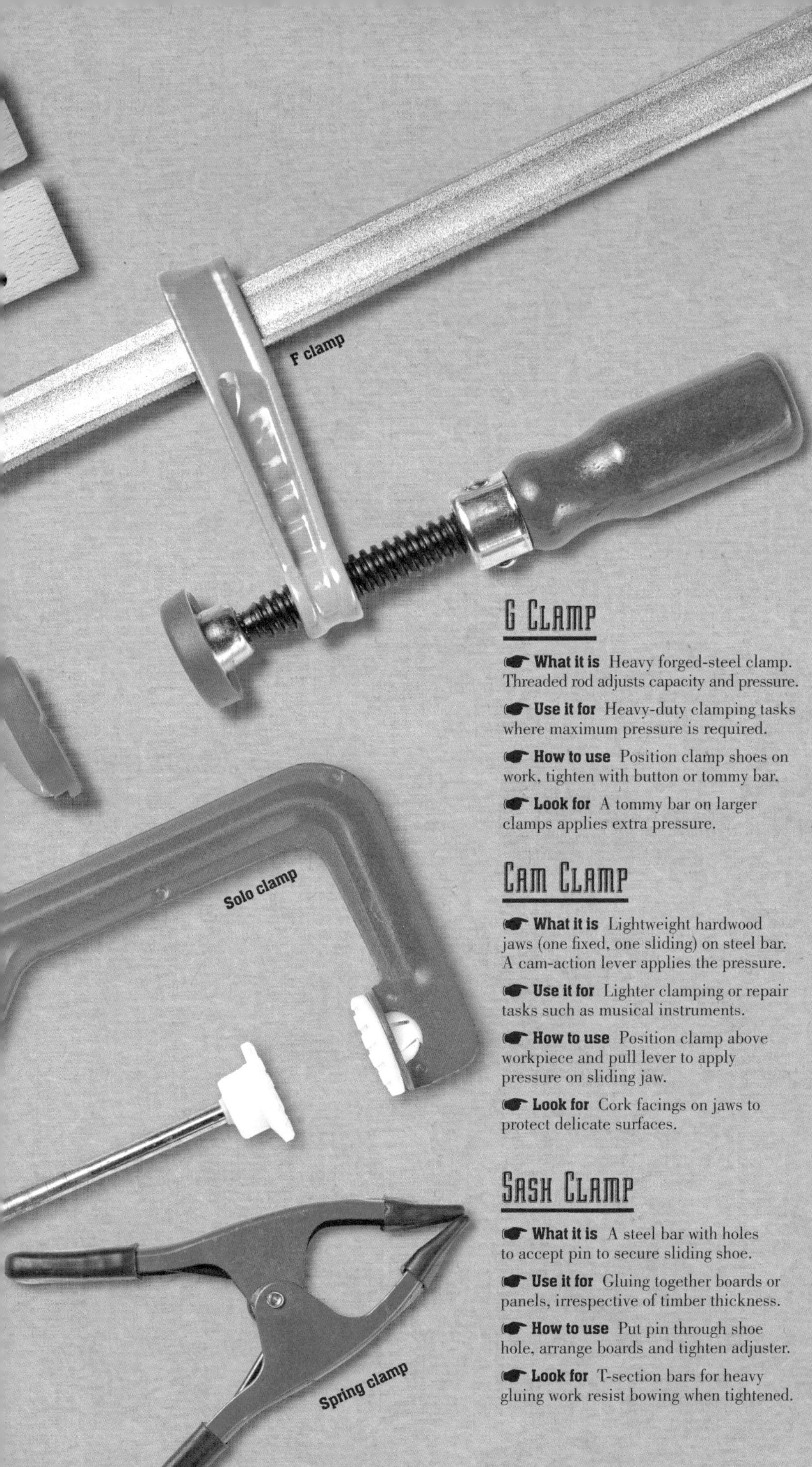

F Clamp

☛ **What it is** Serrated steel bar with one fixed and one sliding jaw, threaded rod with clamping shoe, wood or plastic handle.

☛ **Use it for** General-purpose, heavy clamping tasks. Long bars available, so greater depth capacity than G clamps.

☛ **How to use** Position clamp and slide lower arm up to workpiece. Rotate handle to tighten.

☛ **Look for** Plastic covers on steel shoes to prevent damage to softer surfaces.

Quick-action Clamp

☛ **What it is** Steel bar with fixed, high-density plastic jaw at one end. Opposite jaw slides along bar.

☛ **Use it for** Any situation where single-handed clamping is necessary. Faster to use than G or F clamp.

☛ **How to use** Position clamp around workpiece and squeeze trigger. Reverse jaws on larger clamps to act as spreader.

☛ **Look for** Rubber or plastic shoes on jaws to prevent denting delicate surfaces.

G Clamp

☛ **What it is** Heavy forged-steel clamp. Threaded rod adjusts capacity and pressure.

☛ **Use it for** Heavy-duty clamping tasks where maximum pressure is required.

☛ **How to use** Position clamp shoes on work, tighten with button or tommy bar.

☛ **Look for** A tommy bar on larger clamps applies extra pressure.

Cam Clamp

☛ **What it is** Lightweight hardwood jaws (one fixed, one sliding) on steel bar. A cam-action lever applies the pressure.

☛ **Use it for** Lighter clamping or repair tasks such as musical instruments.

☛ **How to use** Position clamp above workpiece and pull lever to apply pressure on sliding jaw.

☛ **Look for** Cork facings on jaws to protect delicate surfaces.

Sash Clamp

☛ **What it is** A steel bar with holes to accept pin to secure sliding shoe.

☛ **Use it for** Gluing together boards or panels, irrespective of timber thickness.

☛ **How to use** Put pin through shoe hole, arrange boards and tighten adjuster.

☛ **Look for** T-section bars for heavy gluing work resist bowing when tightened.

Solo Clamp

☛ **What it is** Ribbed steel frame with combined handle/lever. Rod runs through frame with plastic clamping shoe attached.

☛ **Use it for** Quick clamping tasks that require the use of only one hand.

☛ **How to use** Position clamp on work, squeeze lever to advance rod and apply pressure. Press small lever to release.

☛ **Look for** Check that the throat capacity is big enough for larger timber.

Spring Clamp

☛ **What it is** Simple steel or plastic jaws hinged together. Pivots with spring action.

☛ **Use it for** Clamping small or lightweight items; temporarily holding items. Only needs one hand to operate.

☛ **How to use** Squeeze ends together to open jaws, position clamp on workpiece and release.

☛ **Look for** Cheaper clamps may not offer sufficient pressure.

"A TOOL IS BUT THE EXTENSION OF A MAN'S HAND, AND A MACHINE IS BUT A COMPLEX TOOL. AND HE THAT INVENTS A MACHINE AUGMENTS THE POWER OF A MAN AND THE WELLBEING OF MANKIND."

HENRY WARD BEECHER

CHOOSING

Pincers or Pliers

Both pincers and pliers operate on the same original principle of two hand-operated levers joined by a fulcrum at one end. Pincers were and are used mainly for holding or levering items such as nails. Pliers have evolved into a multifunction tool, often with a cutting edge, available in a variety of designs for many specialist applications.

> “GRIPPING, TWISTING, CUTTING, OR SHAPING: VERSATILE PLIERS ARE USED IN MANY WAYS”

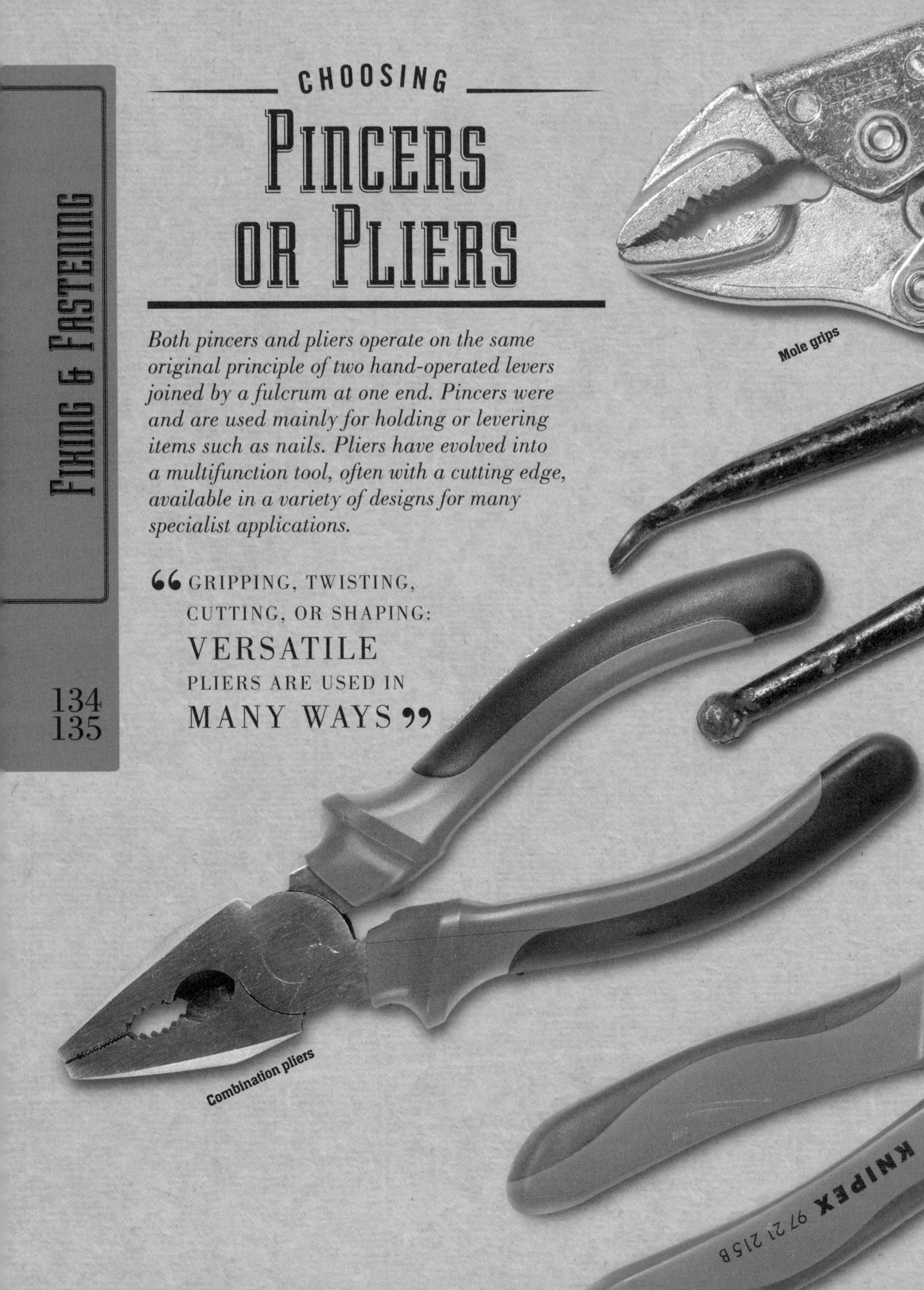

Combination Pliers

What it is Universal handyman's tool with snub-nosed jaws.

Use it for Cutting and bending wire, holding or pulling small items, from bolt heads to pipes.

How to use Hold object using jaw ends; cutting faces are near the fulcrum.

Look for Cutting faces sharp enough to cut cables without causing them to fray.

Side-Cutter Pliers

What it is Pliers designed solely for cutting, crimping, or shaping materials.

Use it for Cutting electrical cables and zip-ties; shaping metal or plastics.

How to use Make precise and delicate incisions or crimps.

Look for Tin snips are side-cutting pliers used to work thin sheets of metal.

Mole Grips

What it is Pliers with over-centred action for holding and gripping. They exert more pressure than conventional pliers and operate with a locking function.

Use it for Exerting a tight grip or lock on anything from a plumbing nut to an item being welded. Can also be used as a temporary handle.

How to use Use screw on handle to set limits of jaw; close handles.

Look for Mole and Vise-Grips are trade names for original locking pliers.

Circlip Pliers

What it is Pliers used for removal and fitting of internal or external wire circlips, often on automotive lines.

Use it for Gently prising apart and moving spring-loaded circlips.

How to use Fit each jaw prong into circlip rings. Open jaws and ease off.

Look for Reversible or adjustable models for internal/external circlip use.

Traditional Pincers

What it is Steel or iron lever, fulcrum, and perpendicular cutting jaws.

Use it for Gripping nail heads and pulling them free.

How to use Grip handles. Once teeth are under nail head, gently lever the nail until it can be pulled out.

Look for A forked-shaped lever on one handle for digging under a nail head.

Crimping Pliers

What it is Pliers with circular cutters and one straight cutter for stripping and cutting coated electrical cables.

Use it for Stripping the ends of electrical cables for wiring to appliances.

How to use Cut cable with flat edge. Use side markings to identify cable size. Clamp cable and pull to strip it.

Look for A crimping function for working on small connectors.

> "Use the RIGHT SIZE pliers for the job, never EXTEND the arms"

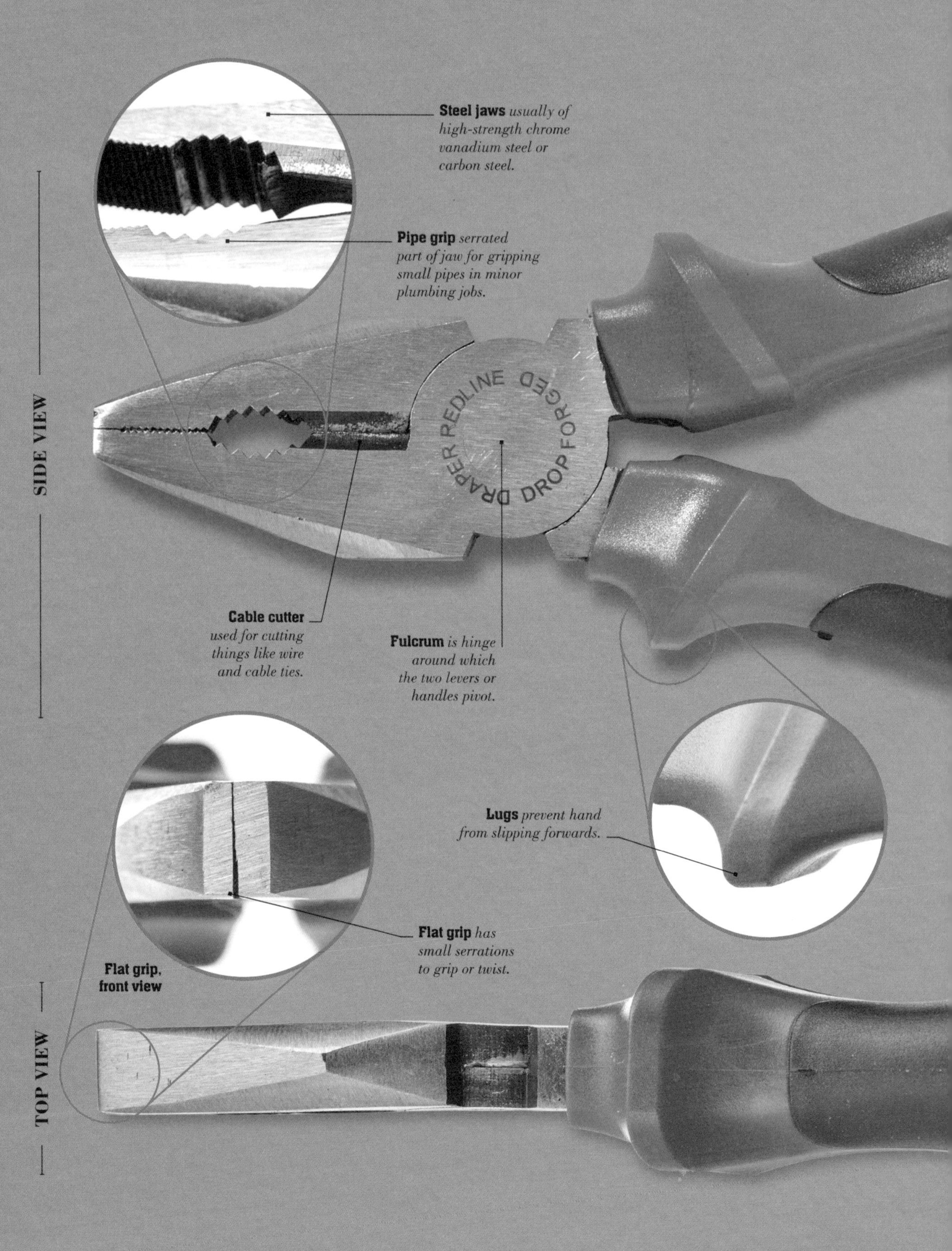
SIDE VIEW
Steel jaws *usually of high-strength chrome vanadium steel or carbon steel.*
Pipe grip *serrated part of jaw for gripping small pipes in minor plumbing jobs.*
REDLINE DROP FORGED DRAPER
Cable cutter *used for cutting things like wire and cable ties.*
Fulcrum *is hinge around which the two levers or handles pivot.*
Lugs *prevent hand from slipping forwards.*
Flat grip *has small serrations to grip or twist.*
Flat grip, front view
TOP VIEW

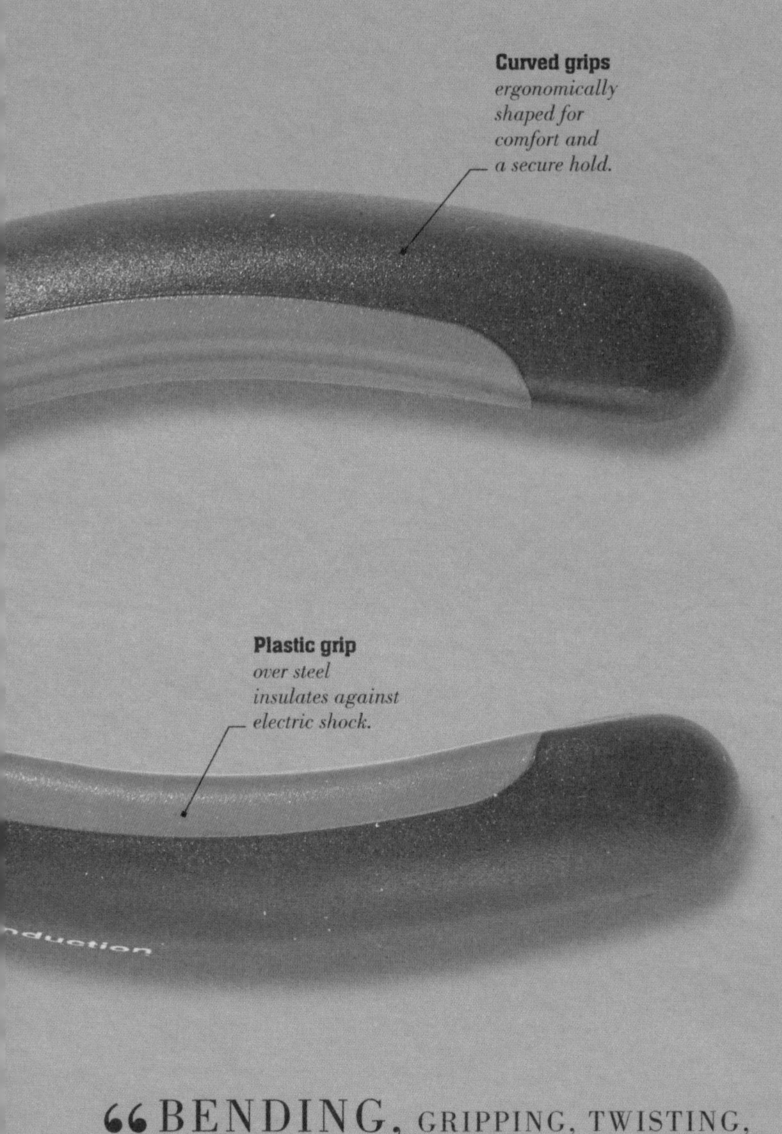

STRUCTURE OF COMBINATION PLIERS

Every toolbox and most multitools contain a pair of pliers that can be used for many minor everyday tasks, such as opening a tight bottle cap, and for more involved jobs like cutting and shaping wire. Hand-sized and universal in design, combination pliers last for years, with virtually no maintenance.

“BENDING, GRIPPING, TWISTING, OR PULLING... COMBINATION PLIERS ARE A TOOLKIT ESSENTIAL”

FOCUS ON...

PLIER TYPES

Combination pliers are one of the go-to tools in the box, but because of their versatility they can often be used incorrectly, either failing to complete the task or even damaging the job. There are many specialist plier types, some suited for multiple applications and some for a single specific task.

Combi pliers
Highly versatile pliers with pipe grip, cutter, and flat grip. Some also have side cutters.

Long-nosed pliers
For delicate and small-scale gripping jobs. Available in various lengths and designs.

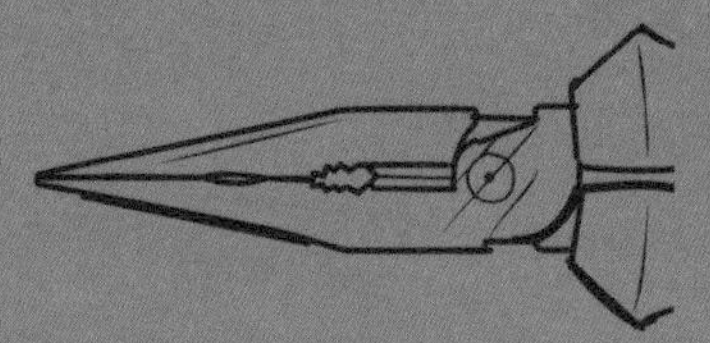

Spring-loaded small pliers
The spring helps open the handles and is very useful for cutting or repetitive tasks.

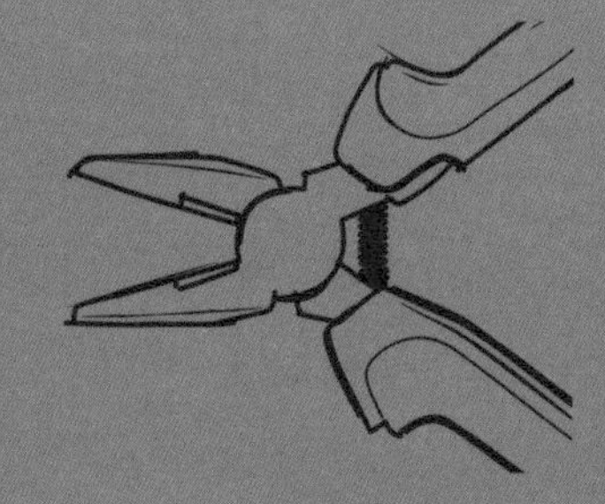

USING

Combination Pliers

Because they are easy to use and are suitable for multiple small tasks, using combination pliers is less about technique and more to do with matching them to the right job – as well as not demanding more than the tool can achieve. With a good grip on the handles and a steady hand, you can use a pair of combination pliers to perform many DIY essentials.

The Process

Before you start

- **Check the task** Is this a job for combination pliers, or would a specialized tool be better?
- **Check the pliers** Make sure the cutting and clamping faces are both in good condition.
- **Another tool?** Often pliers will be used with another tool, so check that the latter is the correct type and size for the job.
- **Protect your eyes** Always wear protective goggles when cutting wire or metal. Even the smallest piece could do a lot of damage to an eye.

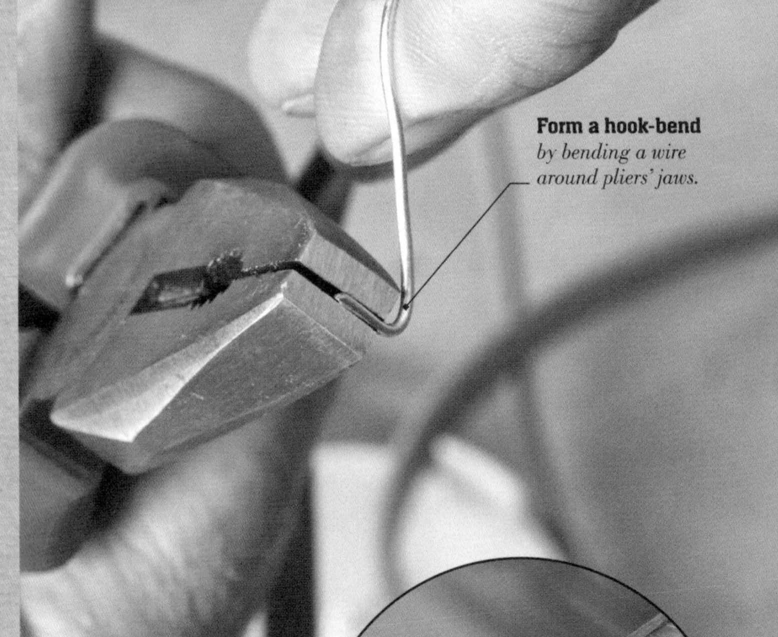

Form a hook-bend *by bending a wire around pliers' jaws.*

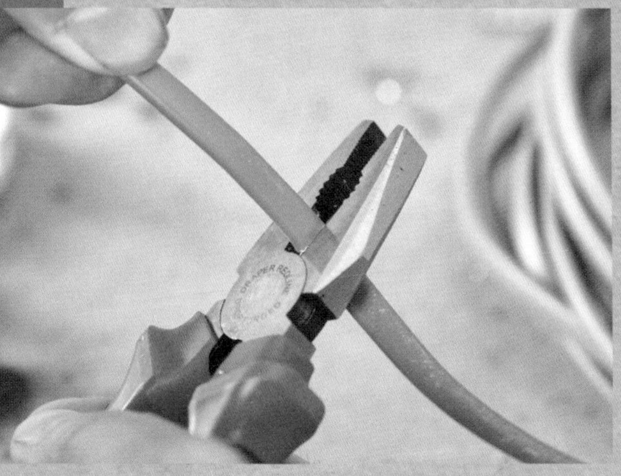

Cutting wire

For cutting that doesn't require a precision finish, grip the pliers and clamp the straight edge onto the wire. Increase the pressure until it cuts through. If that isn't possible, it may still weaken the wire enough to be snapped off.

Bending or splicing wire

Grip the end of the wire in the flat part of the pliers' jaws and gently twist, allowing the cable to wrap itself around the almost-closed jaws. This technique is useful for making a loop or hook. To splice wires, grip with the jaws and twist to form one large, rope-like wire.

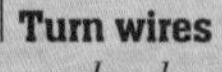

Turn wires *evenly when splicing them so that they twist together.*

FOCUS ON...

Lever Action

Pliers transform a small force applied to two parallel levers (the handles) through a fulcrum (where the handles join) to a greater force in the jaws, providing users with a stronger grip on objects than would be possible by hand. The longer the handles, the greater the force exerted by the jaws. More grip is created closest to the fulcrum, and some pliers have very short jaws to take maximum advantage of this.

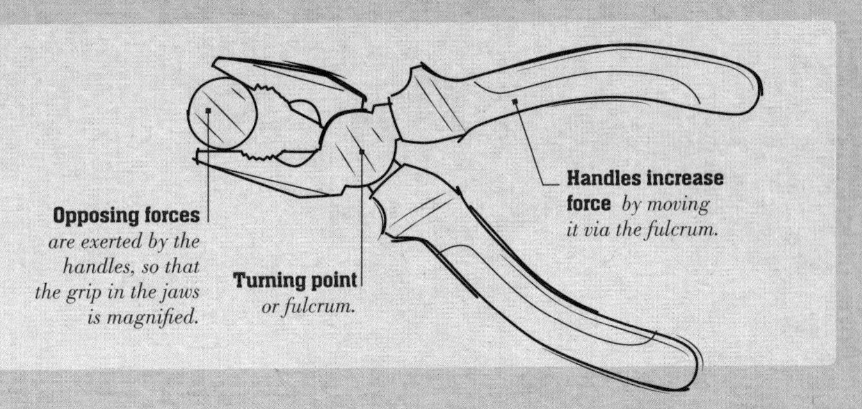

Opposing forces *are exerted by the handles, so that the grip in the jaws is magnified.*

Turning point *or fulcrum.*

Handles increase force *by moving it via the fulcrum.*

Pull a cable *through studs with the pliers' grip.*

Pulling or pushing cables

When rewiring electrics, such as plugs or light switches, you may have to guide cables through walls or studs. Pliers can grip even the smallest of items securely, so are ideal for this type of job. Simply grip the end of the cable in the flat part of the pliers' jaws and gently pull or push it in the desired direction until enough cable is exposed to complete the task by hand.

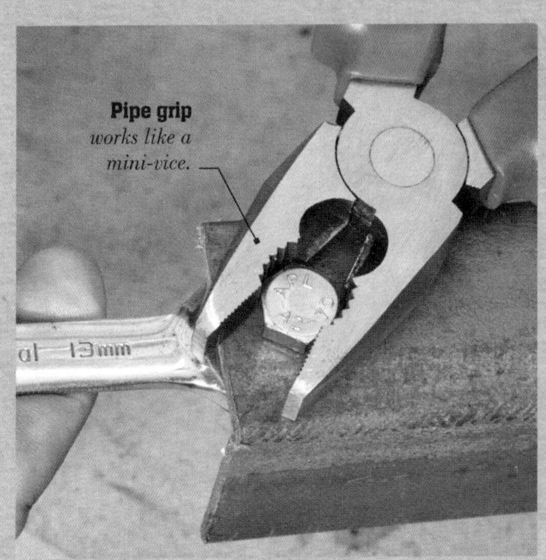

Pipe grip *works like a mini-vice.*

Holding objects with pipe grip

The serrated, semi-circular part of the pliers' jaws is known as the pipe grip, and it makes a useful clamp for bolts, nuts, and small pipes if you don't have another specialized tool available. Close the pipe grip gently around the object and grip it with one hand while you use a wrench or spanner to loosen or tighten the corresponding nut with the other.

After you finish

Inspect your edges Check your pliers over for possible damage, especially on the cutting edges.

Clean your tools Wipe debris or dirt from the pliers' jaws. If necessary, add a drop of oil to the pivot. Store them away carefully.

Maintain tools for

Fixing & Fastening

These tools are simple and robust, needing little maintenance. Keep them free from corrosion and they will last for many years with little upkeep.

CHECKING FOR DAMAGE

With few moving parts and basic design, most fixing and fastening tools are easy to maintain. Inappropriate use and corrosion can cause damage to the tool or part acted upon.

1 Impact damage
Take care not to drop the tool and always use it for the job it was intended for.

2 Rust risk
Steel tools can rust and seize up if left in wet or damp conditions. Store in the dry and wipe with an oily rag.

3 Moving parts
Keep moving parts such as pivots and threaded barrels running smoothly with a few drops of light machine oil.

BATTERY CARE

Many handheld power tools, and especially drills, are powered by rechargeable batteries that have a long service life if looked after. For extensive use it's always worth having a fully charged spare.

Charging
Modern rechargeable battery packs can achieve a full charge in around an hour, and are able to power a drill, saw, or sander where interchangeable. Make sure the battery is fully charged before starting each job.

Storage
After finishing a job, it's a good idea to return the battery to a full state of charge before storage. This ensures that the tool is ready for the next job and prevents the battery discharging if levels are low.

Spare batteries *let you keep working while your other battery charges.*

Tools	Inspection
Screwdrivers	▪ Check tip is not bent or damaged – replace screwdriver if tip or shaft is bent ▪ For cordless screwdriver: check functions work on powered models – ensure battery is fully charged
Wrenches	▪ Check tool is not bent and that the flats that engage with fastener are not damaged or rounded out ▪ Check moving parts, like barrel adjuster or ratchet mechanism, operate smoothly
Drills	▪ Check all functions work on powered models – ensure battery is fully charged ▪ On hand drill check handle turns smoothly and chuck is not damaged ▪ Check drill bits for wear, damage, or bent shaft
Vices	▪ Check closing and opening action of jaws and adjuster handle (tommy bar) ▪ If vice is fitted with quick-release mechanism, check this works smoothly
Clamps	▪ Check jaws slide smoothly. Replace plastic jaw caps if missing
Pincers & Pliers	▪ Check for damage or corrosion to jaws – ensure pivot operates smoothly ▪ Ensure screw and locking mechanism operates smoothly on locking pliers ▪ Check for damaged prongs on circlip pliers

	Cleaning	Oiling	Adjusting	Storage
	▪ Wipe handle, shaft, and tip with dry rag ▪ For cordless screwdriver: keep air vents clean – remove debris with vacuum cleaner and check after any job producing fine dust			▪ Keep in rack or tool box
	▪ Wipe exposed metal with oily rag, or rub down corrosion with wire wool	▪ A drop of machine oil on any moving parts, such as a threaded barrel adjuster, sliding surfaces of jaws, between ratchet collar and spanner head, or on all mechanisms of socket wrench	▪ For torque wrench, refer to service schedule for possible recalibration by trained technician	▪ Hang on hook or in appropriate sized slot in tool box, or drawer in the dry
	▪ Keep air vents clean – remove debris with vacuum cleaner. Check after any job producing fine dust from drilling masonry, timber, etc ▪ Wipe body of tool clean with damp cloth (occasionally)	▪ Apply machine oil to moving parts on hand drill		▪ Keep cordless drill in original plastic case, where provided ▪ Remove battery from tool if unlikely to be used for some time ▪ Keep battery charged
	▪ Wipe dust and debris from threads occasionally – apply grease to concealed threads (e.g. on woodwork vice) if they appear dry or rusted (around every six months)	▪ Spray moving parts with corrosion inhibitor (monthly)		
	▪ Re-grease threads if they appear dry or rusted			▪ Hang on rack or pegs on wall
	▪ Wipe exposed metal with oily rag, or rub down corrosion with wire wool	▪ Apply drop or two of light machine oil to pivots, springs (if fitted), and threads of any locking mechanisms		▪ Keep in tool box or drawer in the dry ▪ It's useful to keep electrical pliers in a small box with electrical spares

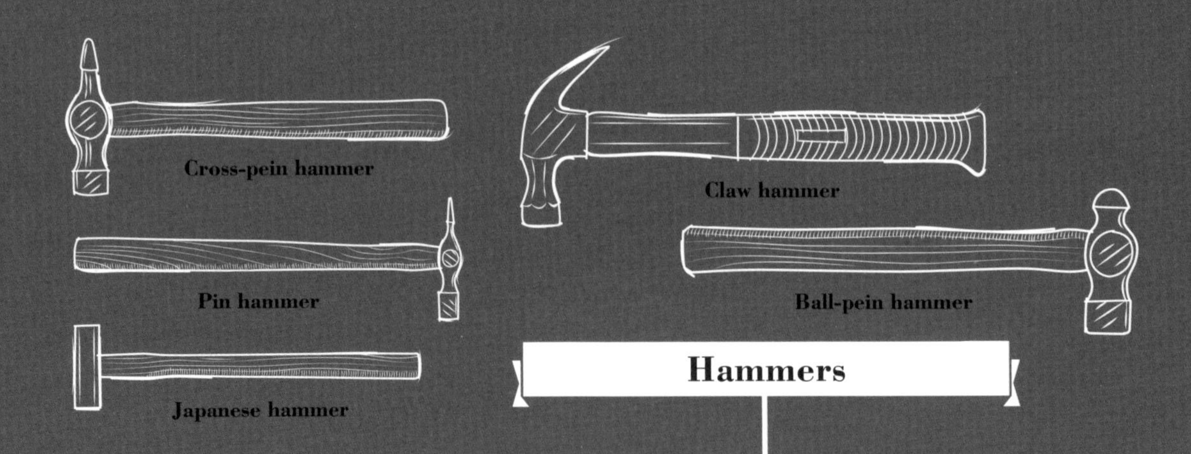

THE TOOLS for STRIKING & BREAKING

These robust tools are surprisingly versatile and can be used for everything from heavy-duty digging and demolition to tapping in small nails and delicately shaping metal.

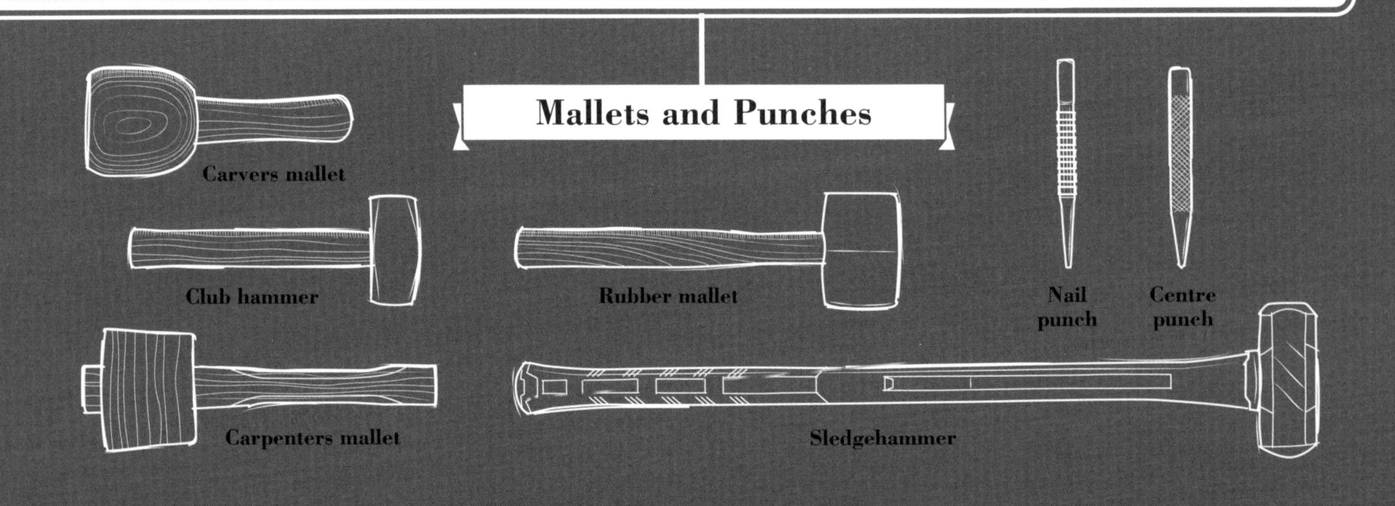

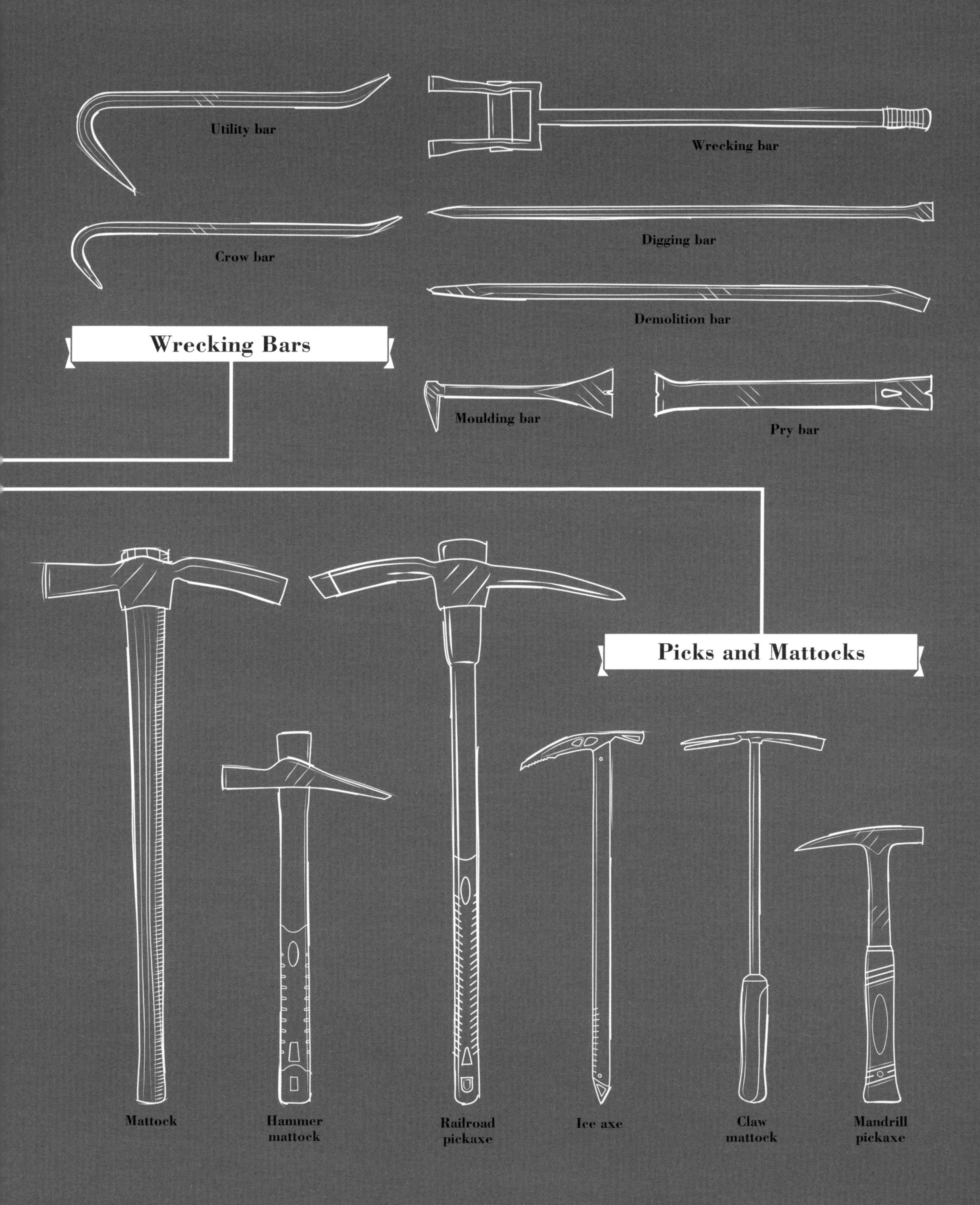

Utility bar
Wrecking bar
Digging bar
Crow bar
Demolition bar
Wrecking Bars
Moulding bar
Pry bar
Picks and Mattocks
Mattock
Hammer mattock
Railroad pickaxe
Ice axe
Claw mattock
Mandrill pickaxe

HISTORY OF

STRIKING & BREAKING

2.6–1.7 MYA — HAMMER STONES AND SOFT HAMMERS

Some of the earliest tools were simply sticks or rocks used for stabbing and crushing. Early hammers were wooden clubs and served many purposes. In the Old Stone Age, "soft hammers" made from pieces of wood, antlers, or bone worked with stone were used to work flint.

Antlers were ideally shaped for hammering.

Pounder or hammerstone.

Early hammer tools

"THE MOMENT MAN FIRST PICKED UP A STONE OR A BRANCH TO USE AS A TOOL, HE ALTERED IRREVOCABLY THE BALANCE BETWEEN HIM AND HIS ENVIRONMENT."

JAMES BURKE

2.6–1.7 MYA — DIGGING STICKS

One of the oldest tools, still used by some subsistence cultures today, is the digging stick. A sturdy stick with a pointed end and handle is the ancestor of many hand tools, including the pickaxe and mattock. It was used for a multitude of tasks, including digging out roots and tubers.

10,000–1900 BCE — EARLY AGRICULTURE

The first handled, or hafted, hammer was used in Neolithic times, possibly as a miner's maul. An oval stone formed the axe head, which was attached to a branch by twisted fibres.

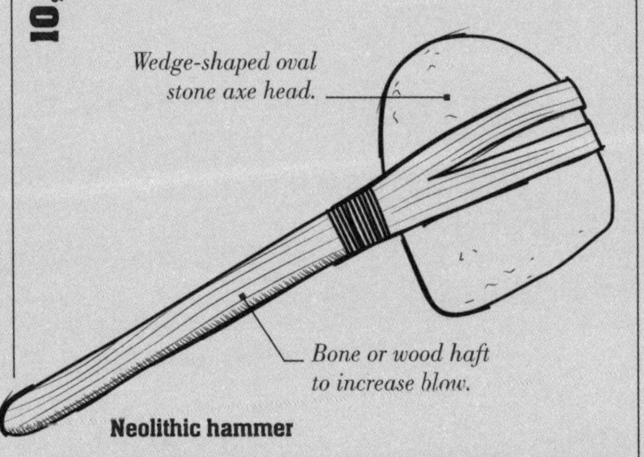

Neolithic hammer

6500 BCE — METAL-AGE HAMMERS

The hammer as we know it was formed, and it was used for metallurgy, nailing, and riveting.

"HAMMER YOUR IRON WHEN IT IS GLOWING HOT."

PUBLILIUS SYRUS
C.85–43 BCE

2300 BCE ANTLER PICK

UK Neolithic flint-mining sites such as Norfolk's Grime's Grave show that the sharp antlers of red deer were used as picks to excavate mines.

Main shaft of antler forms natural handle.

Antler pick

Some shafts here are more than 9m (30ft) deep.

Grime's Grave

London

Neolithic flint mine

3000–1900 BCE SHAFT HOLES

In the Middle East, bronze and copper hammer heads were pierced with a shaft hole, through which wooden handles could be fitted.

3000–1900 BCE MATTOCKS

During the Bronze Age, bronze mattocks were used in ancient Greece, having replaced earlier similar tools made of antler or stone. They were much the same shape as they are today, and the tool has changed relatively little since its creation.

1000 BCE CAST-IRON PICKS

The Iron Age discovery of carburization, where iron absorbs carbon during the smelting process, meant that pickaxe heads became harder, larger, and heavier. Because they were made from iron, their edges also remained sharper for longer than smaller bronze versions, and this durability increased the pace and efficiency of work in activities such as mining.

30KG (66LB)

The weight of the largest stone hammer found at Great Orme, a Bronze Age copper mine in North Wales. It is just one of 2,500 stone hammers of varying weights and sizes found at the mine, which yielded copper of such a high grade that some has been found in bronze made in France and Holland.

50,000 PICKS

The estimated number of red deer antler picks used to create Grime's Grave, a Neolithic flint mine located around 130km (80 miles) northwest of London. The mine, which was begun around 2300 BCE, and covered 34 acres, was worked for over 600 years to exploit a vast seam of flint located beneath chalk beds.

735 BCE–500 CE CLAW HAMMER

The Romans developed the claw hammer and peen hammers (with rounded ends on the opposing face). They also invented the file-maker's hammer with two chisel-shaped heads, which was used to score iron.

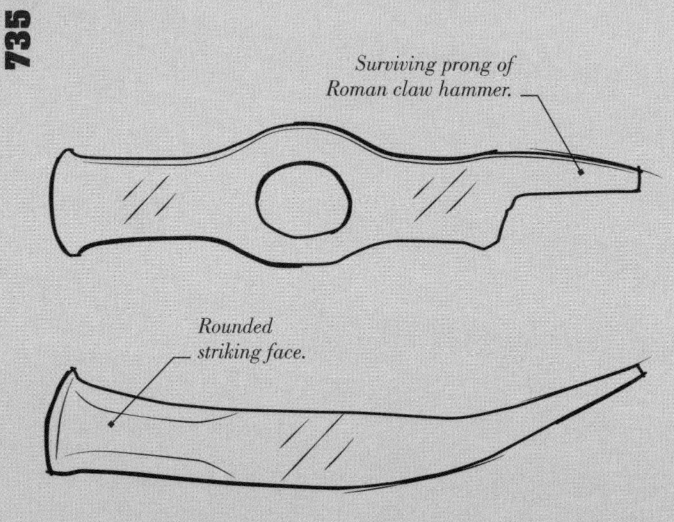

Roman claw hammer head

1000 BCE HEAVY HAMMERS

In Europe, hammer heads began to be pierced with a shaft hole to fit the handle – although this had been the practice in the Middle East for centuries. Iron, however, meant that much weightier iron hammer heads could then be fitted to sturdier wooden shafts, a development that led to the creation of tools such as early sledgehammers and forging hammers for use in blacksmithing. The problem of keeping such heads securely fastened on wooden shafts, though, remains to this day!

Choosing a

Hammer

From driving in small nails to demolition work, hammers are important tools. They vary in size and pattern: some are for specific tasks such as shaping metal; others have a more general purpose. Shafts on smaller tools are traditionally ash or hickory, larger hammers may be steel or carbon fibre, and mallets are often made entirely of hardwood.

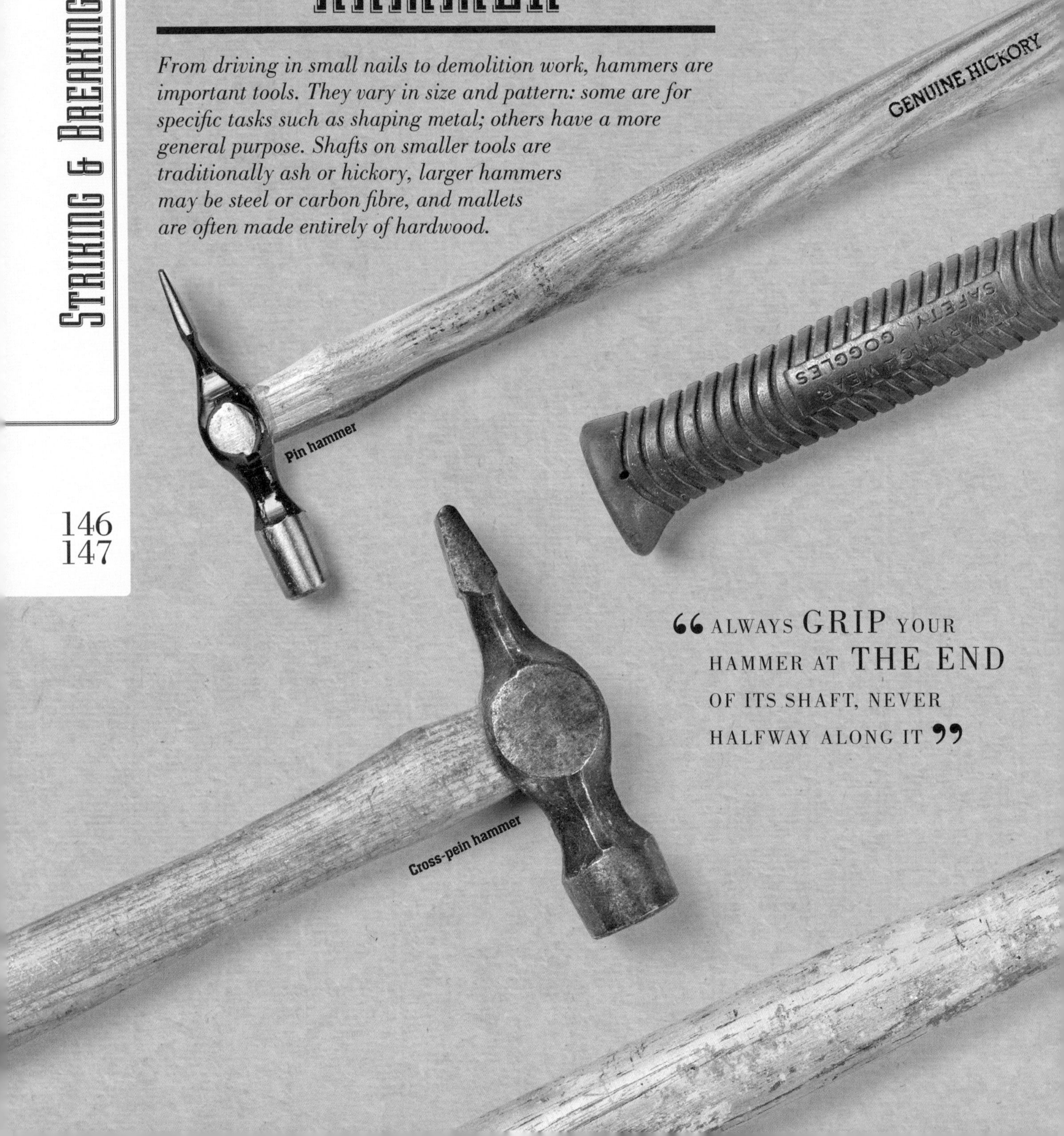

Genuine Hickory

Warning: Wear Safety Goggles

Pin hammer

Cross-pein hammer

“Always GRIP your hammer at THE END of its shaft, never halfway along it”

> “NEVER USE A HAMMER IF THE **HEAD** IS DAMAGED OR HAS WORKED **LOOSE**”

Cross-pein Hammer

What it is Forged-steel head with large striking face, wedge-shaped end, hardwood shaft. Weight: up to 450g (1lb).

Use it for Larger nails. General use, including assembling woodwork joints.

How to use Hold nail between finger and thumb, tap with wedged end. Rotate hammer to finish with striking face.

Look for Striking face should be slightly convex, not dead flat. Medium weight (350g/12 oz) is a good choice.

Pin Hammer

What it is A forged-steel head with small striking face, a wedged end, and a hardwood shaft. Weight: up to 100g (3.5oz).

Use it for Driving panel or veneer pins, tacks, small nails in finer woodwork, upholstery, and hobby work.

How to use Hold pin and tap with wedged end. Once the pin is set, rotate hammer to finish with opposite face.

Look for Check hammer balances well as you swing it. Make sure shaft is firmly wedged into head.

Japanese Hammer

What it is Small steel or bronze head with one flat and one convex face. Slim oak shaft. Weight: up to 375g (13oz).

Use it for Driving smaller nails, striking chisels.

How to use Strike chisels with flat face, nails with convex face.

Look for A head that clearly identifies flat or convex face.

Claw Hammer

What it is Striking face with claw. Steel, carbon-fibre, fibreglass or hardwood shaft. Weight: up to 680g (1.5lb).

Use it for Driving large nails. Removing pins or nails with claw.

How to use Slide V of claw over head of nail, grasp shaft, and exert leverage. Use thin offcut to prevent denting surface.

Look for Fine V tapering of claw to grip smallest pinheads. A shock-resistant shaft (unless hardwood) for comfort.

Ball-pein Hammer

What it is Steel head with ball and flat striking face. Weight: up to 1.1kg (2.4lb). Hardwood, fibreglass, carbon-fibre shaft.

Use it for Shaping/bending metal, setting rivets with ball face.

How to use Tap rivet with hammer to create mushroom head.

Look for A properly hardened, tempered head.

CONTINUED

Centre punch

Carving mallet

Nail punch

Rubber mallet

STANLEY

FATMAX

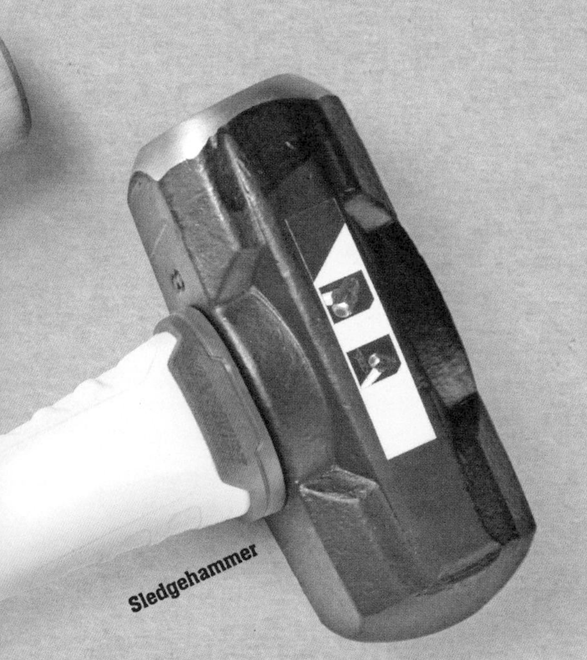

Carving Mallet

What it is Wooden tool with a round head of dense hardwood, specified by weight or head diameter. Turned handle.

Use it for Striking woodcarving chisels and gouges.

How to use Swing mallet so head strikes tool handle square on.

Look for Correct weight. A heavy mallet will be more tiring to use.

Rubber Mallet

What it is Twin-face rubber head on hardwood or fibreglass shaft. Weight: up to 800g (2lb).

Use it for Assembly work where surfaces could be damaged. Driving pegs.

How to use Grip end of shaft and swing mallet to strike workpiece squarely.

Look for Mallets with replaceable screw-in faces (nylon, brass, or copper).

Nail Punch

What it is Steel tool with shaft. Tip comes in sizes to match nail heads.

Use it for Punching nails below timber surface.

How to use Place tip on nail head, tap firmly with hammer until head is flush.

Look for Square section head means punch cannot roll off the workbench.

Sledgehammer

What it is Heavy, twin-faced, steel head on long hardwood or fibreglass shaft. Weight: up to 6.4kg (14lb).

Use it for Breaking concrete, driving fence posts. Use with wedge to split wood.

How to use Hold in both hands due to its weight and swing it like an axe.

Look for Check for damage to shaft and wrap with repair tape if necessary.

Centre Punch

What it is Small steel tool with knurled shaft. One end ground to a point.

Use it for Making small indents in metal or timber to set drill bit.

How to use Position punch tip on pencil mark, tap gently with hammer.

Look for Square section head means punch cannot roll off the workbench.

Club Hammer

What it is Heavy-duty, dual-face steel head. Hardwood or fibreglass shaft. Weight: 1–1.8kg (2.2–4lb).

Use it for Striking cold chisels, general demolition work.

How to use Don gloves and goggles. Strike end of chisel with face of hammer.

Look for Choose a lighter weight hammer, particularly if you are unused to heavy work.

Carpenters Mallet

What it is Hardwood tool with two broad tapered striking faces, flared shaft.

Use it for Striking woodworking chisels or similar.

How to use Hold mallet end. Swing so face strikes chisel handle squarely.

Look for Splits on the striking faces. Re-glue and reshape if necessary.

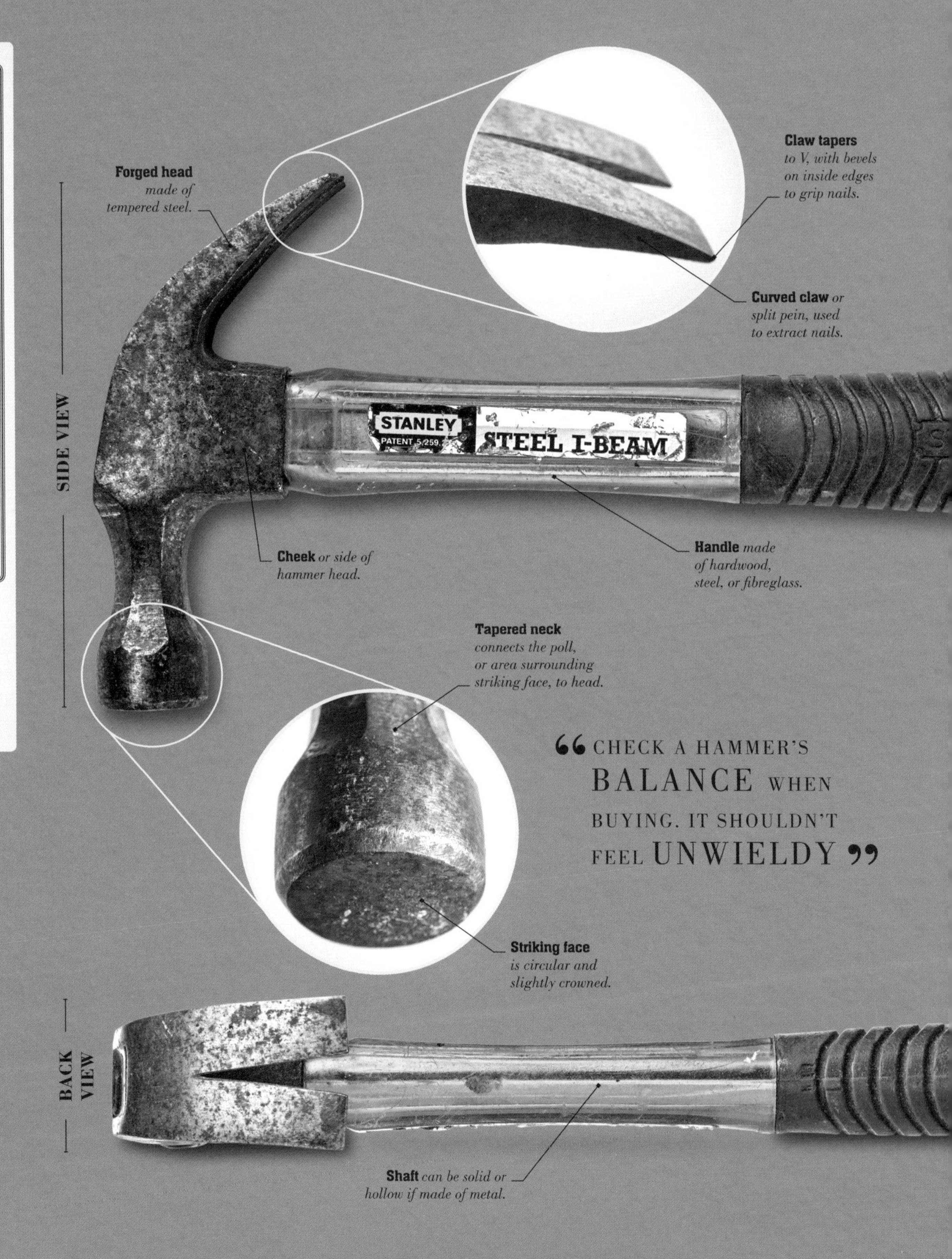
SIDE VIEW
Forged head made of tempered steel.
Claw tapers to V, with bevels on inside edges to grip nails.
Curved claw or split pein, used to extract nails.
STANLEY
PATENT 5,259,
STEEL I-BEAM
Cheek or side of hammer head.
Handle made of hardwood, steel, or fibreglass.
Tapered neck connects the poll, or area surrounding striking face, to head.
“CHECK A HAMMER’S BALANCE WHEN BUYING. IT SHOULDN’T FEEL UNWIELDY”
Striking face is circular and slightly crowned.
BACK VIEW
Shaft can be solid or hollow if made of metal.

“NEVER USE A HAMMER THAT HAS A LOOSE HEAD OR A CRACKED HANDLE”

STRUCTURE OF A CLAW HAMMER

The claw hammer is unique in its capacity to remove nails, as well as drive them into timber and other materials. Nails are removed with a curved or straight claw with a V-shaped groove, which grips the nail head and levers it out. It's a vital general-purpose tool for tasks around the house or workshop.

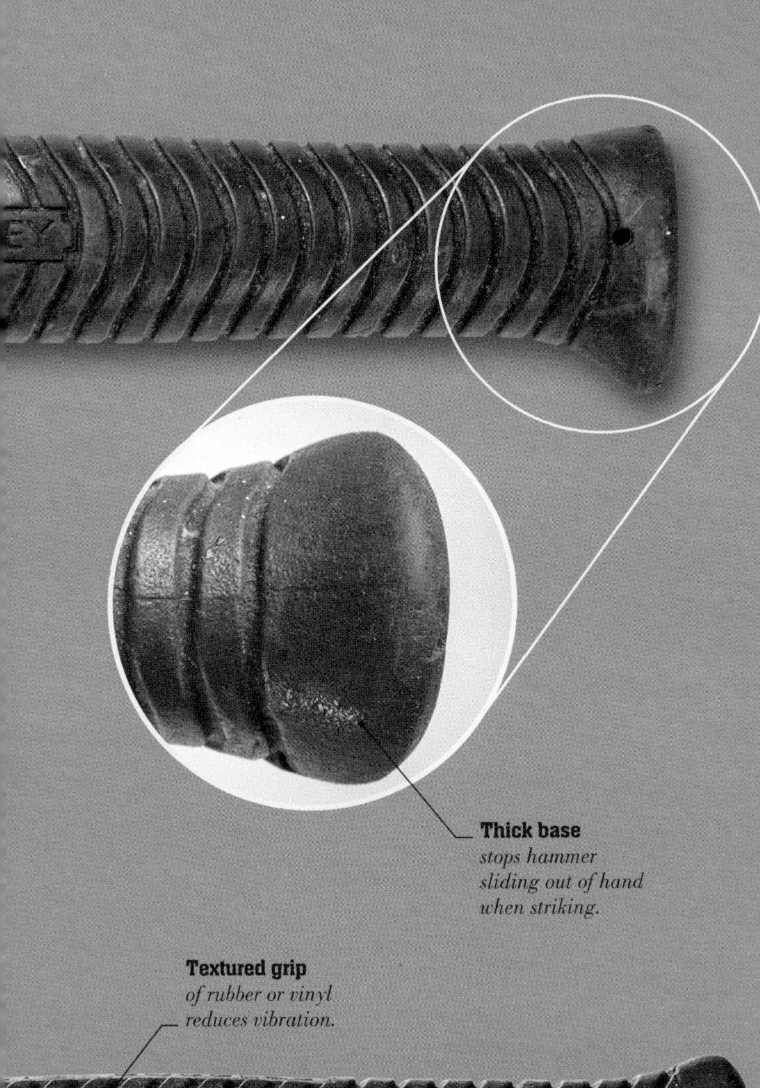

FOCUS ON...

HAMMER HEADS

A hammer is defined by its pattern and head weight, rather than by the weight of the entire tool. A pin hammer typically weighs 100g (4oz), while a ball-pein hammer may be ten times heavier. Some high-tech tools feature anti-vibration pads to isolate the head from its shaft, reducing shock for the user when hammering. A framing hammer has almost straight claws and a large, textured, rather than smooth, striking face, which is designed to prevent or limit skidding off the nail heads.

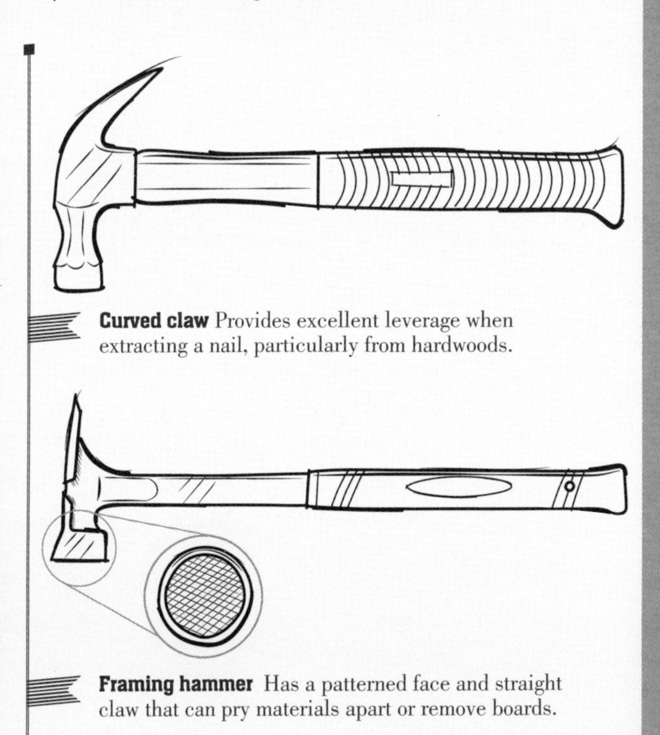

Curved claw Provides excellent leverage when extracting a nail, particularly from hardwoods.

Framing hammer Has a patterned face and straight claw that can pry materials apart or remove boards.

Using a
Claw Hammer

Using a hammer to drive in nails is generally a faster fixing method than driving screws by hand. It's arguably more permanent, though, so ensure that your timber is properly positioned. Bent nails can sometimes be straightened, but you may need to use the hammer's claw to extract them. Use an offcut under the head to avoid damaging surfaces when levering a nail.

The Process

Before you start

- **Work safely** Always wear eye protection when using anything larger than a pin hammer. Even the smallest chip can damage an eye.
- **Inspect hammer** Check the hammer head for chips or flaws. Make sure it's clean to avoid slipping off when striking a nail.
- **Select nail** Always select the correct type and size of nail for the task. Ideally, a nail's length should be three times the depth of the thinnest component when nailing two items together. Otherwise, the fixing may not hold.

1 Choose nail position
Mark the nail position with a pencil, if necessary. Where a nail is to be used close to the end of board, drill a small guide hole first to prevent the timber from splitting, especially if you're using hardwood.

2 Set the nail
Position the point of the nail perpendicular to the mark, gripping it between thumb and forefinger. Holding the nail upright, tap gently with the hammer a few times to get it established.

Nail point *should be held vertically over the mark.*

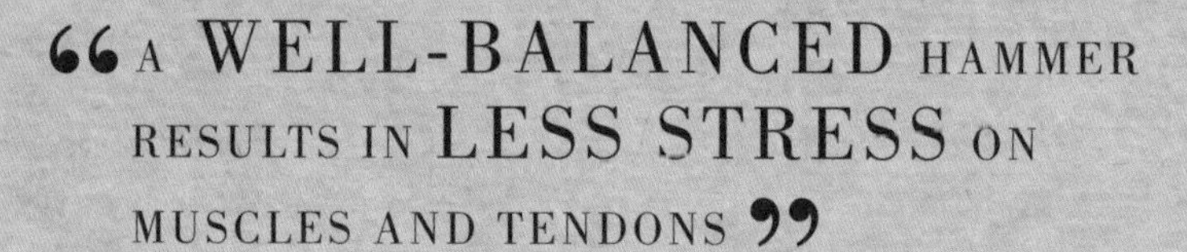

"A WELL-BALANCED HAMMER RESULTS IN LESS STRESS ON MUSCLES AND TENDONS"

FOCUS ON...

The Strike

Using a hammer is rather like operating a lever. Seen as an extension of your arm, the elbow acts as a fulcrum as the tool is gripped and swung downwards. Force is transferred to the hammer head, which then contacts the nail. A well-balanced tool should require little effort to drive any nail completely with just a few firm blows. Always aim to hit the nail head with the hammer's striking face at 90 degrees.

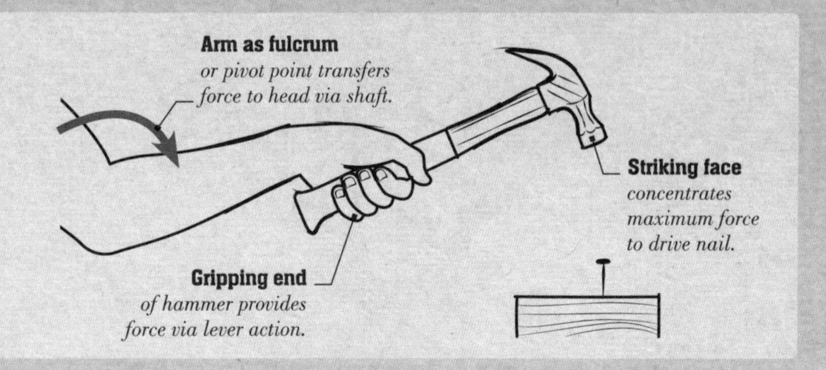

3 Strike the nail

Move your fingers away and strike the nail more firmly, swinging the hammer from the elbow. Be sure the face meets the nail head squarely. Stop hammering when the nail head is just above the wood's surface.

Remove *any poorly placed nails using the claw.*

4 Finish the drive

For the cleanest finish, use a nail punch to drive the nail home. Select a punch that is slightly smaller than the nail head. With your hand resting on the timber to steady the punch, hold it on the nail head. Tap with the hammer until flush or just below the surface.

After you finish

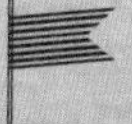

Apply filler If you want to conceal the nail fixing, apply a suitable filler to the hole that matches the colour of the wood.

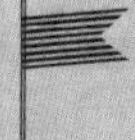

Clean up your tools Keep the striking face of the hammer clean by running it across fine abrasive paper.

THE PHILOSOPHY *of* TOOLS

"A WORKER MAY BE THE HAMMER'S MASTER, BUT THE HAMMER STILL PREVAILS. A TOOL KNOWS EXACTLY HOW IT IS MEANT TO BE HANDLED, WHILE THE USER OF THE TOOL CAN ONLY HAVE AN APPROXIMATE IDEA."

MILAN KUNDERA

CHOOSING

Picks and Mattocks

They may not look like it, but picks and mattocks are highly versatile tools that can be used for digging, chopping, levering, and breaking rock or cement – even as a life-saving anchor in ice. Often confused as the same tool, the mattock's head is different from that of a pickaxe, but both have long handles to give a powerful swing.

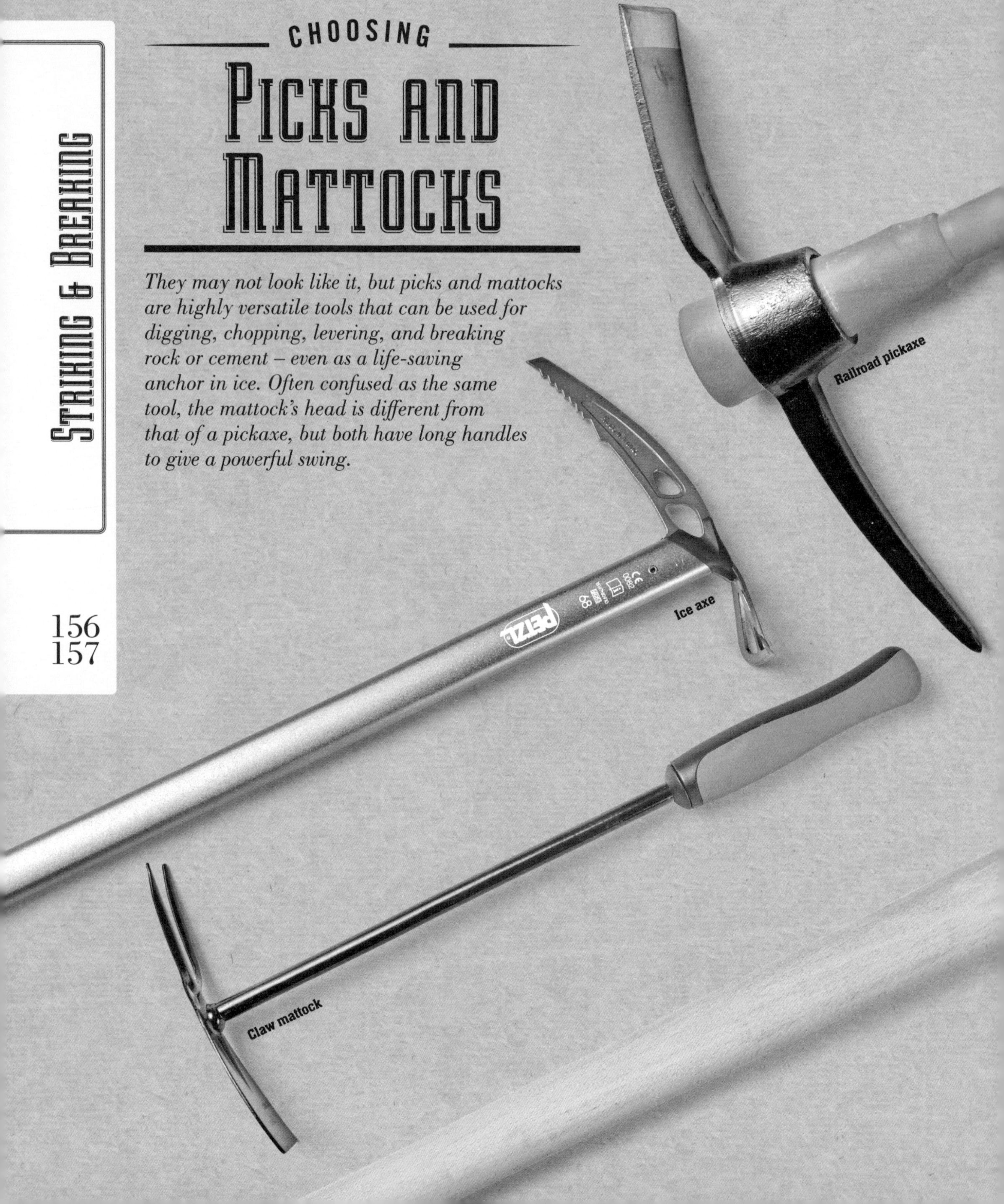

Railroad pickaxe

Ice axe

Claw mattock

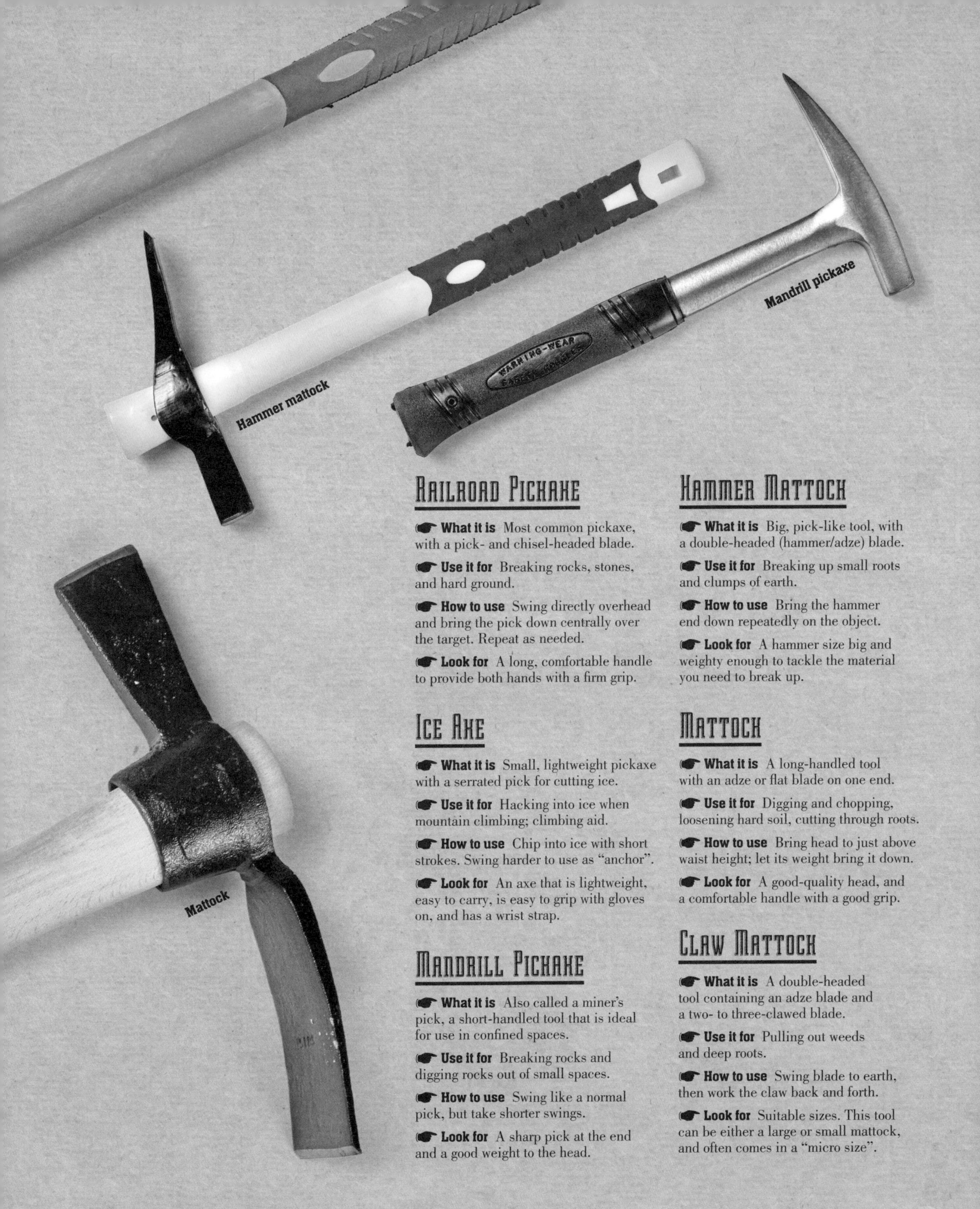

Railroad Pickaxe

What it is Most common pickaxe, with a pick- and chisel-headed blade.

Use it for Breaking rocks, stones, and hard ground.

How to use Swing directly overhead and bring the pick down centrally over the target. Repeat as needed.

Look for A long, comfortable handle to provide both hands with a firm grip.

Ice Axe

What it is Small, lightweight pickaxe with a serrated pick for cutting ice.

Use it for Hacking into ice when mountain climbing; climbing aid.

How to use Chip into ice with short strokes. Swing harder to use as "anchor".

Look for An axe that is lightweight, easy to carry, is easy to grip with gloves on, and has a wrist strap.

Mandrill Pickaxe

What it is Also called a miner's pick, a short-handled tool that is ideal for use in confined spaces.

Use it for Breaking rocks and digging rocks out of small spaces.

How to use Swing like a normal pick, but take shorter swings.

Look for A sharp pick at the end and a good weight to the head.

Hammer Mattock

What it is Big, pick-like tool, with a double-headed (hammer/adze) blade.

Use it for Breaking up small roots and clumps of earth.

How to use Bring the hammer end down repeatedly on the object.

Look for A hammer size big and weighty enough to tackle the material you need to break up.

Mattock

What it is A long-handled tool with an adze or flat blade on one end.

Use it for Digging and chopping, loosening hard soil, cutting through roots.

How to use Bring head to just above waist height; let its weight bring it down.

Look for A good-quality head, and a comfortable handle with a good grip.

Claw Mattock

What it is A double-headed tool containing an adze blade and a two- to three-clawed blade.

Use it for Pulling out weeds and deep roots.

How to use Swing blade to earth, then work the claw back and forth.

Look for Suitable sizes. This tool can be either a large or small mattock, and often comes in a "micro size".

Structure of a Pickaxe

The pickaxe, sometimes called a railroad pickaxe, is an invaluable tool for digging, breaking up hard ground, and chopping up roots. Handles range from around 65–100cm (26–39in) long, while the tool's head is made from forged steel and features a pointed blade on one side with which to break up ground, and a flat chiselled blade on the other, which also doubles as a lever.

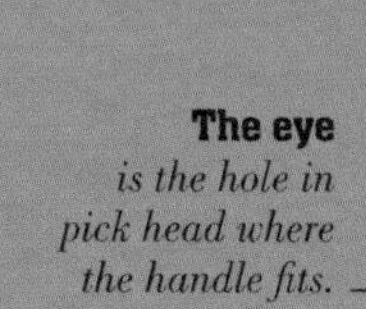

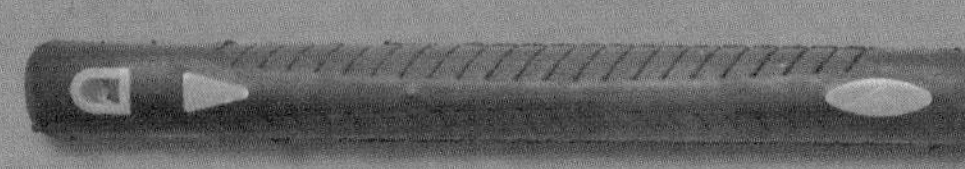

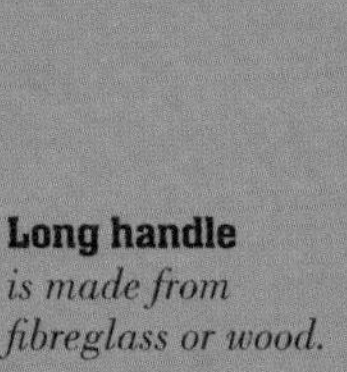

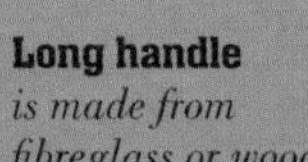

The eye *is the hole in pick head where the handle fits.*

Top of handle *often formed into tapered wedge.*

OVERALL VIEW

“THE RAILROAD PICKAXE WAS SO NAMED DUE TO ITS EXTENSIVE USE DURING THE CONSTRUCTION OF AMERICAN RAILROADS”

Long handle *is made from fibreglass or wood.*

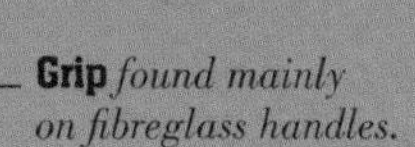

Grip *found mainly on fibreglass handles.*

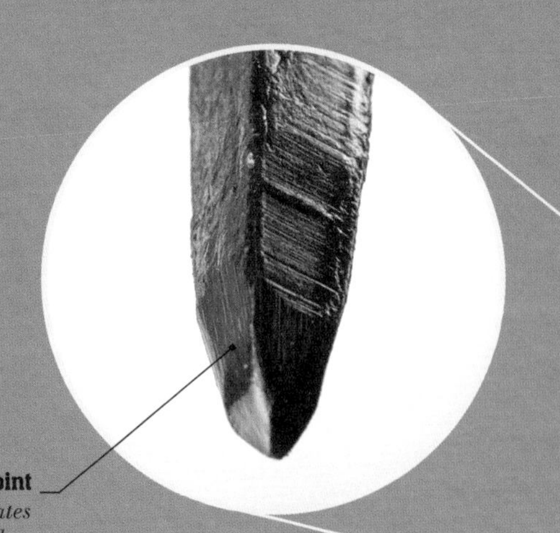

Pick point *concentrates force of blow into small area.*

FOCUS ON...

Striking Action

Pickaxe heads are deliberately curved so that when they make contact with the ground or rock, the impact of the strike occurs at an angle. This not only makes the strike more effective at breaking up a surface, it also prevents debris from flying directly upwards into the face of the user, as well as reducing the likelihood that the tool head bends due to the impact. The point concentrates the force of a strike into a small area, while the sharp chisel spreads it into a cutting face.

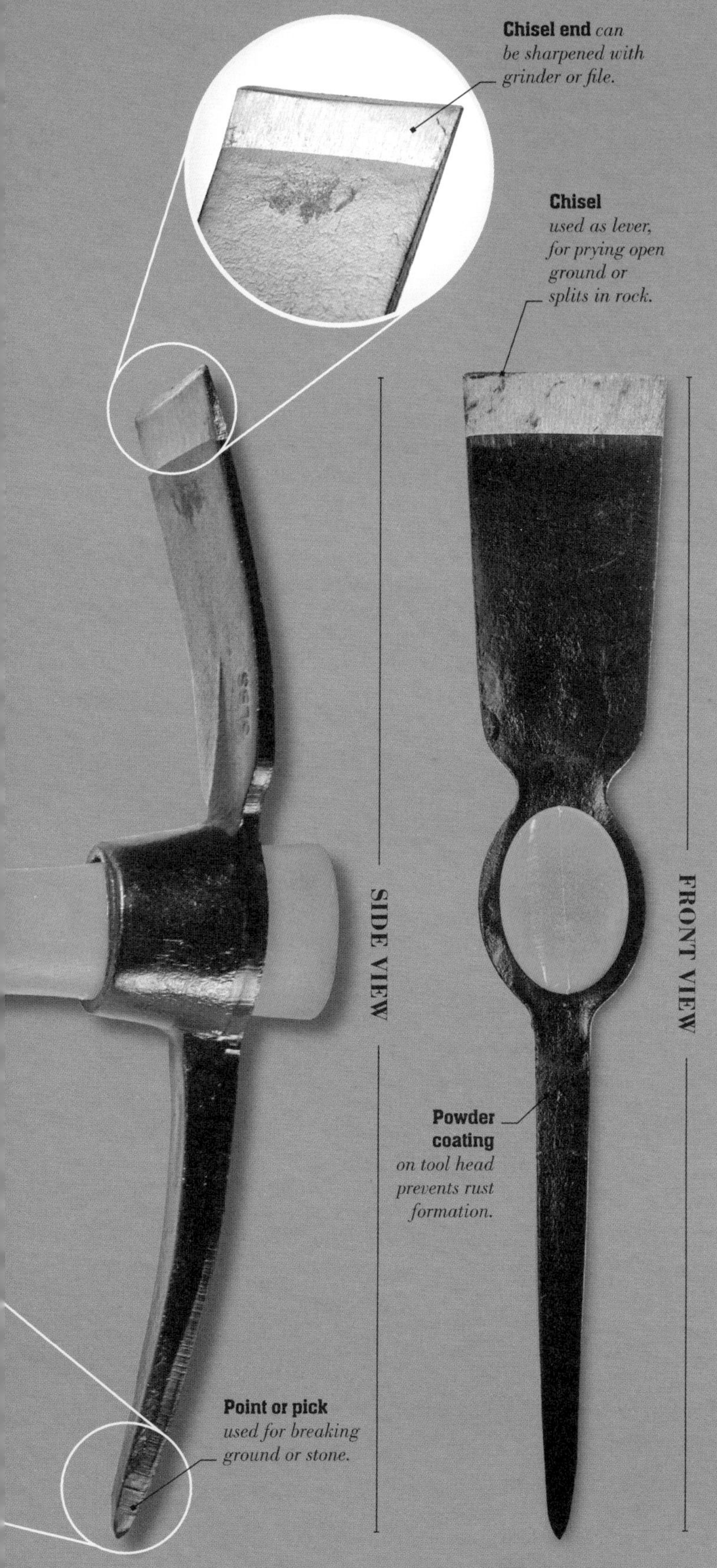

Chisel end *can be sharpened with grinder or file.*

Chisel *used as lever, for prying open ground or splits in rock.*

SIDE VIEW

FRONT VIEW

Powder coating *on tool head prevents rust formation.*

Point or pick *used for breaking ground or stone.*

USING A

PICKAXE

When trying to break up rock or hard ground, use the pick end of the pickaxe. To use the chisel end, place it into a crack that needs to be split and rock the head back and forth, like a lever prising the object open.

The Process

Before you start

- **Practise the swing** If you aren't used to dealing with the weight of a pickaxe, start slowly and practise pulling it over your head.
- **Stay safe** Work in a clear area with nothing behind you. Wear gloves with extra grip, and safety goggles to stop fragments of earth or rock flying up into your eyes.

1 Position yourself
Stand with your feet slightly apart and your dominant leg forward. The object you need to hit should be slightly in front of you.

2 Grip the pick
Place your non-dominant hand towards the bottom of the grip, and put your other hand slightly further up the handle, with a space between the two hands.

3 Make the swing
Bending at the waist and keeping your knees flexed, lift the pickaxe over your head, or if new to using this tool, swing it from just over the shoulder. Bring the axe down in an arc, keeping your arms extended. Keep your eye on the object you need to hit as you bring the tool down. Grip tightly as you make contact with the object so the pickaxe doesn't slip.

After you finish

- **Clean the tool** Wipe the handle and head of the pickaxe clean of any debris.
- **Check the handle** If using a wooden-handled tool, check the handle for splinters. Smooth them away with sandpaper if need be. Cracked handles should be replaced.

Choosing a
Wrecking Bar

Wrecking bars of many types have been around for centuries. Their strength and length provide excellent leverage, while a wide variety of grooves and points make them ideal for prying apart items, pulling out tight fasteners, or breaking down rocks – all essential for demolition projects. Demolition may seem daunting but with the right tools for the job, it will take much less effort.

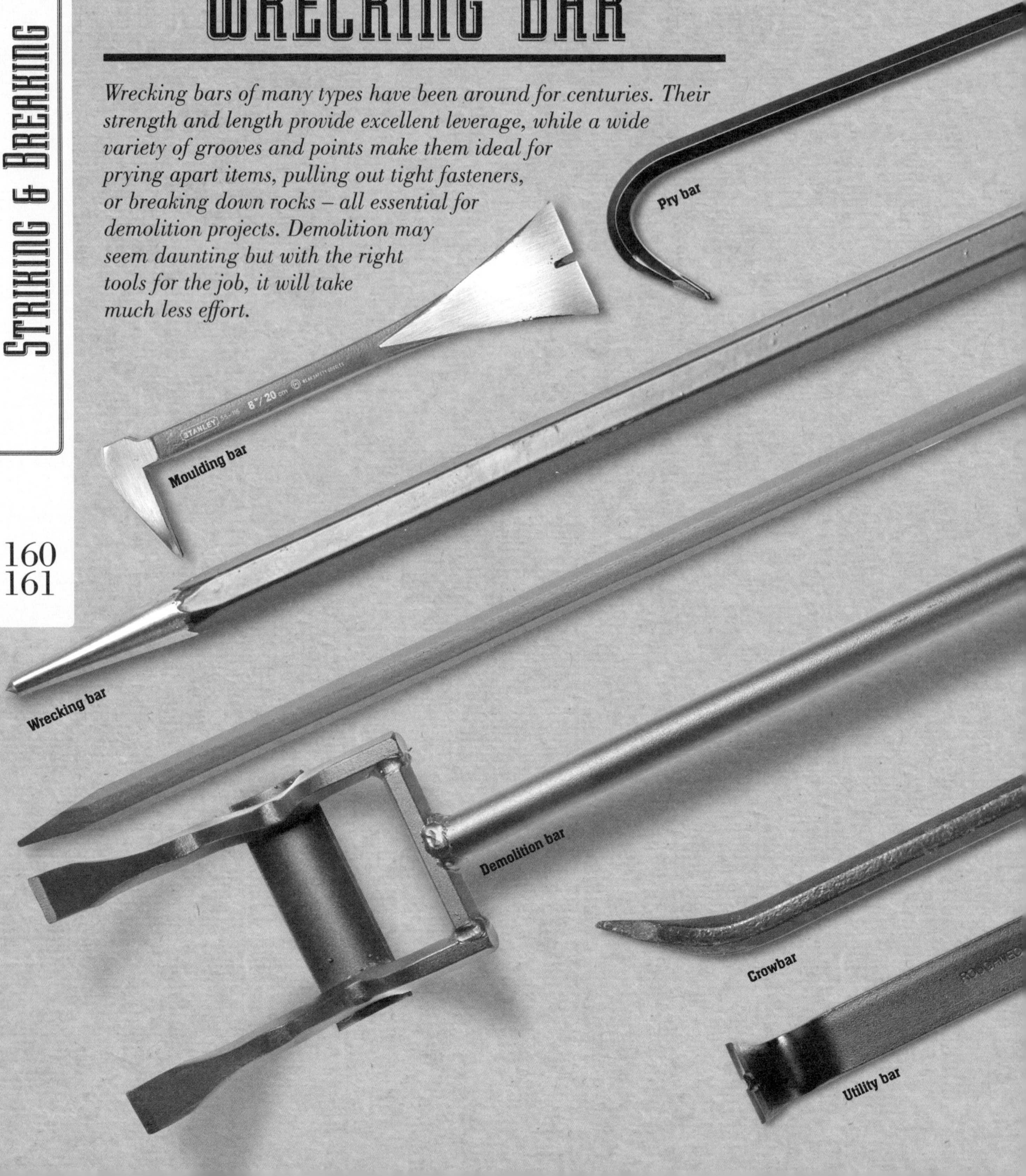

Pry bar

Moulding bar

Wrecking bar

Demolition bar

Crowbar

Utility bar

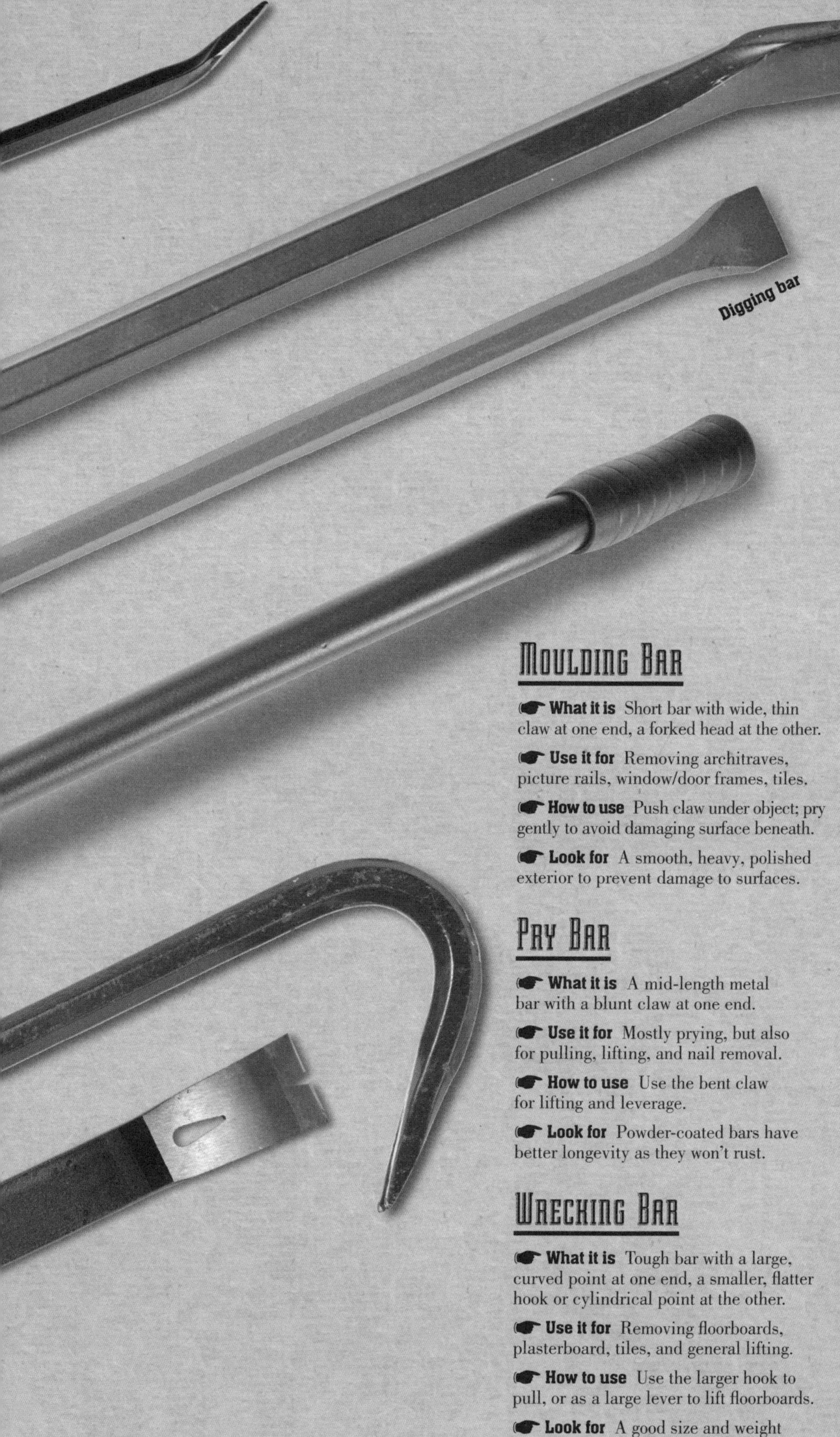

Digging Bar

☛ **What it is** A long metal bar with a chisel at one end and a point at the other.

☛ **Use it for** Digging post holes, breaking up hardened or frozen ground, digging out tree roots; also used as a lever.

☛ **How to use** Drive pointed end into the ground in a rotational motion to make a hole.

☛ **Look for** Longer and thicker bars will tackle harder surfaces.

Demolition Bar

☛ **What it is** A long bar with a bent end that has two prongs like a fork, with a chisel at the other end.

☛ **Use it for** General demolition, breaking, removal, lifting heavy objects.

☛ **How to use** Put the forked section under the object that needs to be removed and lever the bar.

☛ **Look for** Make sure the bar is long enough to use standing up. A rubberized grip on the handle is a must.

Moulding Bar

☛ **What it is** Short bar with wide, thin claw at one end, a forked head at the other.

☛ **Use it for** Removing architraves, picture rails, window/door frames, tiles.

☛ **How to use** Push claw under object; pry gently to avoid damaging surface beneath.

☛ **Look for** A smooth, heavy, polished exterior to prevent damage to surfaces.

Crowbar

☛ **What it is** A long tool with a chisel-like end and either a claw or point at the other end.

☛ **Use it for** Breaking rock, general demolition, lifting, and levering.

☛ **How to use** For breaking, drive the point or chisel end down directly over the item with a firm hit.

☛ **Look for** End type. For breaking, a pointed end is better, but for lifting, use a bent forked end.

Pry Bar

☛ **What it is** A mid-length metal bar with a blunt claw at one end.

☛ **Use it for** Mostly prying, but also for pulling, lifting, and nail removal.

☛ **How to use** Use the bent claw for lifting and leverage.

☛ **Look for** Powder-coated bars have better longevity as they won't rust.

Wrecking Bar

☛ **What it is** Tough bar with a large, curved point at one end, a smaller, flatter hook or cylindrical point at the other.

☛ **Use it for** Removing floorboards, plasterboard, tiles, and general lifting.

☛ **How to use** Use the larger hook to pull, or as a large lever to lift floorboards.

☛ **Look for** A good size and weight to provide leverage and endurance.

Utility Bar

☛ **What it is** A flat, short bar, with one bent and one flatter, slightly curved end.

☛ **Use it for** Pulling up flooring, pulling out nails, and removing tiles.

☛ **How to use** Insert flatter end under item to be removed; rock bar back and forth to loosen. Use bent end for pulling.

☛ **Look for** Choose size that works well in the space you have. Some have a small hole in the bar to remove nails easily.

“YOU CAN BURY A LOT OF TROUBLES DIGGING IN THE DIRT.”

ANONYMOUS

Maintain Tools for Striking & Breaking

If looked after well your tools should last many years. Ensure you clean excess dirt off after use and store them in a dry place to prevent rust.

KEEP TOOLS CLEAN

Cleaning is the key to keeping tools in good shape for longer. Avoid using abrasive chemical cleaners and stick to soap and water, or even sugar soap, for stubborn dirt.

1 Clear dry dirt
Knock any dry dirt off your tools when you have finished using them.

2 Wash tools
Use warm water and a rag to clean off any really stubborn dirt.

3 Dry and store
Dry your tools thoroughly and store them in a dry place to stop rust forming.

Dry dirt *can simply be knocked off digging tools.*

STORAGE

Store your tools safely and in an order that makes sense to you, so that they are easy to find when you need them. Tools with sharp edges should be hung safely or stored so the sharp edge cannot fall easily – for example pickaxes should be stored with the head resting on the floor.

Get organized
If you have a place for each of your tools to live, they are easier to find when you need them. Digging into a large box or bag that has tools randomly thrown into it can be dangerous and can also cause damage to the tools.

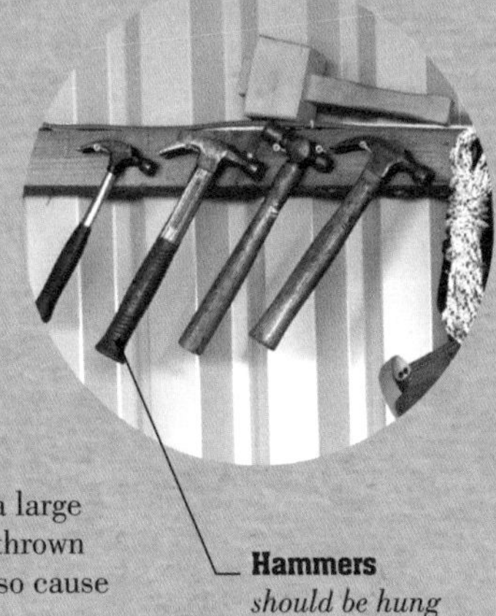

Hammers *should be hung up securely.*

Tools	Inspection
Hammers	▪ Check for any damage after use, ensure that the hammer heads are securely fitted and don't wobble ▪ Check handles for grip, any damaged or peeling of coated handles could make them slippery.
Picks & Mattocks	▪ Check for splinters in handles ▪ Check for loose heads before use
Wrecking Bars	▪ Check for any bends and chips in tools when cleaning them

Cleaning	Repair	Tips	Storage
▪ Wipe down after use	▪ A broken shaft on a hammer with a traditional hardwood handle can be replaced: remove broken wood from eye, then, using the old shaft as a guide, whittle a new shaft to size, testing fit with hammer head as you go. Set head into new shaft with a rubber mallet. Remove wedge from old shaft and tap it into slot sawn across the end to fix head firmly in place.	▪ Treat wooden handles with linseed oil to preserve the wood and maintain a smooth but comfortable grip	▪ Store in a cool, dry place to stop wood from swelling and metal heads from rusting
▪ Clean excess dirt off after job ▪ For dry dirt use a wet cloth	▪ Splits in fibreglass handles cannot be repaired, but small splits or splinters in wooden handles can be sanded away	▪ If head becomes loose during a job, soak the head in water for about 30 minutes to swell wood and make head fit – this is only a temporary fix though, and not a long-term solution to a loose head	▪ Store in a cool, dry place to stop wood from swelling and metal heads from rusting
▪ Clean with WD40 to lift dirt and protect corrosive elements	▪ Breaker bars are hard to break so won't need any repairs. If they bend they can't be used again so need to be replaced	▪ If you feel like your bar is bending during use stop and use a heavier weight bar	▪ Ensure all tools are dry before storage ▪ Hang on hooks or store flat in a safe box or bag

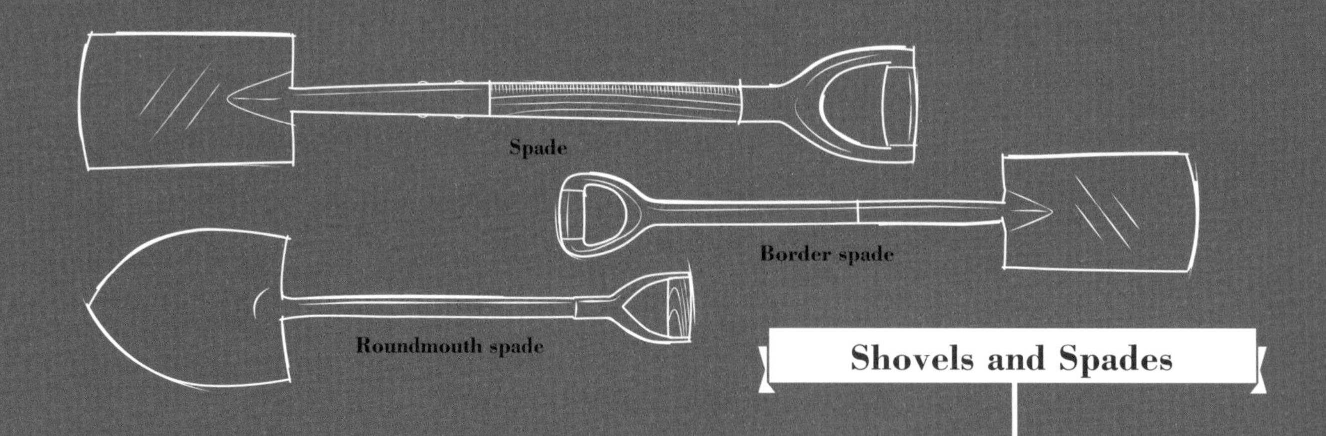

Shovels and Spades

THE TOOLS *for* DIGGING & GROUNDWORK

Preparing soil, planting, clearing, and digging all require tools suited to the task. From the humble spade to the specialized cultivator, the right tool can make light work of groundwork.

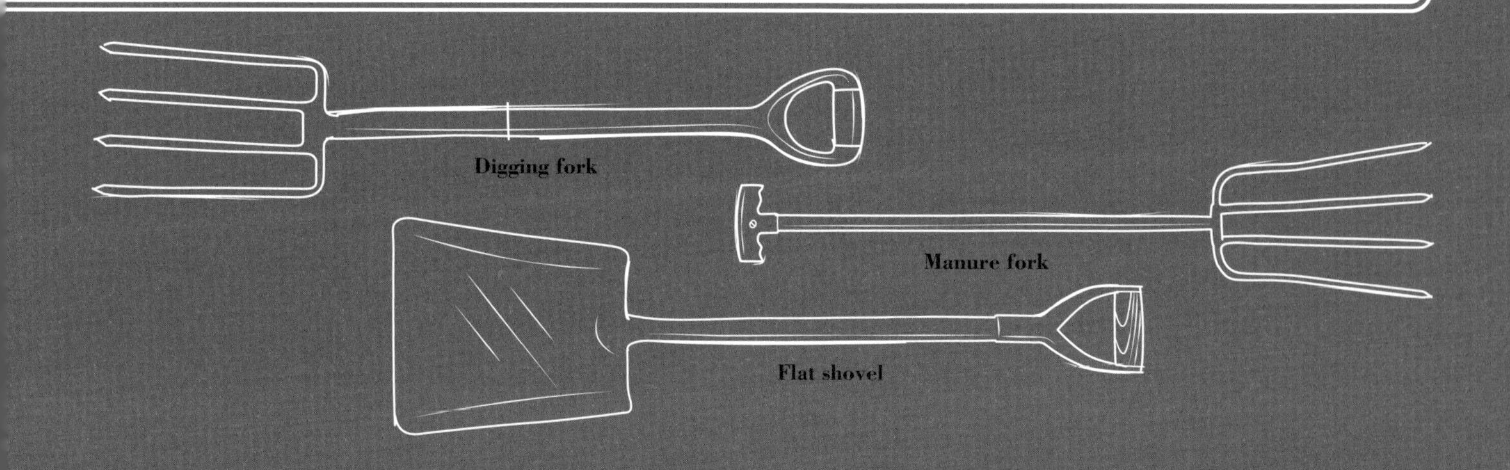

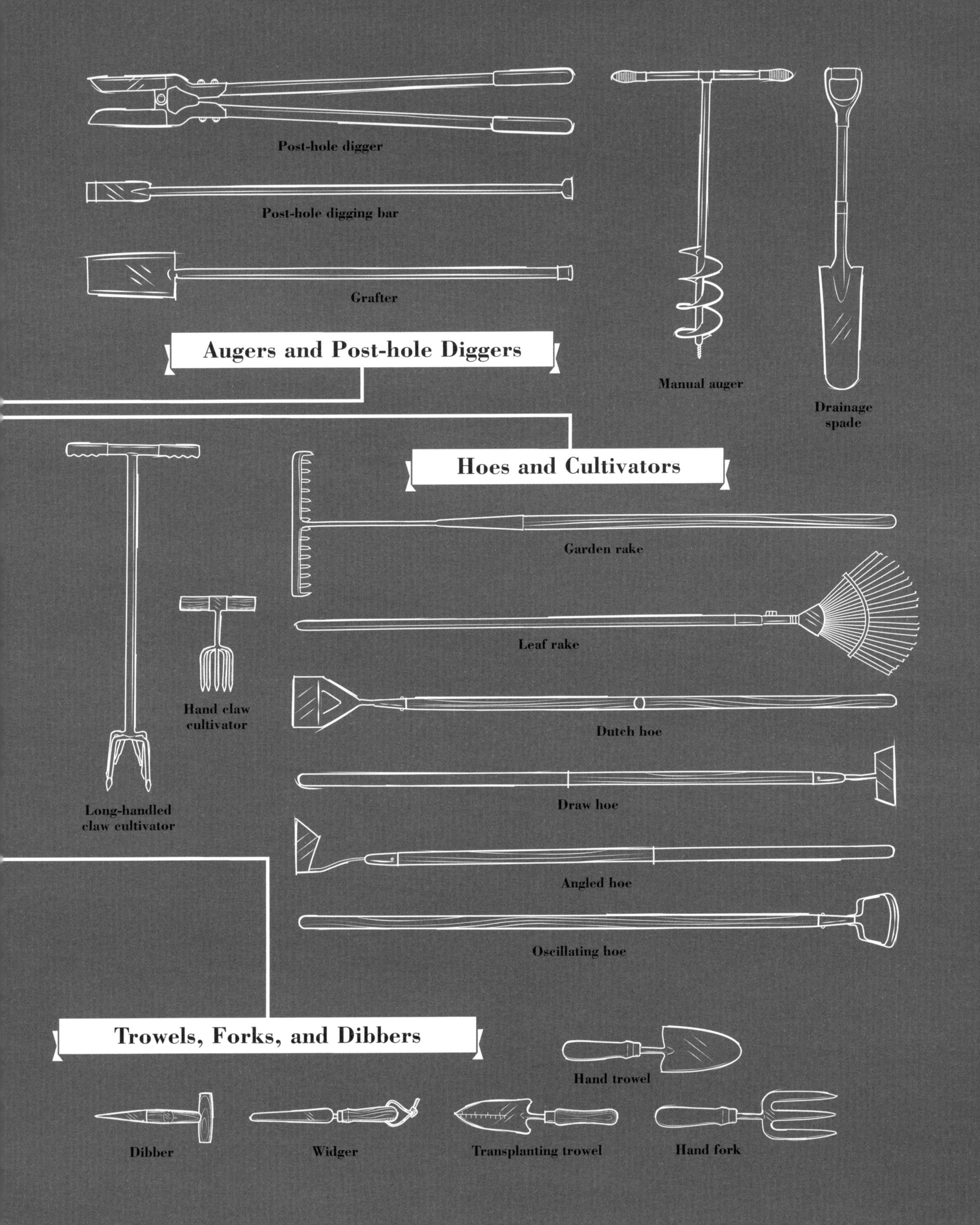
Post-hole digger
Post-hole digging bar
Grafter
Augers and Post-hole Diggers
Manual auger
Drainage spade
Hoes and Cultivators
Garden rake
Leaf rake
Dutch hoe
Draw hoe
Angled hoe
Oscillating hoe
Hand claw cultivator
Long-handled claw cultivator
Trowels, Forks, and Dibbers
Hand trowel
Dibber
Widger
Transplanting trowel
Hand fork

HISTORY OF

Digging & Groundwork

2.6–1.7 MYA

DIGGING STICKS

The basic but highly versatile digging stick is the ancestor of tools such as the spade and the hoe. It was used mainly for digging out underground food, such as tubers and roots, as well as flushing out burrowing animals or accessing insects within anthills.

"A WISE MAN WILL MAKE TOOLS OF WHAT COMES TO HAND."

THOMAS FULLER

10,000 YA

EARLY TROWEL

In Neolithic times, the shoulder blades of large mammals were used to dig out soil and rocks, especially when mining for flint. The bones of oxen were put to use in much the same way as a garden trowel is used today.

5000 BCE

EARLY HOES

Cave paintings depict hoe-like tools being used by ancient peoples. These resemble long, fork-like sticks and are believed to have been used to weed farmland, chop up plant matter, and create furrows for planting.

1800 BCE

WOODEN TROWELS

Ancestors of Native Americans belonging to the Katzie First Nation used wooden tools similar in design to modern garden trowels. The tools were carved with broad, rounded tips and were found at a site near Vancouver, Canada. Used for cultivation of wild potatoes known as wapato, they provide some of the earliest evidence of wild food gardening on the continent.

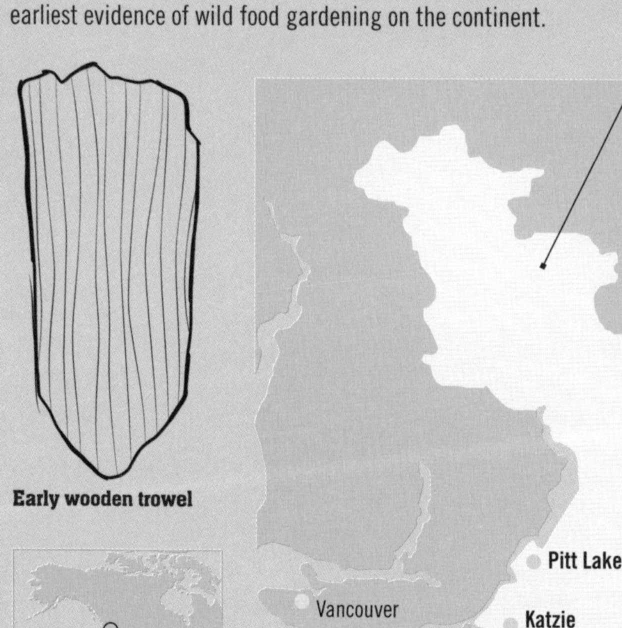

Early wooden trowel

Region of Katzie First Nation Territory

"ENOUGH SHOVELS OF EARTH – A MOUNTAIN. ENOUGH PAILS OF WATER – A RIVER."

CHINESE PROVERB

1750 BCE

SHOVELS

Artefacts found at copper-mining sites in the UK show that wooden shovels were used to dig mines.

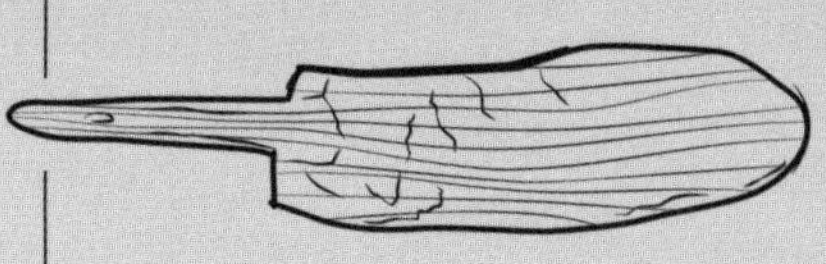

Preserved wooden shovel

500 BCE–500 CE

ROMAN GARDENING TOOLS

The Romans developed many of the gardening tools still in use today. The *pala* was the forerunner of the spade, the *sarculum* was equivalent to a hoe or weeding hook, and the *bidens* resembled a modern rake.

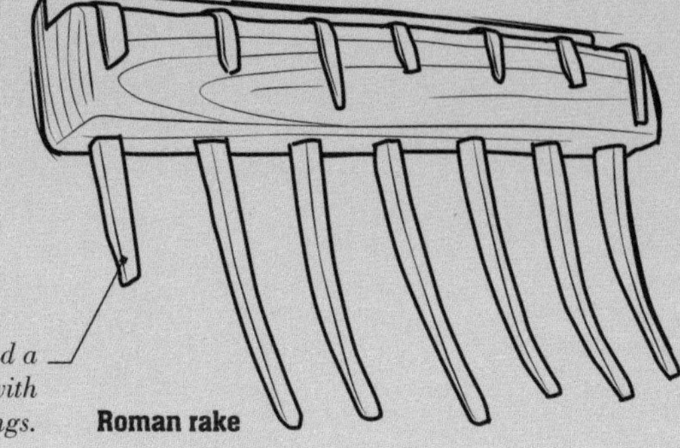

The Roman bidens *had a wooden head inserted with several wooden tines or prongs.*

Roman rake

c.55 BCE

BRONZE SHOVELS

Romans designed and used bronze shovels. Examples, including versions for shovelling incense, have been found at various sites.

140–165 CE

ROMAN SPADES

The Romans added a cutting edge to reinforce wooden spades in the form of "spade shoes" or iron rims. The spades and shovels used today owe much to Roman improvements to the tools, especially the development of forged iron.

> "IF YOU HAVE A GARDEN AND A LIBRARY, YOU HAVE EVERYTHING YOU NEED."
>
> CICERO

BRONZE

Bronze is an alloy of copper and tin, although sometimes other metals are added, including zinc and nickel. Historically, the composition varied widely, with craftsmen using whatever metal scraps were available.

12%
Tin

88%
Copper

The main part of most bronze blends is copper.

Bronze alloy

1300s

LIGHTWEIGHT TOOLS

In the Middle Ages, iron smelting led to more lightweight tools that required less effort to use, with more precisely crafted shapes.

1600s

GARDENER'S KIT

By the mid-1600s, contemporary illustrations show cultivating forks and trowels as part of the gardener's wide-ranging toolkit.

1774

CAST-IRON SHOVEL

The first cast-iron shovel in America was forged by Captain John Ames. The Ames company went on to modify the shovel further by introducing a back-strapped model in 1817, for use by soldiers in wartime, and then adding a wooden handle in 1824.

The cast-iron blade was more durable than wrought iron.

Socket-sleeve allowed for fitting of wooden or metal handles.

Ames shovel

CHOOSING A

Shovel or Spade

Spades, shovels, and forks vary in size, shape, and length, meaning they are suited to specific tasks. Though the terms are often used interchangeably, shovels and spades do not do the same job. A shovel's angled head makes it useful for scooping, while the straighter spade is for digging.

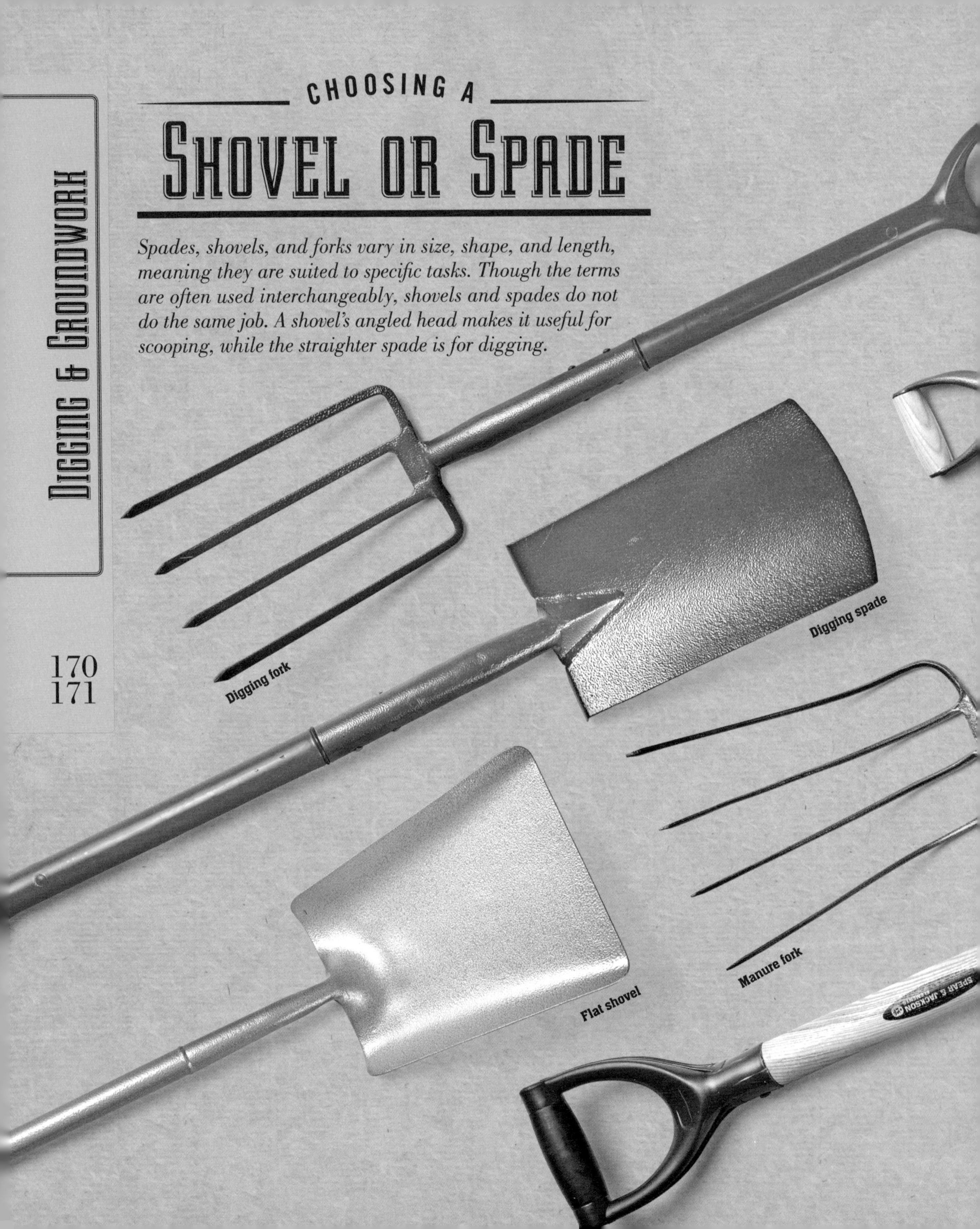

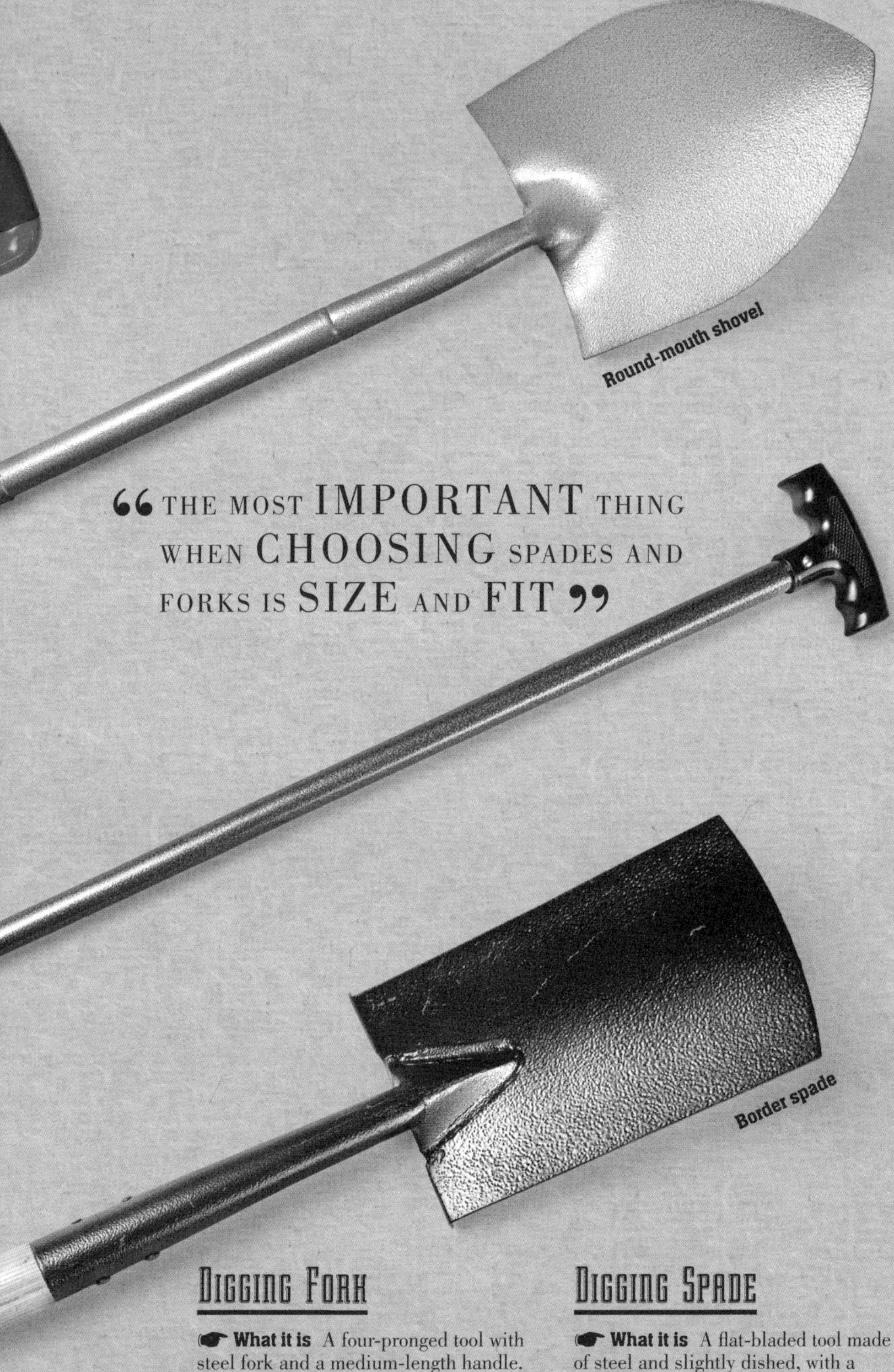

“THE MOST IMPORTANT THING WHEN CHOOSING SPADES AND FORKS IS SIZE AND FIT”

Round-mouth Shovel

What it is Shovel with a medium-length handle and a large, scooped, rounded head (mouth).

Use it for Scooping and moving large quantities of loose earth, sand, or gravel. Not for digging.

How to use With one hand at the far end and one near the mouth end, scoop in a sweeping motion.

Look for Strong but lightweight to ease effort. Wood or composite-fibre handles are strong.

Flat Shovel

What it is Shovel with a flat mouth, square end, and raised edges.

Use it for Scooping and moving large volumes of loose material. Good for scraping flat surfaces.

How to use A consistent motion makes loading and scooping more efficient. Don't overload the mouth.

Look for A strong join between the mouth and the steel shaft leading to the handle. Lightweight build.

Manure Fork

What it is A medium-length fork with wide, slender tines that have sharp, pointed tips.

Use it for Moving, clearing, and loading loose material, including manure, hay, and weed piles.

How to use Push the tines into bulkier material or piles that bind well, such as leaves and grass clippings.

Look for Some versions have a long handle, which is useful for spreading manure and loading trailers.

Digging Fork

What it is A four-pronged tool with steel fork and a medium-length handle.

Use it for Turning over, breaking up cultivated soil. Filtering roots from soil.

How to use Put the fork onto the soil and push down with one foot. Lever the handle back and turn soil over.

Look for Strong fork tines with well-pointed or sharp tips and a good, sturdy handle without play.

Digging Spade

What it is A flat-bladed tool made of steel and slightly dished, with a medium-length handle.

Use it for Digging holes in soil, turning soil over, such as when planting trees, and large shrubs.

How to use Push the blade (spit) into the soil with your foot, lever back and forth, and turn over,

Look for Blade size (small or large) and handle length to suit your height.

Border Spade

What it is Much like a digging spade but with a much smaller blade or head.

Use it for Lighter gardening jobs, including border work in tight spaces and loading soil into large pots.

How to use Use like the digging spade, but it may not need to be pushed with the foot.

Look for A good fit, and a smooth blade surface, with an edge that is as sharp as possible.

Structure of a
Spade

If you want to do some digging in the garden, you'll always need a spade, which makes it one of the most fundamental tools in the gardening shed. There is little variation in design, although handle shape and angle, and blade size or moulding do vary.

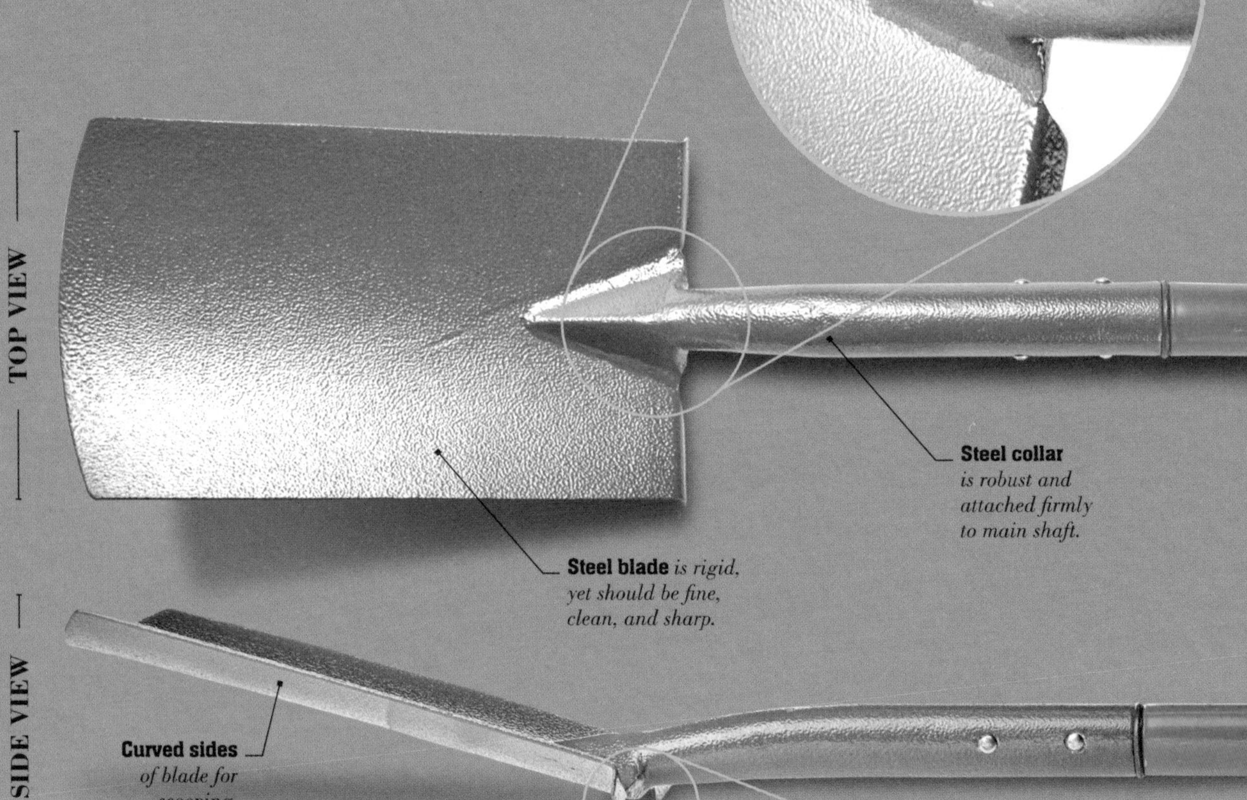

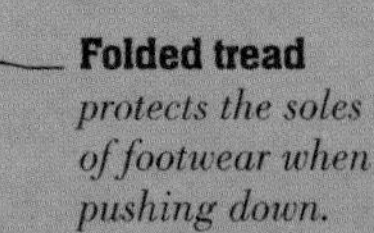

FOCUS ON...

HANDLE SHAPES

Handle grips are either D- or T-shaped and come in a range of materials. People with large hands can find a D-grip restrictive. All grips need to be well finished to avoid snags or splinters, and can sometimes be sanded. A slight angle reduces back strain and increases leverage.

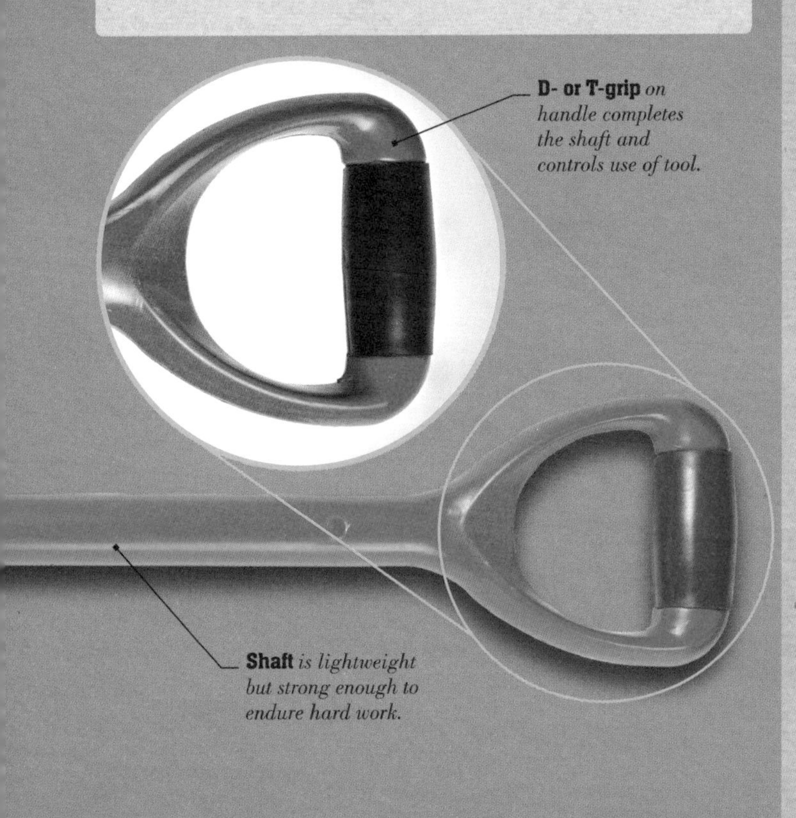

Angled grip *helps reduce bending, saving back from strain.*

> “A SPADE IN USE BECOMES AN EXTENSION OF THE ARMS, SO ITS DESIGN SHOULD FIT THE USER PERFECTLY”

USING A SPADE

Spades are very versatile, and a well-chosen one is pleasing to use. Working slowly and methodically, digging manageable volumes, and straightening the back is important, as is selecting the right spade for the task.

The Process

Before you start

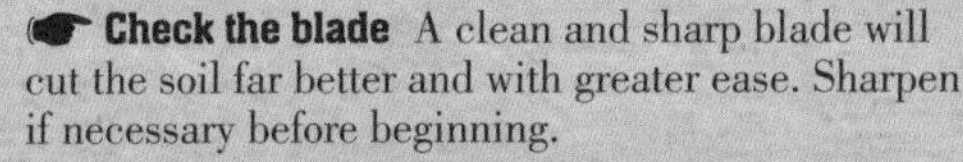

- **Check the blade** A clean and sharp blade will cut the soil far better and with greater ease. Sharpen if necessary before beginning.
- **Inspect the shaft** Check for play in the shaft and grip. Spades with wooden shafts are better stored in an unheated shed to avoid drying out.

1 Place and push
Position the spade vertically on the surface of the ground and give it a slight push with your hands. Place one foot on the blade tread and push firmly into the ground.

2 Mark out the hole
Repeat this action several times when digging holes to determine the shape all the way around before removing any soil. Gently rock or lever the spade when in the soil if the ground is hard.

3 Lever the spade
Bending your knees and keeping your back straight, lever the spade handle back towards you, either turning the soil over when cultivating, or lifting it out if digging a hole. Repeat until the job is done.

After you finish

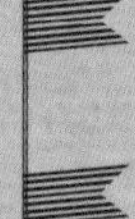

- **Clean the blade** Clean blade and shaft with a cloth. Give uncoated steel blade a light coat of general-purpose oil to prevent rust. Rub wooden shafts with linseed oil.
- **Check for play** If wooden shaft develops slight play, soak it in water for 24 hours as a short-term solution to rehydrate wood and allow it to swell to fit.

CHOOSING A

Post-hole Digger

Digging holes for posts can be very hard work, especially as the job often has to be performed in a tight spot, and sometimes in stony ground. The holes must be deep, vertical, and narrow, leading to the need for specialist tools. Post-hole tools have long handles and long, narrow digging heads, unlike most other spades and shovels.

Drainage spade

“POST HOLES ARE EASIER TO DIG WHEN YOU HAVE THE RIGHT TOOLS”

Post-hole digger

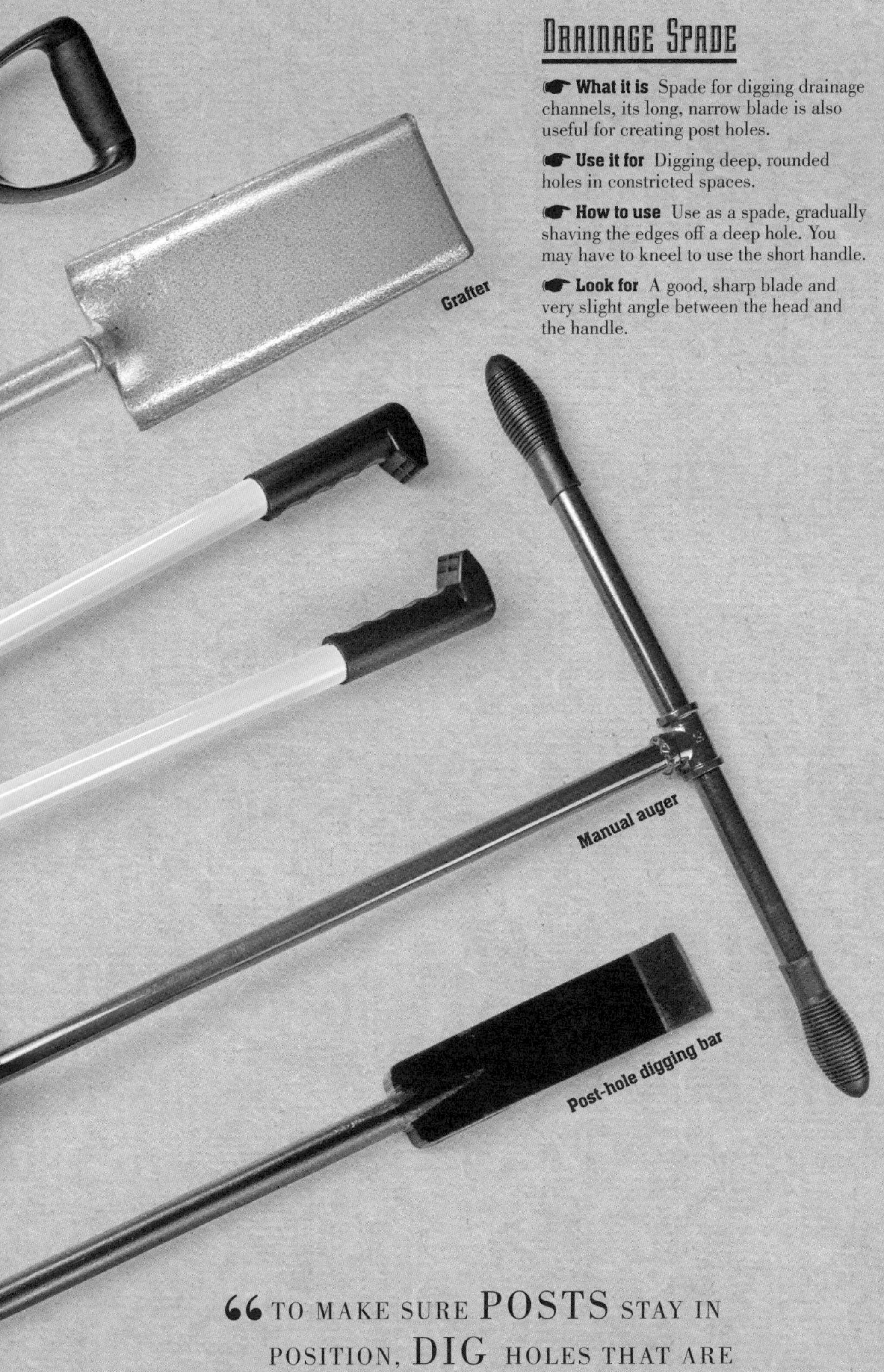

Drainage Spade

What it is Spade for digging drainage channels, its long, narrow blade is also useful for creating post holes.

Use it for Digging deep, rounded holes in constricted spaces.

How to use Use as a spade, gradually shaving the edges off a deep hole. You may have to kneel to use the short handle.

Look for A good, sharp blade and very slight angle between the head and the handle.

Grafter

What it is A narrow, flat-bladed spade with a very long, heavy handle.

Use it for Digging deep, narrow holes in combination with post-hole diggers.

How to use Push blade into ground with foot. Use blade to shave the sides of the holes and loosen bottom for scooping.

Look for A sharp and narrow head, with a strong and weighty handle.

Post-hole Digger

What it is Essentially a twin-bladed spade, with a scissor action and long handles for reach.

Use it for Digging post holes in tandem with the post-hole digging bar. Scooping loosened material from holes.

How to use Holding both handles, drop the head into the loosened soil, pull handles apart to grip, lift out and empty.

Look for A well-made and durable scissors mechanism. Long handles.

Manual Auger

What it is A very large screw thread on a metal shaft turned via a long, T-shaped handle.

Use it for Making deep, circular holes in smooth, stone-free soil such as clay.

How to use Position the auger and rotate the thread clockwise to make hole. Remove periodically to empty soil.

Look for A tough, sharp auger thread and a very strong T-handle.

Post-hole Digging Bar

What it is A heavy, solid-iron bar with a chisel-shaped iron head and a long handle.

Use it for Breaking up hard ground or stony surfaces when digging.

How to use Use the blade end to break up the ground in sections, a few centimetres at a time before scooping out hole with post-hole digger. Loosen any stones or obstructions as you go.

Look for Long handle with a solid-iron core for extra weight.

“TO MAKE SURE POSTS STAY IN POSITION, DIG HOLES THAT ARE DEEP BUT NOT TOO WIDE”

Choosing a

Hoe or Cultivator

Choosing the right hoe or cultivator can make a considerable difference to the ease and effectiveness of your work in the garden, and it can save a lot of other tasks further down the line. Frequent use of a hoe deals with young weeds easily, which in turn avoids the need to dig older ones out later on. A sturdy, well-made rake with the right kind of tines will make surface preparation a joy.

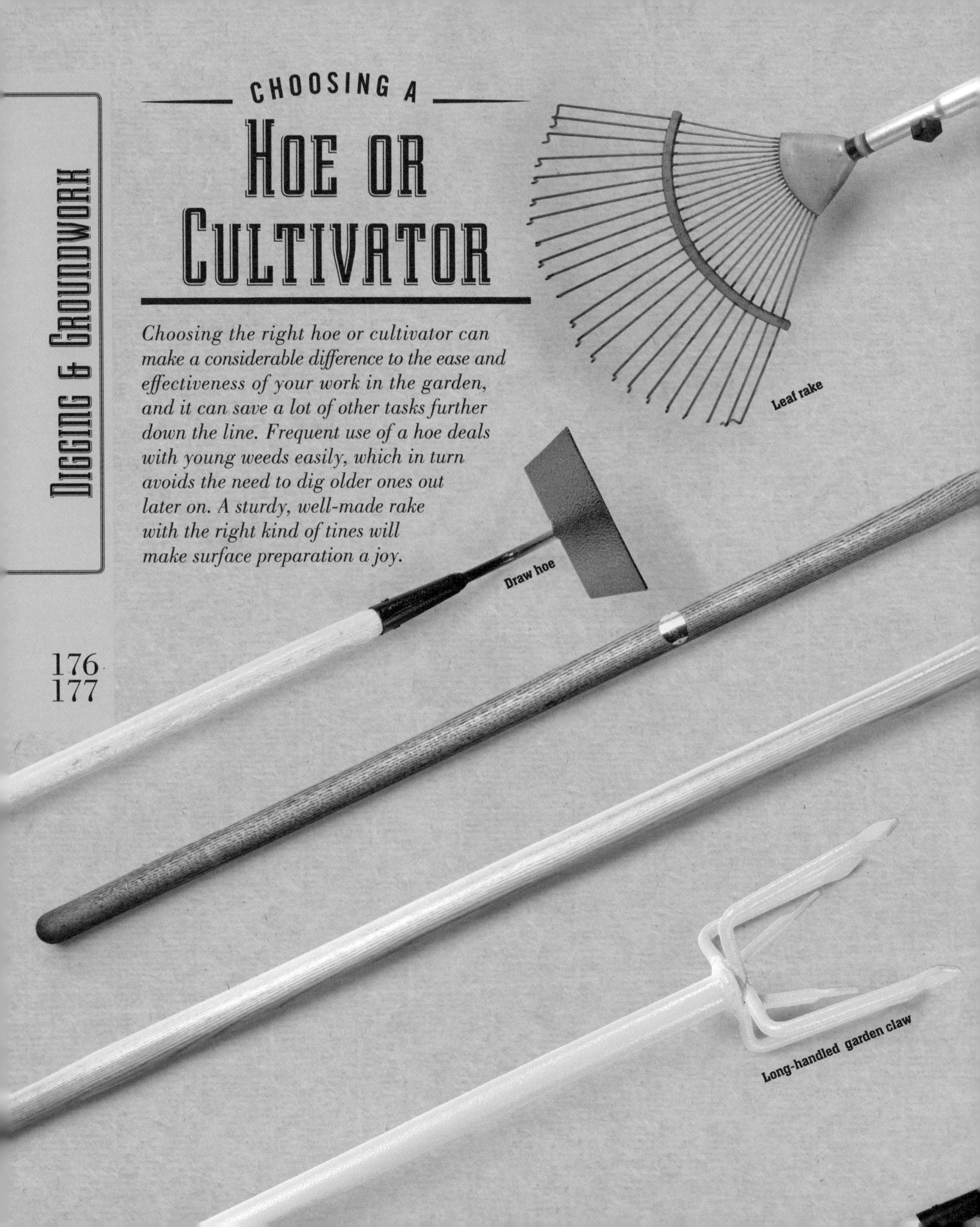

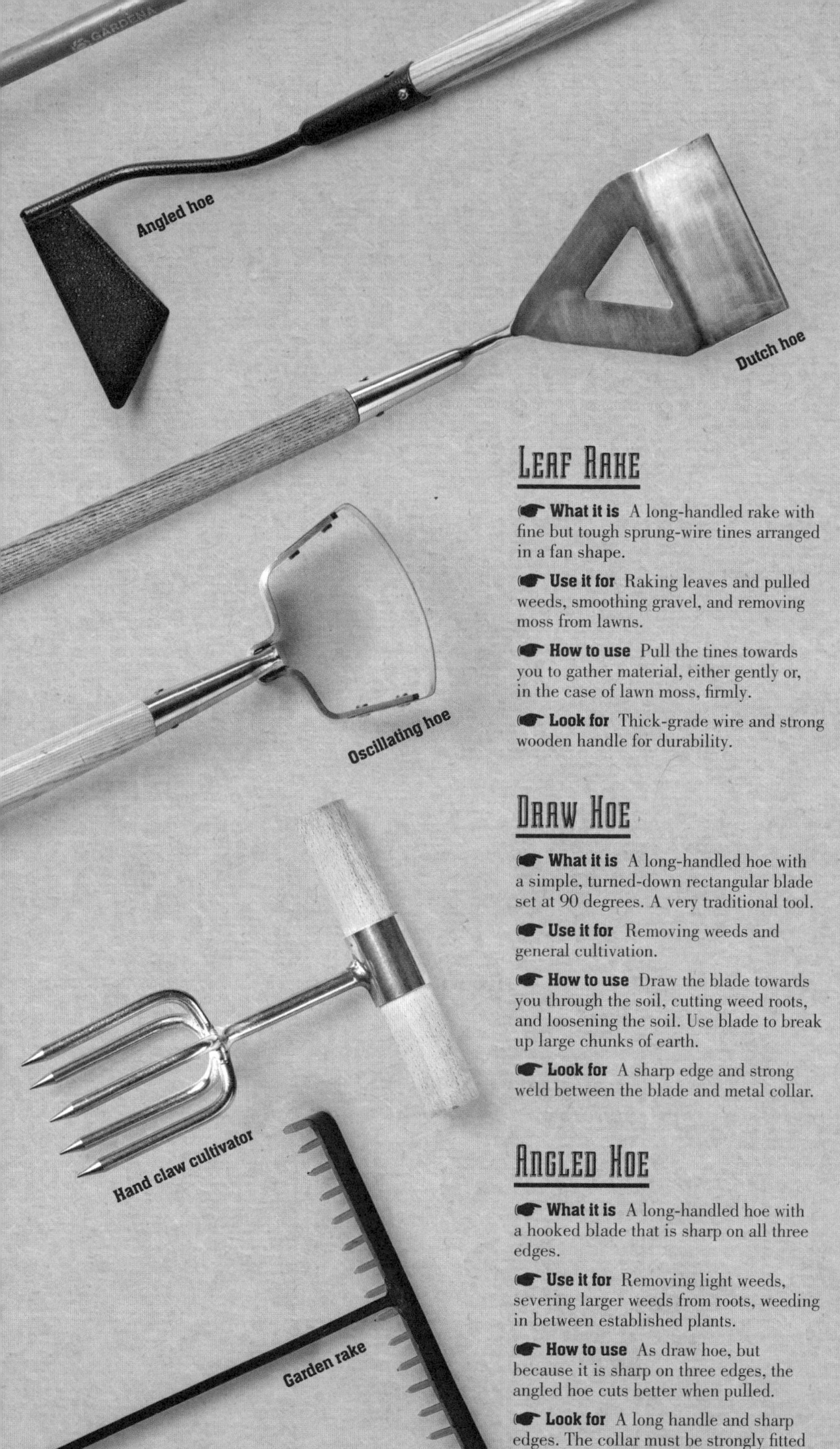

Dutch Hoe

What it is Traditional hoe with a flat, D-shaped head, sharp on the front edge.

Use it for Removing weeds, cultivating soil in more open ground.

How to use Use a push and pull motion in between plants and seedlings.

Look for A sharp edge to blade, a long handle suitable for your height.

Oscillating Hoe

What it is Hoe with flexing, stirrup-shaped head and sharp, curved blade.

Use it for All hoeing tasks, from light weeding to thick weeding in gravel.

How to use Push and pull; the blades cut in both directions, oscillating slightly.

Look for Different head sizes and handle lengths.

Long-handled Garden Claw

What it is Medium-height tool with T-handle. Four short prongs, often twisted, are arranged in a square.

Use it for Cultivating and loosening soil, removing weeds, turning compost.

How to use Push prongs into the soil and rotate the handle.

Look for A comfortable fit and strong prongs and fixings.

Hand Claw Cultivator

What it is Similar to long-handled claw, but with short handle for close work.

Use it for Cultivating worked soil, working around established plants.

How to use Push into ground and twist the tines repeatedly with one hand.

Look for A smooth handle for comfort when using repeatedly.

Garden Rake

What it is Long-handled rake, with a wide metal head holding many short tines.

Use it for Levelling seed beds, landscaping, spreading gravel or mulches.

How to use Work backwards and forwards to level or sculpt materials.

Look for A long handle and a heavy head to carry it through surface materials.

Leaf Rake

What it is A long-handled rake with fine but tough sprung-wire tines arranged in a fan shape.

Use it for Raking leaves and pulled weeds, smoothing gravel, and removing moss from lawns.

How to use Pull the tines towards you to gather material, either gently or, in the case of lawn moss, firmly.

Look for Thick-grade wire and strong wooden handle for durability.

Draw Hoe

What it is A long-handled hoe with a simple, turned-down rectangular blade set at 90 degrees. A very traditional tool.

Use it for Removing weeds and general cultivation.

How to use Draw the blade towards you through the soil, cutting weed roots, and loosening the soil. Use blade to break up large chunks of earth.

Look for A sharp edge and strong weld between the blade and metal collar.

Angled Hoe

What it is A long-handled hoe with a hooked blade that is sharp on all three edges.

Use it for Removing light weeds, severing larger weeds from roots, weeding in between established plants.

How to use As draw hoe, but because it is sharp on three edges, the angled hoe cuts better when pulled.

Look for A long handle and sharp edges. The collar must be strongly fitted to the blade.

"TO FORGET HOW TO DIG THE EARTH AND TO TEND THE SOIL IS TO FORGET OURSELVES."

MAHATMA GANDHI

Structure of a
Hoe

The hoe is one of the best tools for the modern gardener, and can even be well deployed to achieve no-dig gardening. Because hoeing vastly reduces labour, the right choice of hoe is important. The professionals' choice is the oscillating hoe, which has a head that uses a swivel action. In all cases, hoes must have a long handle for comfort, and be kept very sharp.

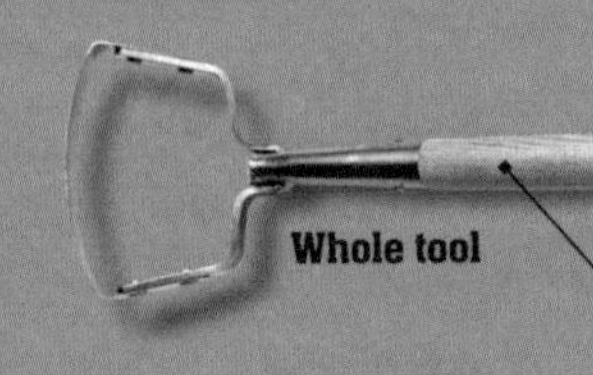

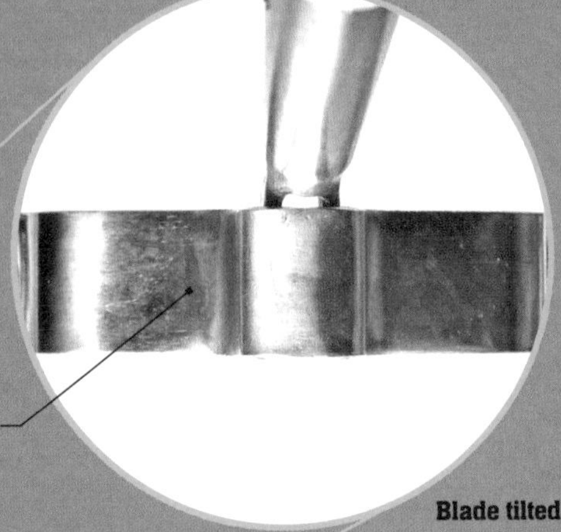

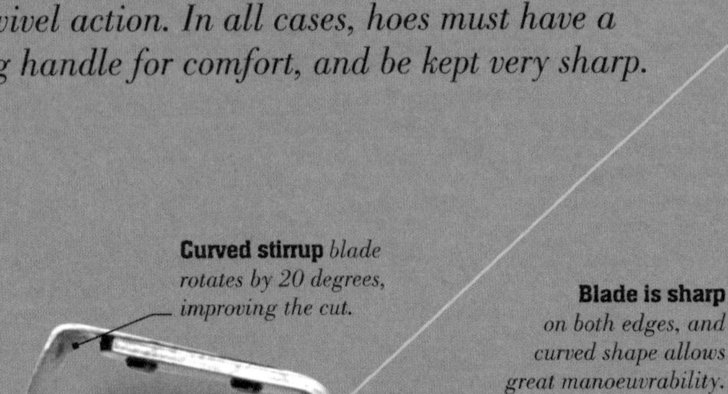

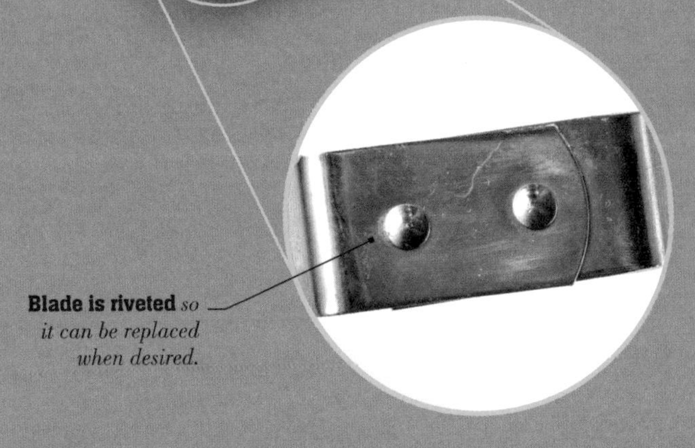

FOCUS ON...
Head Types

Hoes come in many shapes and sizes, and because so many hoe heads are available, choice is important. Heads that cut in both directions are sharp on each edge, which can halve the time needed for each task, making them more efficient. The edge must be slightly angled into the soil for best results, which is why the oscillating hoe is a good bet. Dutch hoes are harder to work, and less effective.

Handle length *gives extended reach and saves the back.*

“NO GARDENER SHOULD BE WITHOUT A HOE. USED CONSISTENTLY, IT CAN GREATLY REDUCE THE CHORE OF KEEPING THE GARDEN WEED-FREE”

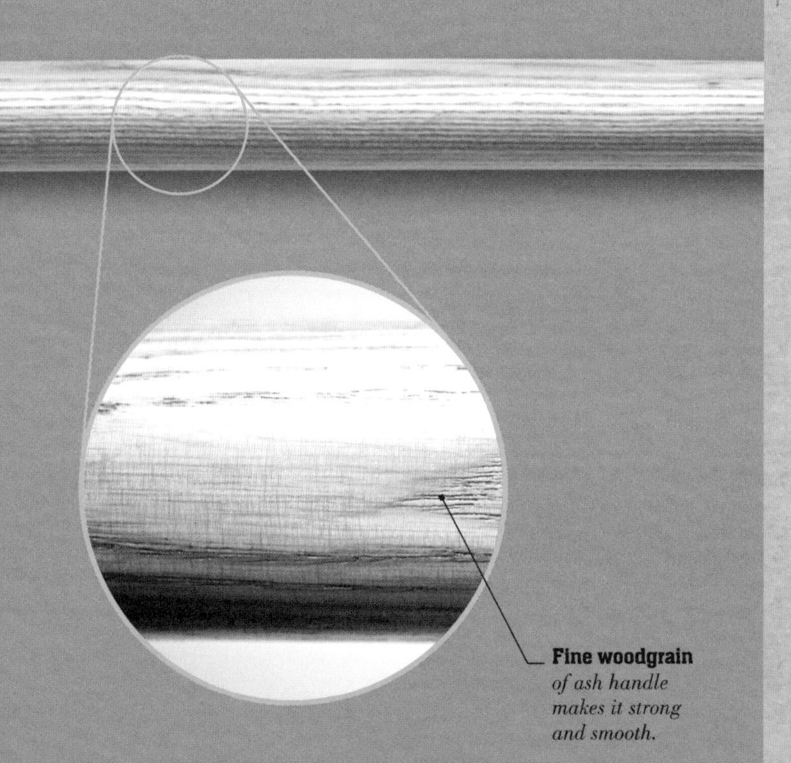

Fine woodgrain *of ash handle makes it strong and smooth.*

USING A HOE

Keeping on top of weeds by regular hoeing can reduce digging, as well as limiting the surface compaction involved. This retains and builds healthy soil structure and improves the overall plant health.

The Process

Before you start

- **Plan a hoe-friendly space** Gardens that are designed for hoeing leave space between plants for the hoe, so be sure to plan ahead when you're planting and sowing.
- **Check blade** Ensure the hoe blade is sharp before you start; sharpen with a file if necessary. Some blades, such as the oscillating hoe, are said to be self-sharpening.

1 Check the weather

Choose a dry, preferably hot day to hoe. Hoe when weeds are tiny seedlings for maximum effect, not when they are established. This saves a lot of time later.

2 Work in rows

Choose a position near the area that reduces surface compaction, limiting the amount you walk over the cultivation surface. Hoe up and down rows, methodically and not randomly.

3 Skim smaller weeds

Push and pull the hoe gently through the soil, although not too deeply, skimming the surface to cut the roots of seedling weeds. Larger, perennial weeds can be beaten by frequent hoeing too, as the front of the blade can be used to cut through their roots and lever them out of the soil.

After you finish

- **Clear up the weeds** If you have a compost heap that generates enough heat to kill seeds, rake up and dispose of the weeds there. If not, either bag them or burn them.
- **Clean the tool** Wipe down both hoe blade and shaft. Sharpen the blade if necessary, giving it a light coating of oil, preferably vegetable oil, before storing.

Choosing a

Trowel, Fork, or Dibber

Trowels and forks come in all shapes and sizes, and while it's useful to have a selection, two well-chosen items – a standard hand trowel and fork – will cover almost every job. Dibbers and transplanting trowels are useful if you sow and plant seedlings frequently. Choose the best quality tools you can afford, and take good care of them.

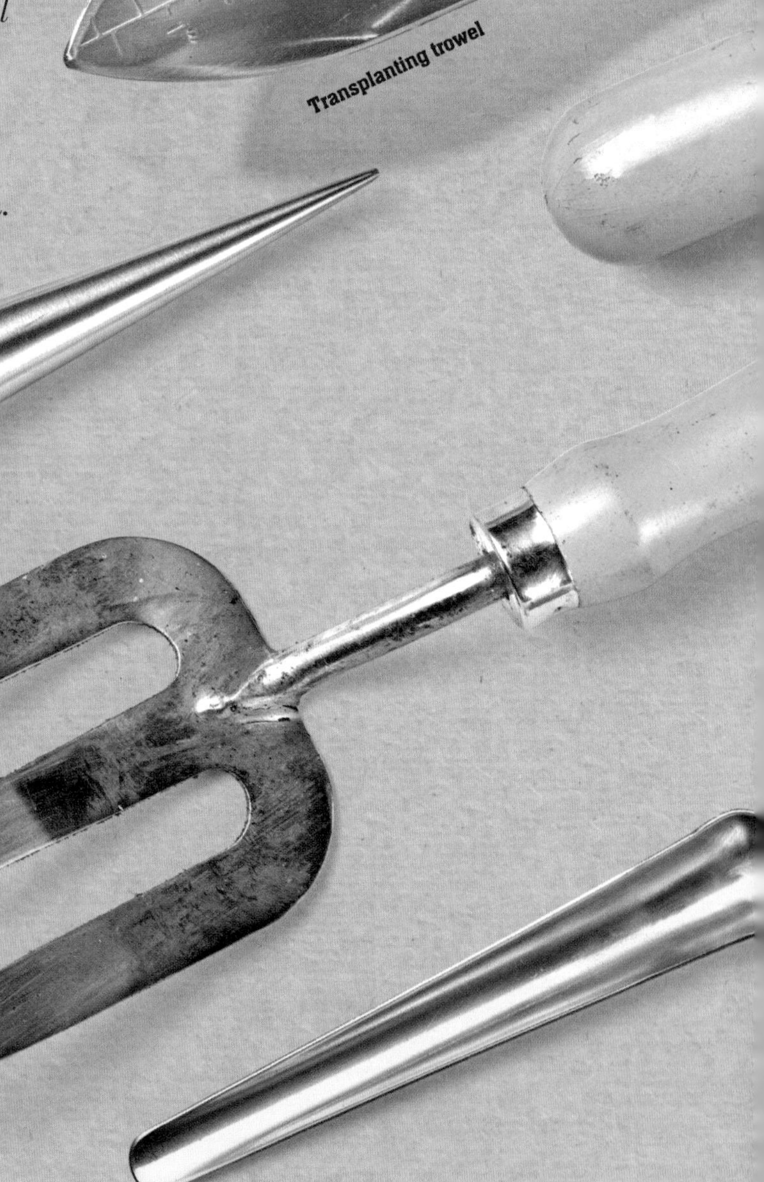

Transplanting trowel

Dibber

Hand fork

Widger

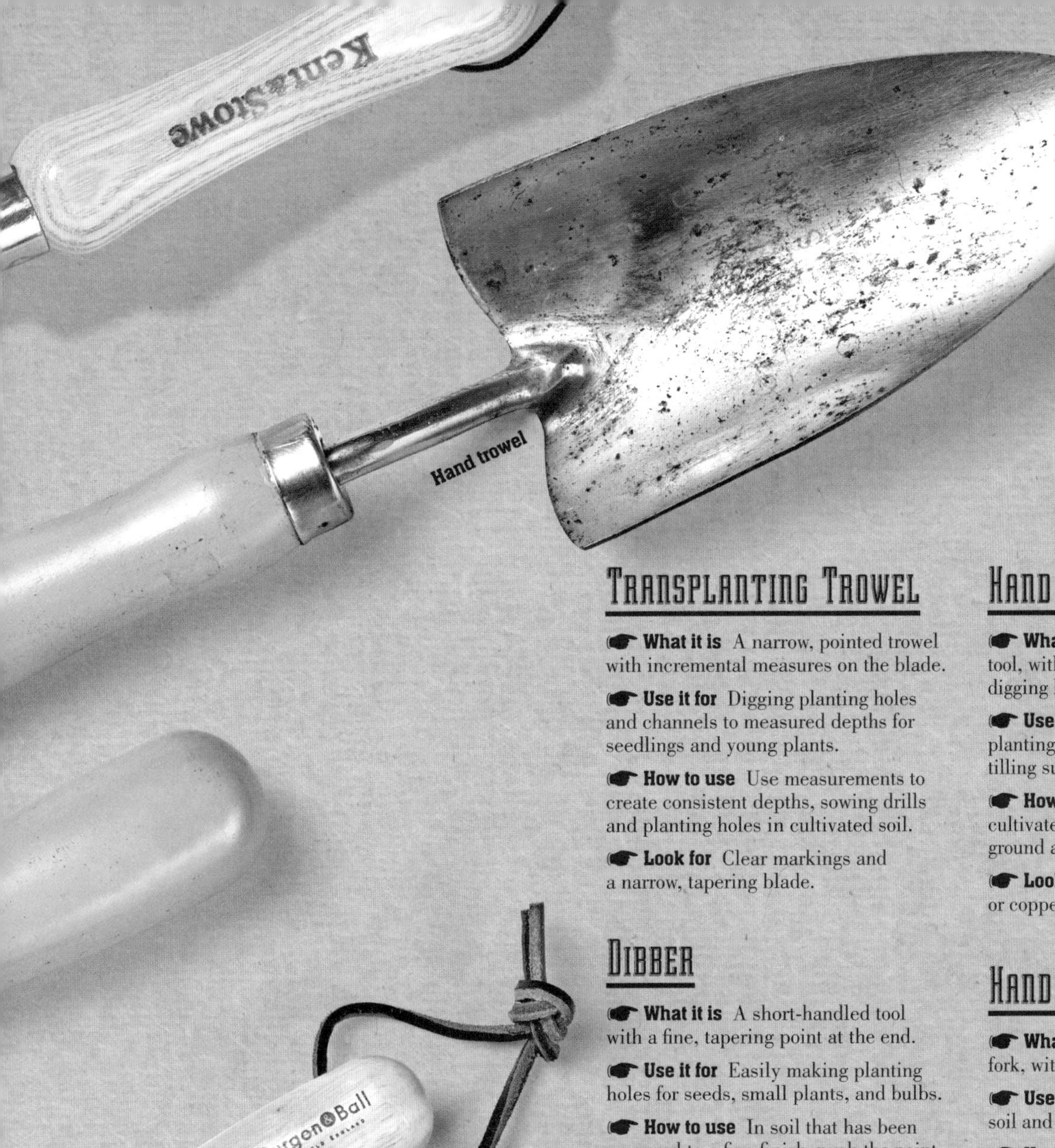

Transplanting Trowel

What it is A narrow, pointed trowel with incremental measures on the blade.

Use it for Digging planting holes and channels to measured depths for seedlings and young plants.

How to use Use measurements to create consistent depths, sowing drills and planting holes in cultivated soil.

Look for Clear markings and a narrow, tapering blade.

Dibber

What it is A short-handled tool with a fine, tapering point at the end.

Use it for Easily making planting holes for seeds, small plants, and bulbs.

How to use In soil that has been prepared to a fine finish, push the point in vertically to the required depth.

Look for A smooth finish to the point, like metal, to avoid any pulling of soil back out of the hole.

Hand Trowel

What it is An essential gardening tool, with a short handle and scooped digging blade.

Use it for General gardening: planting small plants, uprooting weeds, tilling surface soil, scooping compost.

How to use Push blade into loose, cultivated soil. Avoid use in very hard ground as this might bend the blade.

Look for Best quality, with a steel or copper blade and a strong handle.

Hand Fork

What it is A miniature garden hand fork, with short handle and three tines.

Use it for Mainly weeding through soil and loosening soil surface in borders.

How to use Working close to the ground, push into the soil and lever or turn to cultivate or lift out weeds.

Look for Solid, strong tines that will not easily bend. A comfortable handle.

Widger

What it is A very long and narrow hand tool with a scooped head.

Use it for Transplanting seedlings, making seed drills and holes; also weeding in tight spots.

How to use Slip blade down the length of seedling roots to lift them out gently, or push into gaps between stones to pull out weeds.

Look for A tapering blade with reasonably sharp edges and a very robust handle.

“A QUALITY TROWEL OR FORK CAN BE A HIGHLY SATISFYING TOOL TO OWN, HOLD, AND USE”

Structure of a

Trowel

In its design and purpose, a trowel is really a miniature spade. It is a must-have tool for smaller planting and maintenance operations in the garden. With its small blade and handle, it's also ideal for single-handed use.

Blade *may rust over time, but can be treated with wire wool and oil.*

Carbon-steel *blade is strong but prone to rust, so requires light coat of oil for protection.*

Blade *made of carbon steel, stainless steel, aluminium, or plastic.*

Curved sides *strengthen the blade and allow easy scooping.*

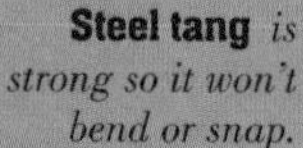

Steel tang *is strong so it won't bend or snap.*

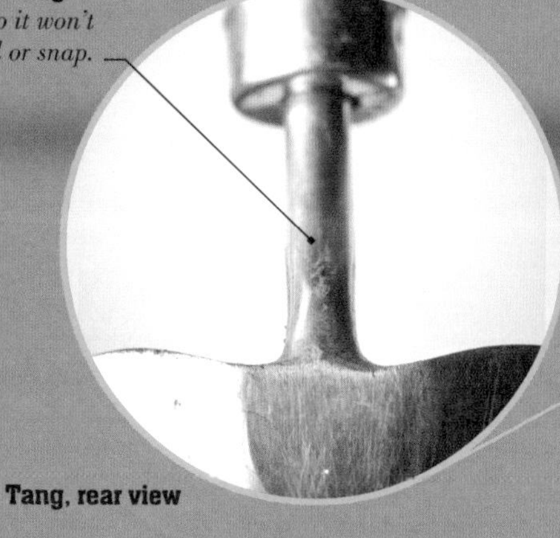

Tang, rear view

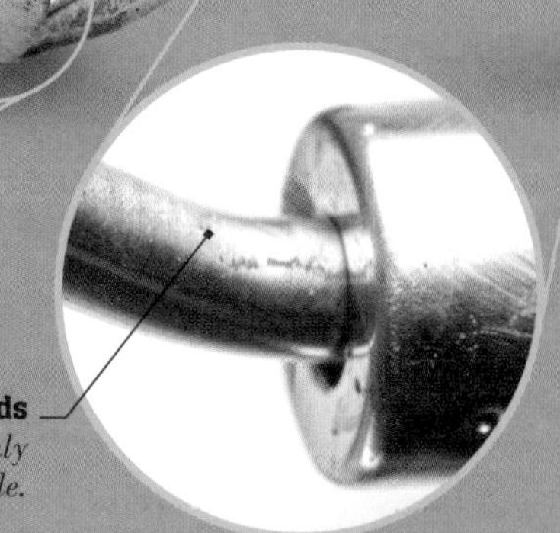

Tang embeds *the blade firmly in the handle.*

FOCUS ON...

Head Shapes

Trowel head shapes vary greatly, and are made from a range of materials. Narrow and pointed heads are good for weeding or planting small plants; digging trowels are very wide and almost triangular. Plastic or very thin steel budget trowels are easy to use but don't last long and break easily. Forged stainless-steel or copper heads, mounted into wooden handles, are the longest-lasting, and easy to maintain.

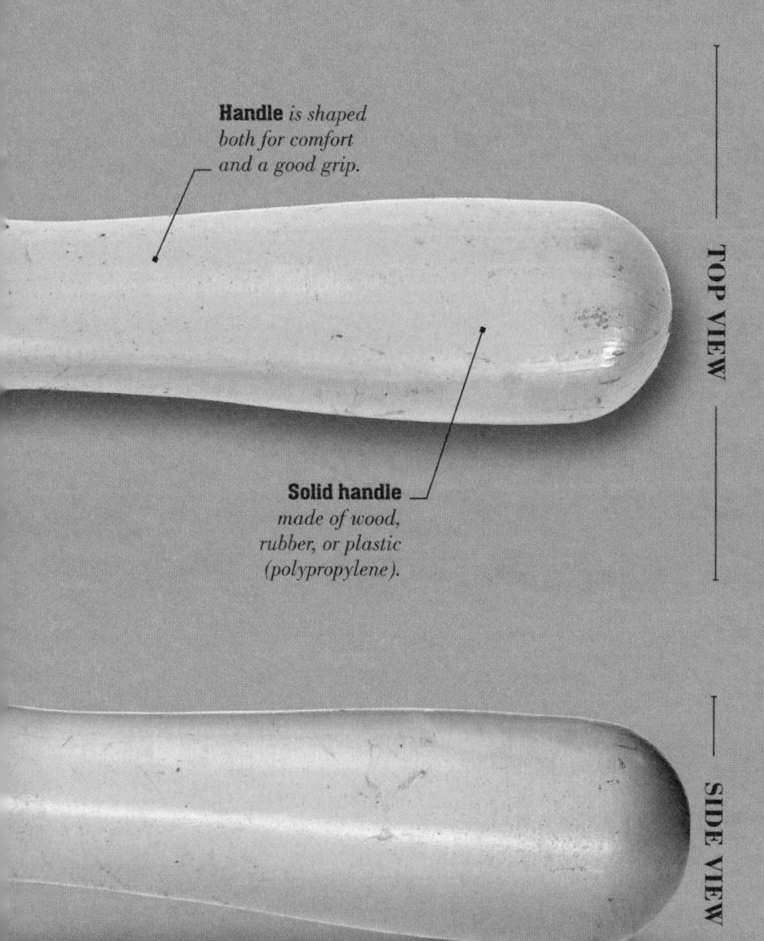

"A VITAL TOOL FOR USE IN BORDERS, CONTAINERS, OR VEGETABLE GARDENS"

Using a Trowel

A trowel is involved in many small gardening tasks, and like all tools, a good fit for you will make it more enjoyable to wield. Make sure you like the feel of the handle, as well as the size and shape of the head.

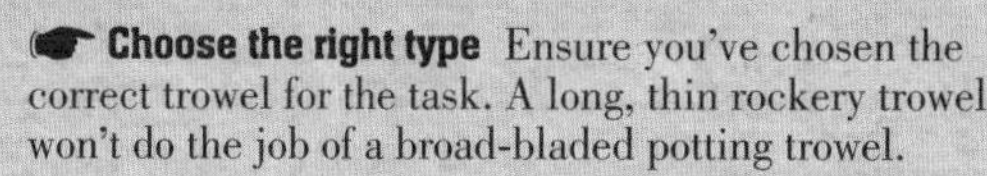

The Process

Before you start

Choose the right type Ensure you've chosen the correct trowel for the task. A long, thin rockery trowel won't do the job of a broad-bladed potting trowel.

Inspect the tool Check that your trowel is clean, the blade isn't bent, the edges are in good condition, and the handle is sound.

1 Prepare the ground

Working hard, compacted soil with a trowel is difficult, so work over any soil first with a border fork prior to using your trowel to plant. Planting in well-worked and composted soil is easy, so preparation helps.

2 Dig the hole

Push the trowel vertically into the soil, and do this several times to form the shape of the desired planting hole in firm soil. Apply backward pressure while removing the soil to capture and remove it. Use one hand only, and don't overwork the tool.

3 Hold back the soil

In loose soil, or when planting into potting compost, the soil often falls back into a planting hole. Use the trowel to pull the soil away, then leave it in place at the edge to hold the soil back while planting.

After you finish

Remember your tool Be careful not to throw trowels away. They often end up in a bucket, mixed in among the weeds, and then get flung onto the compost heap!

Keep it clean Make sure to clean off the trowel after use, give it a light coat of general-purpose oil, if needed, then store safely.

THE PHILOSOPHY *of* TOOLS

"WHEN I GO INTO THE GARDEN WITH A SPADE, AND DIG A BED, I FEEL SUCH AN EXHILARATION AND HEALTH THAT I DISCOVER THAT I HAVE BEEN DEFRAUDING MYSELF ALL THIS TIME IN LETTING OTHERS DO FOR ME WHAT I SHOULD HAVE DONE WITH MY OWN HANDS."

RALPH WALDO EMERSON

Maintain tools for Digging & Groundwork

Often shoved in the shed, grubby and rusty, with wobbling handles, digging tools may be robust, but they need care like any other tools.

SHARPENING EDGED BLADES

A good, sharp hoe will reward with ease of use, lessening many labour-intensive chores. Regular maintenance keeps it performing well at all times.

1 Check edges

All hoes have a cutting edge – some have one; others two or three. Check all edges for damage.

2 Clamp and file

Clamp the hoe in a vice or on a workbench, and, using a flat file or stone, sharpen one side of each edge only. Follow the sharpening angle already present in the edge.

Sharp edge *of an oscillating hoe.*

3 Keep it sharp

The edge need not be razor-sharp, but good enough to cut roots on young plants. For the best results, sharpen with each use.

TREAT WITH CARE

All tools benefit from maintenance, which in some cases is only a check over, or perhaps a quick clean off after use. Checking tools means they don't fail you when really needed.

Wash and oil

Untreated metals will rust, and the pitting this causes leads to resistance and collects dirt. Be sure to clean all tools with water, including coated, synthetic, or rust-resistant metals. Coat any untreated metals with light vegetable oil, which is better for the soil than some alternatives.

Care for wood

Many tools, both new and old, have wooden handles, and they are often the most pleasing to use. Storage in hot sheds or glasshouses can dry out the wood and cause play in the joints. Soak wooden handles in water to revive them and then store in cool shade.

Tools	Inspection
Shovels & Spades	▪ Check for play in joints of handle and shaft before use – wobbling handles can pinch skin ▪ Check spade edges are sharp and face clean before work and after finishing
Post-hole Diggers	▪ If used infrequently, take extra care to check that your post-hole digger is in good condition before storing ▪ Check scissor action, usually a nut-and-bolt arrangement, is working well – on many models this can work loose – and adjust if necessary
Hoes & Cultivators	▪ Check tools are structurally sound, with any special mechanisms working smoothly ▪ Tools with cutting edges need to be sharp, and hoes are often overlooked – we expect them to cut, yet push them through harsh, blunting material
Trowels, Forks, & Dibbers	▪ Check for stress damage, such as cracks at top of shafts of forks and trowels, or bent areas ▪ Look for rust and deep pitting, as texture of thickset rust will attract soil and therefore builds-up a resistant surface

	Cleaning	Protection	Adjustment	Storage
	▪After use, wash with water and hand brush if necessary – using a rain-fed trough is best, rather than a hose, as it leaves two hands free ▪Clean soil away when wet, as baked-on dirt can be very tough to get off, hardening like fired clay	▪These tools need little, if any, oiling when stored in a cool, dry shed or room – seasoned wood does not need protection and neither does good steel ▪If metalwork does need oil, such as in prolonged or damp storage, then use vegetable-based chainsaw oil, which will not harm your soil	▪Fix play in wooden handles by soaking in water for 24 hours – man-made handles can require new rivets, or may be beyond repair ▪Sharpen spade edge with a flat file to give it a good, fine angle	▪Store tools with wooden handles in cool and dry sheds or rooms – a shed in sun or a glasshouse can get very hot, drying out handles ▪Avoid damp conditions, which can lead to rusting tool heads
	▪After use, wash with water and hand brush if necessary – preferably using rainwater ▪Clean soil off when wet	▪Little, if any, oiling needed – should metalwork need attention use vegetable-based chainsaw oil	▪Adjust scissor action via nut and bolt – it relies upon a shaft running through the centre, with a threaded end – ideally nut will have a nylon locking core to keep it accurate, replace if necessary	▪Make sure tools are clean and well-adjusted, as often they are needed for an emergency repair and must be ready when required
	▪After use, wash with water and hand brush if necessary – preferably using rainwater ▪Clean soil off when wet	▪Little, if any, oiling needed – should metalwork need attention use vegetable-based chainsaw oil	▪Keep mechanisms, such as that of oscillating hoe, clear of compressed mud ▪Sharpening hoe edges is important as they need to sever roots of weeds cleanly – use a flat file and solid vice, clamp, or workbench to grip tool head, sharpen one or both edges to tight angle that's sharp to touch	▪Store in a tool bin or on specific hook if awkwardly shaped
	▪Wash every time you use them, or simply brush dirt off with gardening glove as you finish your work		▪Overworked tools may get bent – clamp tool and apply gentle pressure to bend metal back to shape ▪Take great care when bending a hand tool, as wrong or excessive force may simply snap it	▪Keep these tools at hand, perhaps in a grab bag, garden trug, bucket, or similar – it is good for efficiency, but put them in there as you hope to find them ▪Do not "store" your favourite hand tools on a compost heap – they will not be the same after a year of compost action

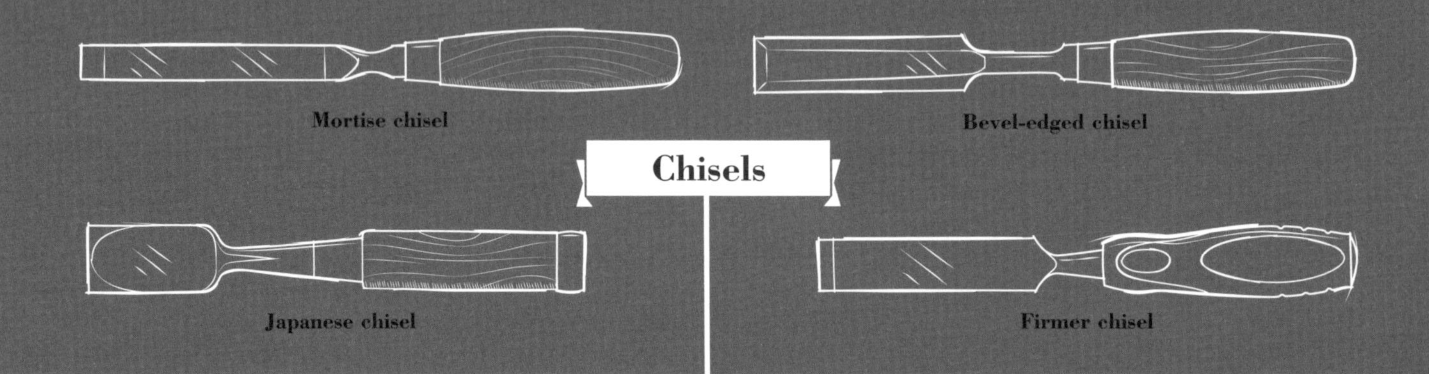

THE TOOLS *for* SHAPING & SHARPENING

A carpenter's toolkit includes a range of chisels, planes, gouges, and files for shaping wood. Sharpness is key for fine woodwork and sharpening stones are vital for blade maintenence.

Files and Rasps

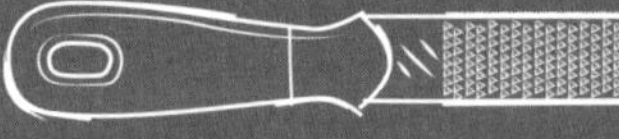
Rasp

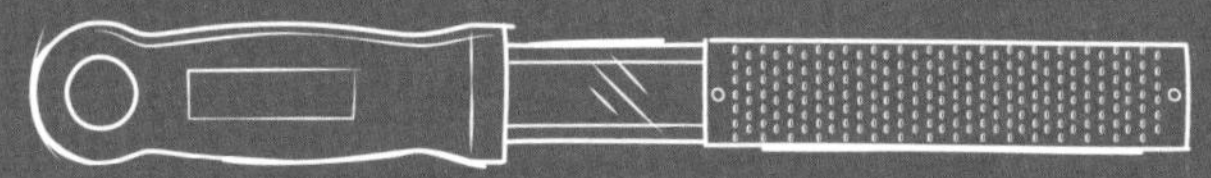
Microplane

File

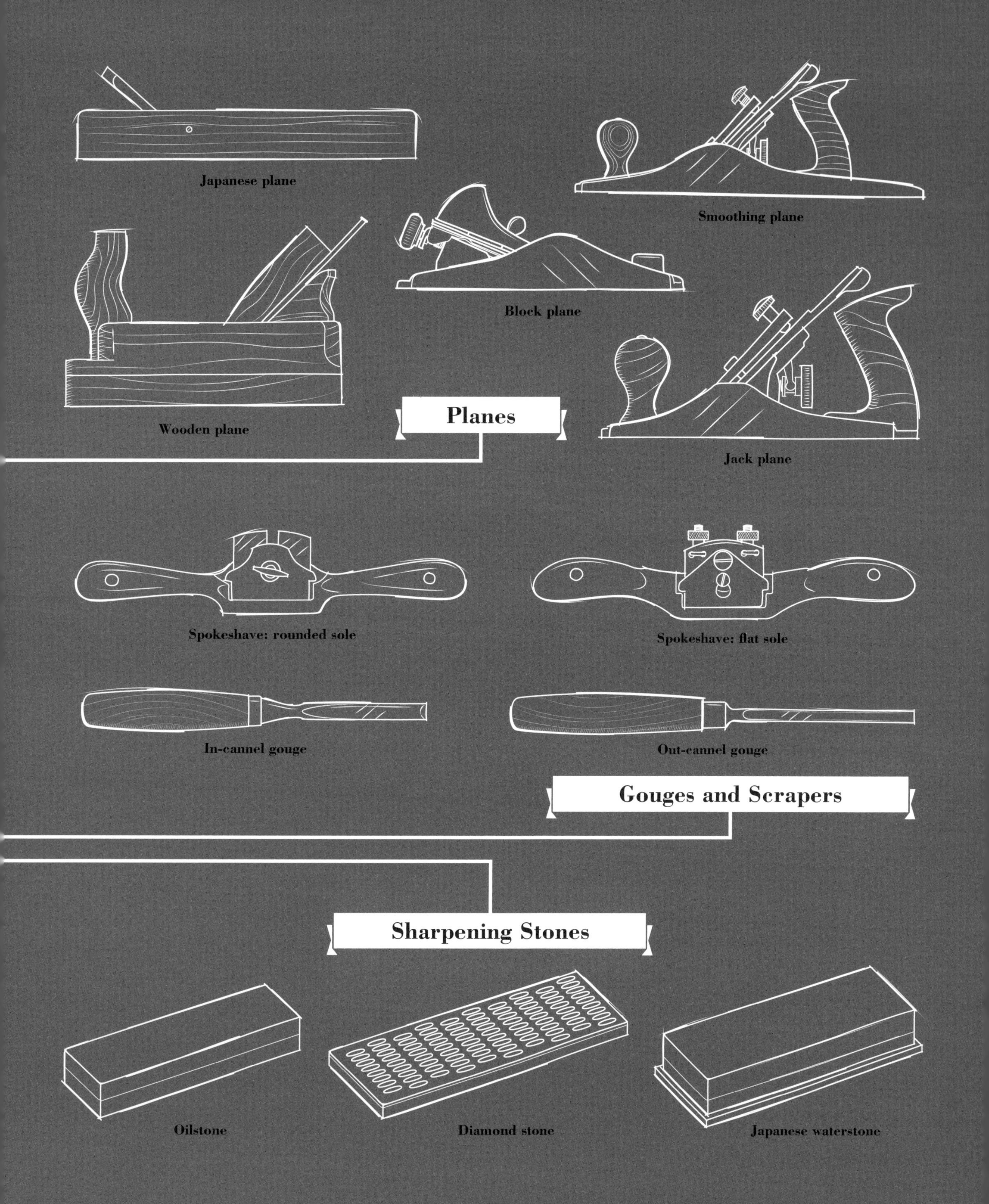
Japanese plane
Smoothing plane
Block plane
Wooden plane
Planes
Jack plane
Spokeshave: rounded sole
Spokeshave: flat sole
In-cannel gouge
Out-cannel gouge
Gouges and Scrapers
Sharpening Stones
Oilstone
Diamond stone
Japanese waterstone

HISTORY OF SHAPING & SHARPENING

8000 BCE — FIRST CHISELS

Long, chisel-like stone tools made of flint appeared around this time. These were further developed in the late Neolithic period by grinding the flint.

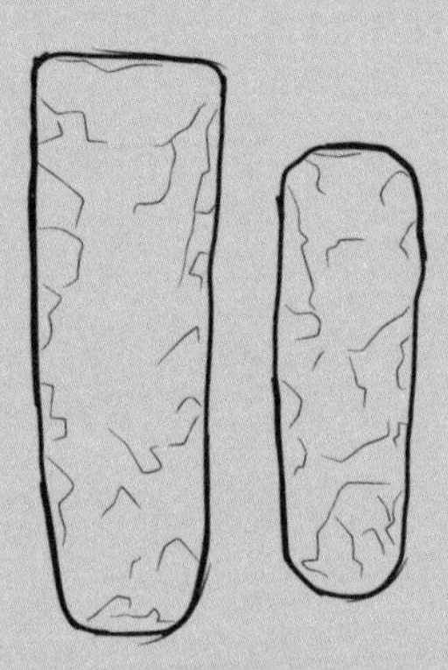

Paleolithic chisels

7000 BCE — EARLY GOUGES

Chisels and gouges were made from ground and polished stones, such as jadeite, diorite, and schist. All last longer than flint, a type of quartz that fractures easily.

"IT IS WELL WITH ME ONLY WHEN I HAVE A CHISEL IN MY HAND."

MICHELANGELO

MOHS SCALE

In 1812, German mineralogist Friedrich Mohs developed a way to identify minerals based on their resistance to scratching by ten reference minerals. Here, flint scores higher than diorite, but diorite is actually more durable.

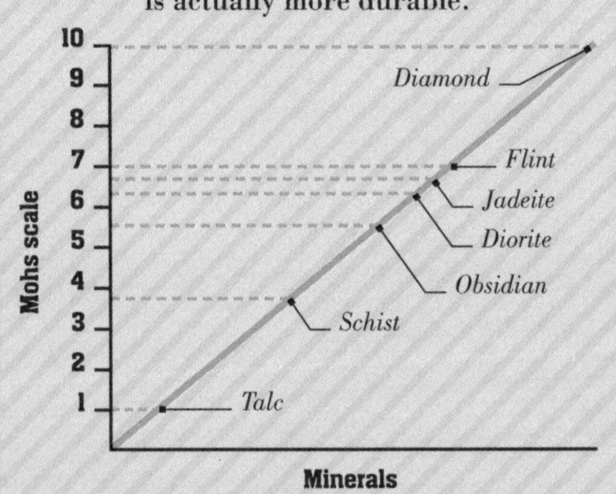

c.3000–1900 BCE — EARLY BRONZE CHISELS

When smelting and casting techniques were developed, the first bronze chisels were made. They initially consisted of one solid piece of metal with no attachable handle, and could be used to cut and shape soft rocks such as sandstone and limestone, as well as wood.

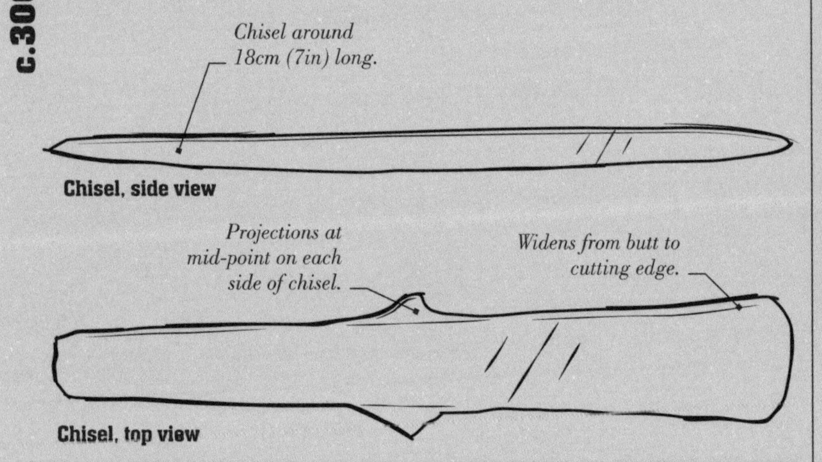

Chisel, side view

Chisel, top view

c.1500 BCE — EGYPTIAN FILES AND CHISELS

Ancient Egyptians used flat bronze files and iron as well as bronze chisels. Some were cast with tangs – rods pushed into a wooden handle – or with sockets, into which wooden handles were set.

1200–900 BCE — OBSIDIAN TOOLS

Chisels and knife-chisels made out of obsidian, a type of volcanic glass, were used to shape softer stone, as seen in the highly intricate sculptures of pre-Columbian Central America. Gouges, basically chisels with concave sections, were also used around this time to scoop out hollows or create holes with curved, instead of straight, sides.

735 BCE–500 CE

ROMAN CARPENTRY

Roman carpenters used a wide range of woodworking tools, including a variety of files, chisels, and gouges. A combined round and flat file made from bronze was developed and was widely used in the Iron Age.

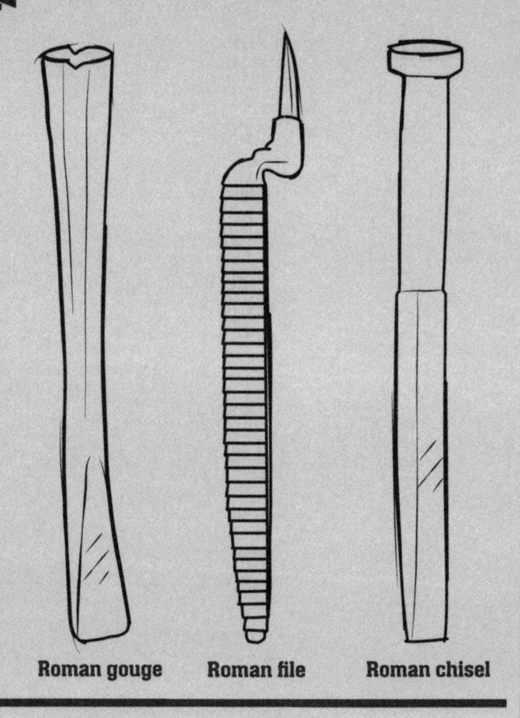

79 CE

FIRST PLANE

The earliest-known planes were Roman, with examples of the tool discovered at Pompeii. The Roman tool worked in much the same way as the modern plane. Roman planes came in a variety of sizes ranging from about 20cm (8in) long up to 43cm (17in) long.

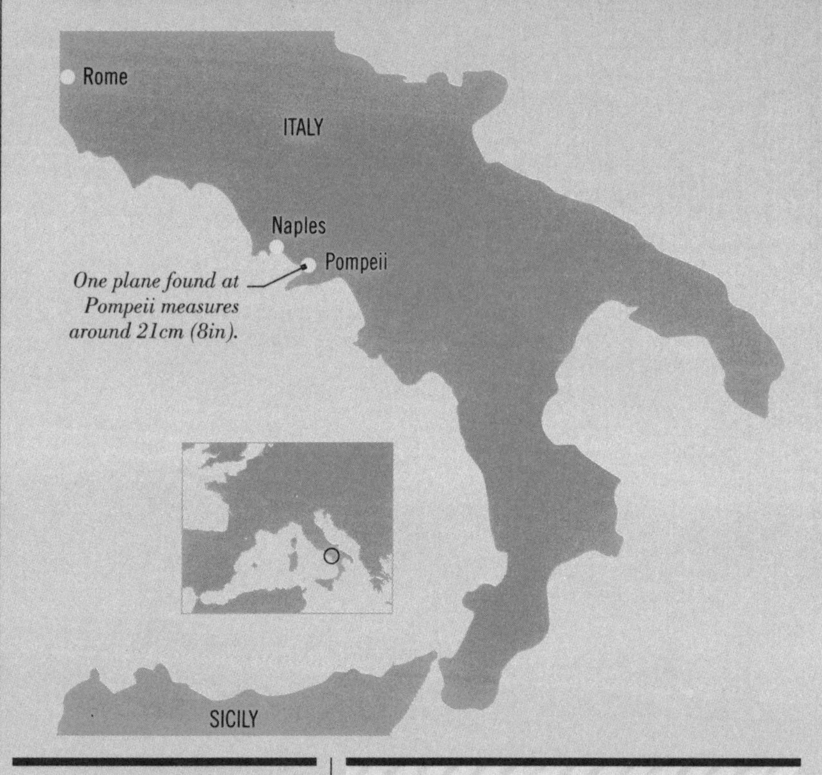

One plane found at Pompeii measures around 21cm (8in).

1100 CE

FILE SHAPES

Files made of carburized steel were available in a variety of shapes, including square, triangular, and round. The tools were hardened after being cut to the desired shape and length by a sharp chisel and hammer.

1600s

BENCH PLANE

A widely used carpentry tool, the bench plane was used for smoothing wood and straightening edges in furniture and house construction.

RARE IVORY

Roman woodworking planes made with ivory are rare. The Goodmanham plane, found in 2000 in Yorkshire, northern England, is one of the most complete examples with a solid-ivory stock.

"THERE ARE ALSO FILES MADE OF PURE STEEL ... FOUR-CORNERED, THREE-CORNERED, AND ROUND."

THEOPHILUS PRESBYTER

c.1890

PLANES

The screw and lever adjustor for a plane's iron was created, replacing the wedge that was hammered in place. The plane has changed little since.

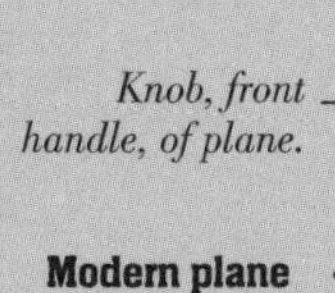

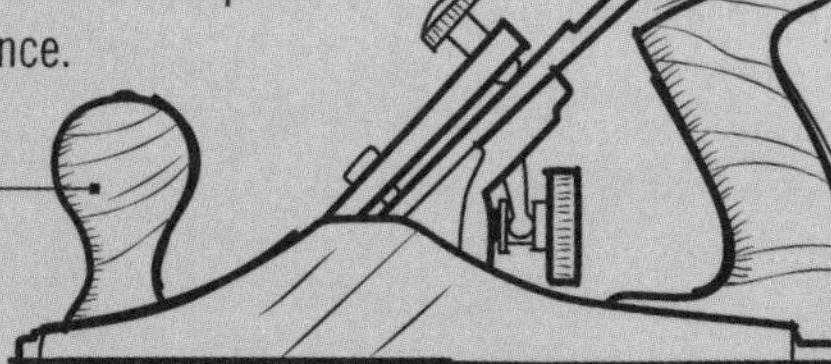

Knob, front handle, of plane.

Modern plane

Choosing a
Chisel

Because chisels are such fundamental woodworking tools, it's important to select the correct type for each task. A bevel-edged chisel is designed for cutting dovetails and finer woodwork, but its blade is not sturdy enough to chop out a mortise. Conversely, a mortise chisel is too cumbersome for most cabinetmaking jobs.

Mortise chisel

Japanese chisel

> "A BLUNT CHISEL REQUIRES MORE EFFORT TO CUT WOOD CLEANLY THAN A SHARP ONE"

Mortise Chisel

What it is Heavy, square-edged steel blade with broad neck. Hardwood handle with steel hoop prevents splitting.

Use it for Chopping mortises and levering out waste material without the risk of the blade snapping.

How to use Hold the tool upright to start chopping and use a mallet to strike the handle.

Look for Models with a leather washer between the blade and handle to absorb mallet blows.

Firmer Chisel

What it is Tool with a rectangular-section carbon-steel blade and a hardwood or polypropylene handle.

Use it for Carpentry and joinery work, particularly framing. Also general construction and DIY projects.

How to use Either grip the tool and push with both hands, or use a mallet for striking the handle.

Look for You may need to buy secondhand, as this tool is much less common nowadays.

Japanese Chisel

What it is Laminated blade (soft and hard steel), with a hollow ground on back. The oak handle has a steel hoop to withstand hammer blows.

Use it for Cutting joints and fine woodwork generally. Heavier versions are specifically for mortise cutting.

How to use Strike with Japanese hammer or use with two hands, like Western chisels.

Look for Requires single bevel (not two), when sharpening. Hollow back must be maintained as steel wears.

Bevel-edged Chisel

What it is Parallel steel blade with shallow bevelled edges. Has a hardwood or polypropylene handle.

Use it for Cutting pins and tails on dovetail joints. Horizontal/vertical paring, lightweight cuts, getting into corners.

How to use Grip tool with both hands or use a mallet for striking handle.

Look for Boxwood handles can be quite elaborate in shape, including octagonal or bulbous.

"SHARPEN A CHISEL CORRECTLY BY USING A HONING GUIDE"

TOP VIEW

Cutting edge *ground and honed to specific angles.*

Leather washer *to absorb shock (on registered pattern chisel).*

SIDE VIEW

Stout forged-steel *blade with square edges and precision-ground flat back.*

Blade end *angle is bevelled to 25 degrees.*

Chisel tip *usually sharpened to 30 degrees.*

"ALWAYS STRIKE A HARDWOOD CHISEL HANDLE WITH A MALLET, NEVER WITH A HAMMER"

Turned end *of handle reduced to accept hoop.*

Steel hoop *reinforces the top end of handle.*

STRUCTURE OF A

MORTISE CHISEL

A mortise chisel's blade has square edges, rather than bevelled. This makes it much stronger, which is important when levering out waste material from a mortise to prevent the tool from breaking. The handle is sturdier and traditionally made from ash or hornbeam. It's fitted with a steel hoop to prevent splitting when striking with a mallet.

Steel hoop *fitted tightly to handle to prevent wood splitting.*

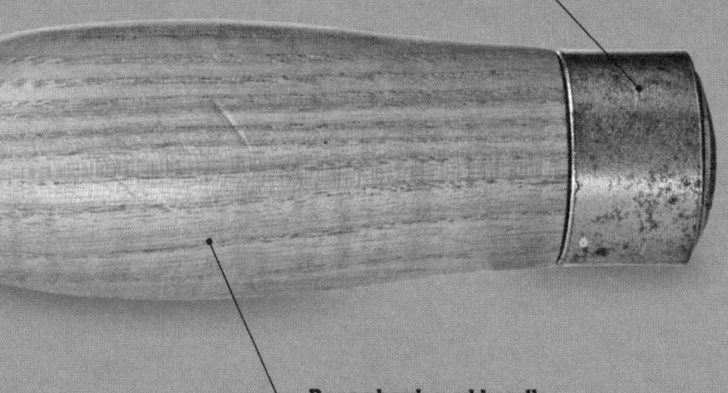

Dense hardwood handle *commonly ash, which is flexible and absorbs mallet blows readily.*

FOCUS ON...

CHISEL BLADES

Chisel blades are forged from carbon steel and come in various widths, from 3mm (0.12in) on bevel-edge tools up to 50mm (2in) on framing chisels. Lighter blades can be used to cut freehand or with gentle mallet blows, although heavier chisels are simply struck. Timber-framing chisels feature sturdier blades, which are virtually unbreakable.

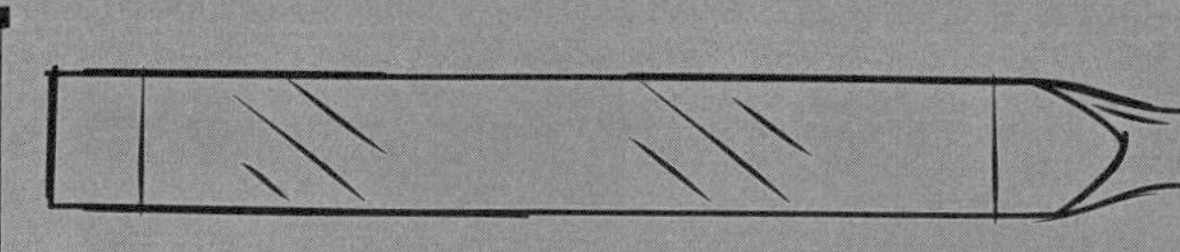

Mortise Tempered and hardened steel blade has square edges for chopping mortises cleanly. Designed to be struck with a mallet.

Bevel-edge Finely ground bevels on edges enable tool to undercut and cut dovetails. Widths down to 3mm (0.12in). Likely to have boxwood handle.

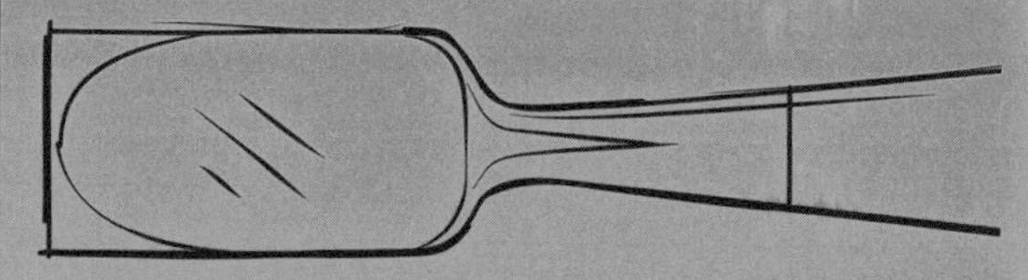

Japanese Laminated steel blade (hard back, softer front) with hollow back. Creates extremely sharp edge. Should be reshaped with a fine hammer occasionally.

USING A

Mortise Chisel

For chopping square or rectangular holes in timber, remember that the blade of a mortise chisel is heavier and more suitable for the task than that of a bevel-edged or even a firmer chisel. Always chop a mortise first before cutting the matching tenon. It's much easier to trim the tenon to fit, if necessary, than vice versa.

The Process

Before you start

- **Check the blade** Check the chisel blade is the correct width to match the mortise and that the edge is sharp.
- **Mark the mortise** Mark mortise precisely on the wood with correctly adjusted gauge plus a try square.
- **Mark the waste** With a pencil, cross-hatch the waste to be removed from mortise and tenon. If not, you could cut either the wrong way.

1 Clamp and position
Clamp the workpiece to the benchtop, close to a leg to transmit mallet blows down to the floor. With the chisel held vertically, place the blade edge about 3mm (0.12in) in from the end of the mortise, bevel facing inwards.

Strike chisel *the same number of times at each cut to maintain equal depth.*

2 Make the cut
Strike the chisel firmly with a mallet. Make a series of cuts about 3mm (0.12in) apart along the mortise to the same depth each time, keeping the chisel vertical. Count the number of strikes at each position to repeat and match the depth. Reverse the chisel when you reach the opposite end.

Keep blade *bevel-side down to remove the first layer of waste wood.*

FOCUS ON...

Chisel Bevels

The end of a chisel blade is always ground to an angle of 25 degrees; this is known as the primary bevel. A tiny secondary bevel of 30 degrees is created by sharpening (honing) the blade on a stone. This small additional bevel strengthens the blade edge because it forces the fibres of the timber apart during cutting, which reduces the force needed by the blade to move through the wood.

Bevel *forces wood fibres apart when cutting.*

Wood splits *as it is pushed apart by chisel bevel.*

Waste sections *at ends can be used for leverage, pushing the rest of the pieces out.*

3 Clear out the waste

Clean out the first layer of waste wood, keeping the blade bevel-side down. Use the waste section at each end of the mortise for leverage. Continue chopping across the next layer of the mortise and repeat the process, removing the waste at the end of the section as before.

4 Complete the mortise

When you reach just over half the timber depth, remove the waste at each end of the mortise by moving the chisel just inside pencil line and chopping downwards, keeping the blade vertical. Turn the timber upside down, tip out the chips, and repeat the process from opposite face.

“KEEP BOTH HANDS BEHIND THE CUTTING EDGE TO PREVENT INJURY”

After you finish

- ☛ **Check the cut** Hold the blade of a small try square inside the mortise to check the end depths match. If not, clean it up with the chisel.
- ☛ **Mark the tenon** Mark out a matching tenon before resetting the mortise gauge.

Choosing a Plane

Most home workshops need one or two bench planes. Defined by their size, the most versatile are metal jack and smoothing planes. A longer plane is better for getting boards straight, removing higher areas to produce a flat surface. A shorter plane tends to ride the contours on lengthy boards, although it's better for joints and finer work. You can still buy traditional wooden planes, although these are trickier to adjust.

Smoothing plane

Wooden plane

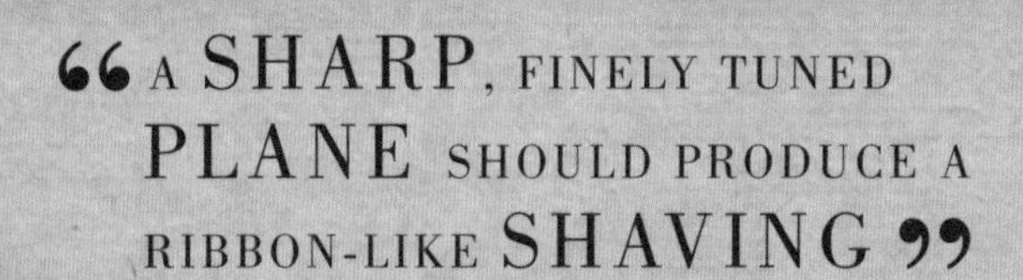

“A SHARP, FINELY TUNED PLANE SHOULD PRODUCE A RIBBON-LIKE SHAVING”

Jack plane

Japanese plane

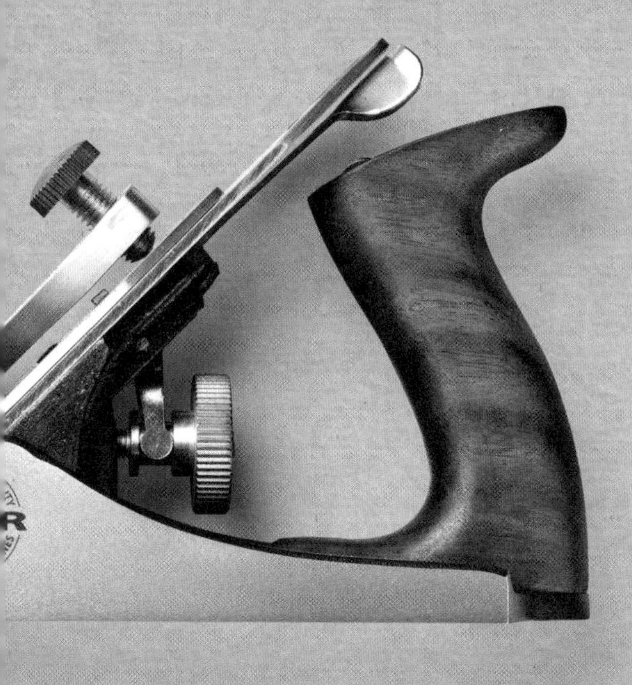

Smoothing Plane

What it is Iron body with carbon-steel blade, adjusts for cutting depth via thumbwheel. Hardwood or plastic handle.

Use it for Trimming joints, final cleaning of surfaces before sanding, planing smaller components to size.

How to use Set blade depth with the thumbwheel. Sight down the sole to check that the projecting blade is even.

Look for No 4 is the most common size. No 4½ a slightly wider and heavier tool.

“THE LONGER THE PLANE’S BODY, THE MORE EVENLY IT PREPARES A SURFACE”

Wooden Plane

What it is Body of dense hardwood (beech or hornbeam), with cutout for the steel blade, which is wedged in place.

Use it for General planing tasks depending on model, from preparing rough-sawn timber to final finishing.

How to use Gently tap upper edge of blade with small hammer to increase cutting depth. Tap rear of body to reduce it.

Look for Wooden planes with blade adjustment tend to be more expensive than equivalent metal versions.

Jack Plane

What it is Iron body, carbon-steel blade, adjusts cutting depth via thumbwheel. Hardwood or plastic handle.

Use it for Preparing rough-sawn timber to size. Hanging doors, general carpentry and joinery work.

How to use Set blade depth with the thumbwheel. Sight down the sole to check that the projecting blade is even.

Look for No 5 is the most common size, No 5½ a slightly wider, heavier tool. Hardwood handles are most comfortable.

Japanese Plane

What it is Simple body from oak. Sloping cutout to accommodate steel blade, held in place with wedge.

Use it for Smaller tools for fine finishing, longer ones for preparing timber. Specific planes for chamfers and rebating.

How to use Japanese planes cut on the pull stroke. Tap the end of the blade to increase cutting depth.

Look for Laminated steel blades have hollow backs, so a special hammer is needed for eventual reshaping.

Block Plane

What it is Small iron tool that can be used in one hand. Has a carbon-steel blade, with depth and lateral adjusters.

Use it for Planing end grain, narrow edges, chamfers, trimming joints, and fine detailed work.

How to use Adjust blade for fine cut, place palm over top of tool. Using two hands will increase pressure.

Look for More sophisticated block planes have adjustable throats for finer control of shavings.

Block plane

TOP VIEW
SIDE VIEW
Front handle made of dense hardwood or plastic screwed to the body.
Brass screw tightens screw cap, cap iron, and blade assembly against frog.
Screw, side view
Frog screwed to body, seats blade at 45 degrees.
Brass screw cap iron with locking screw.
Cast-iron body and sole accurately machined to lie completely flat.
SUPERIOR TRADE QUALITY
AXMINSTER
RIDER
5½
WOODWORKING PLANES
Toe of plane where downward pressure is applied at the start of cut.
Steel blade projects through mouth in sole.
Blade on sole

STRUCTURE OF A

JACK PLANE

A jack plane is a traditional bench plane used for preparing rough-sawn timber to size. In terms of plane size, the No 5 jack plane makes a versatile all-rounder for general carpentry and woodwork. Bodies of metal planes are generally made of very strong cast iron, although more upmarket models may be made of bronze.

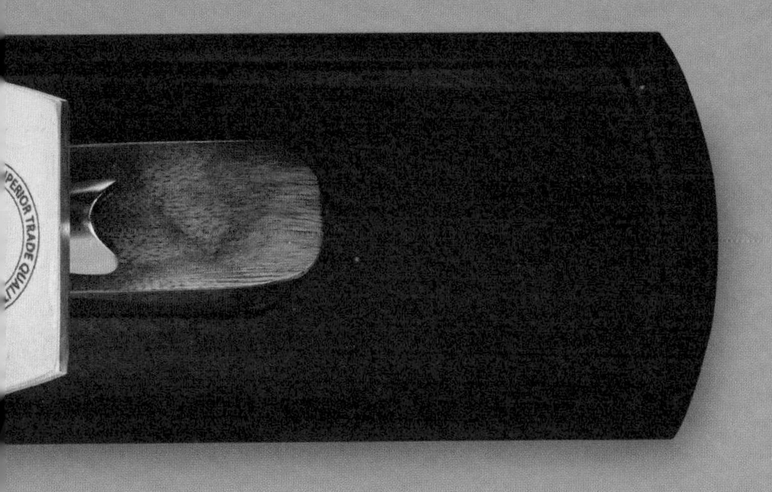

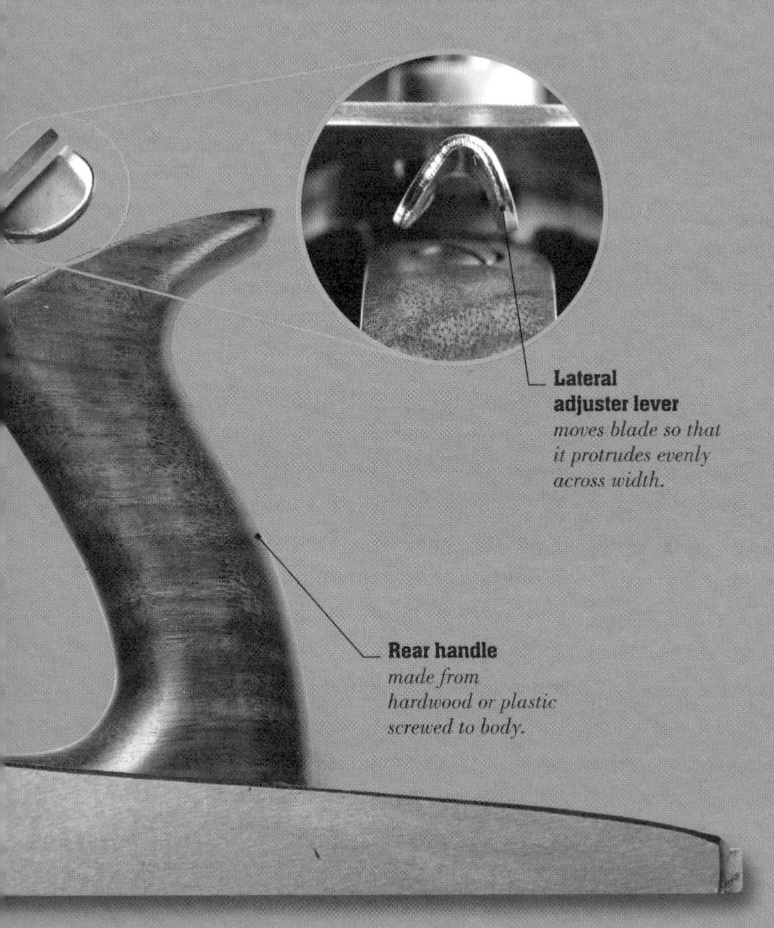

Lateral adjuster lever *moves blade so that it protrudes evenly across width.*

Rear handle *made from hardwood or plastic screwed to body.*

FOCUS ON...

PLANE SIZES

Bench planes vary in length and width; some have specific jobs. Each is defined by a number, such as No 4, as well as by a common name: smoothing, jack, fore, etc. Blades (or plane irons) vary in size, depending on tool width, but the most common sizes are 50mm (2in) and 60mm (2.25in). Shorter planes are better for finishing timber, while longer ones are perfect for flattening undulating surfaces.

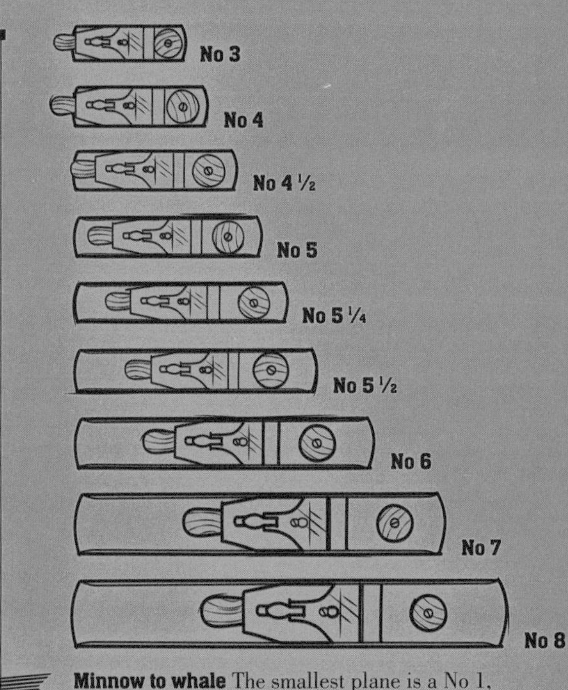

Minnow to whale The smallest plane is a No 1, though it's hard to find. Smoothing planes include No 3 through to No 4½, while Nos 5 and 5½ are known as jack planes. Even longer are the No 6 (fore) and Nos 7 and 8 try or jointer planes. Superb for truing up long boards, they are unwieldy and difficult to handle.

> "RUBBING THE SOLE OF THE PLANE WITH CANDLE WAX WILL REDUCE THE EFFORT REQUIRED WHEN PLANING"

Using a Jack Plane

The No 5, commonly known as a jack plane, is a good all-rounder when it comes to choosing a plane for timber preparation. Used for planing wood square as well as reducing material to exact dimensions, it's long enough to true up the edge of a door while not being too big to store in a tool box.

The Process

Before you start

- **Check the blade** Inspect the blade and sharpen it, if required, on a suitable stone. Remove any oil from the blade before using.
- **Secure the workpiece** Either position the timber securely in a vice or secure it to the benchtop with a suitable clamp.

1 Adjust the blade

Hold the plane upside down to check the blade is projecting evenly across its mouth. If it's uneven, adjust the lateral blade level. Turn depth-adjuster wheel for a fine cut.

Twist depth *adjuster wheel.*

2 Plane in position

Standing with feet apart, grasp the rear handle with your index finger pointing down the side of the frog. Applying downward pressure on the front knob, push the plane forwards along the wood, following the grain. As you reach the end of the workpiece, transfer the pressure from the toe to the heel of the tool.

FOCUS ON...

Shaving Action

A correctly adjusted and sharpened plane will create a thin shaving of wood as it moves across the surface. As the blade severs the wood fibres, the convex cap iron (or chip-breaker) forces a shaving up through the mouth, which then curls backwards in a string of small cracks. In most bench planes the blade is seated on the frog at 45 degrees, with the bevel-side down.

3 Square it off

Using a try or combination square, check that the face edge is at 90 degrees to the adjacent planed surface (the face side). If the edge is not square, adjust the lateral lever of the plane slightly to compensate and plane the edge again. Keep checking until the edge is perfectly square.

Face edge *should be at 90 degrees to surface.*

Look for light *beneath the straight edge that indicates gaps and unlevel surfaces.*

4 Check for level

The planed edge must also be flat, so check this by tilting the plane on its edge and sighting along it. With longer timber it's best to use a steel straight edge or even a long spirit level. Always use a marking gauge for accuracy when planing to width or thickness.

After you finish

- **Clean the plane** Brush away debris from the tool and remove any shavings jammed under the cap iron.
- **Store it safely** If stored in an unheated workshop, wipe the sole with an oily rag and retract the blade.

"THE FEEL AND BEAUTY OF FINELY CRAFTED WOOD... THE REFRESHING SMELL OF YOUR WORKSHOP... THE ABSORBING JOY OF CUTTING AND JOINING THAT MAKES THE HOURS RACE BY... THESE ARE THE REASONS YOU LOVE WOODWORKING."

JACK NEFF

Choosing a
File or Rasp

Used to shape both metal and wood, files and rasps come in varying sizes and shapes: flat, half-round, square and circular. Only files can be used on metal, while traditional rasps and modern Microplane blades are designed for efficient use on timber. Needle files, with a range of profiles, are miniature versions used for detailed work.

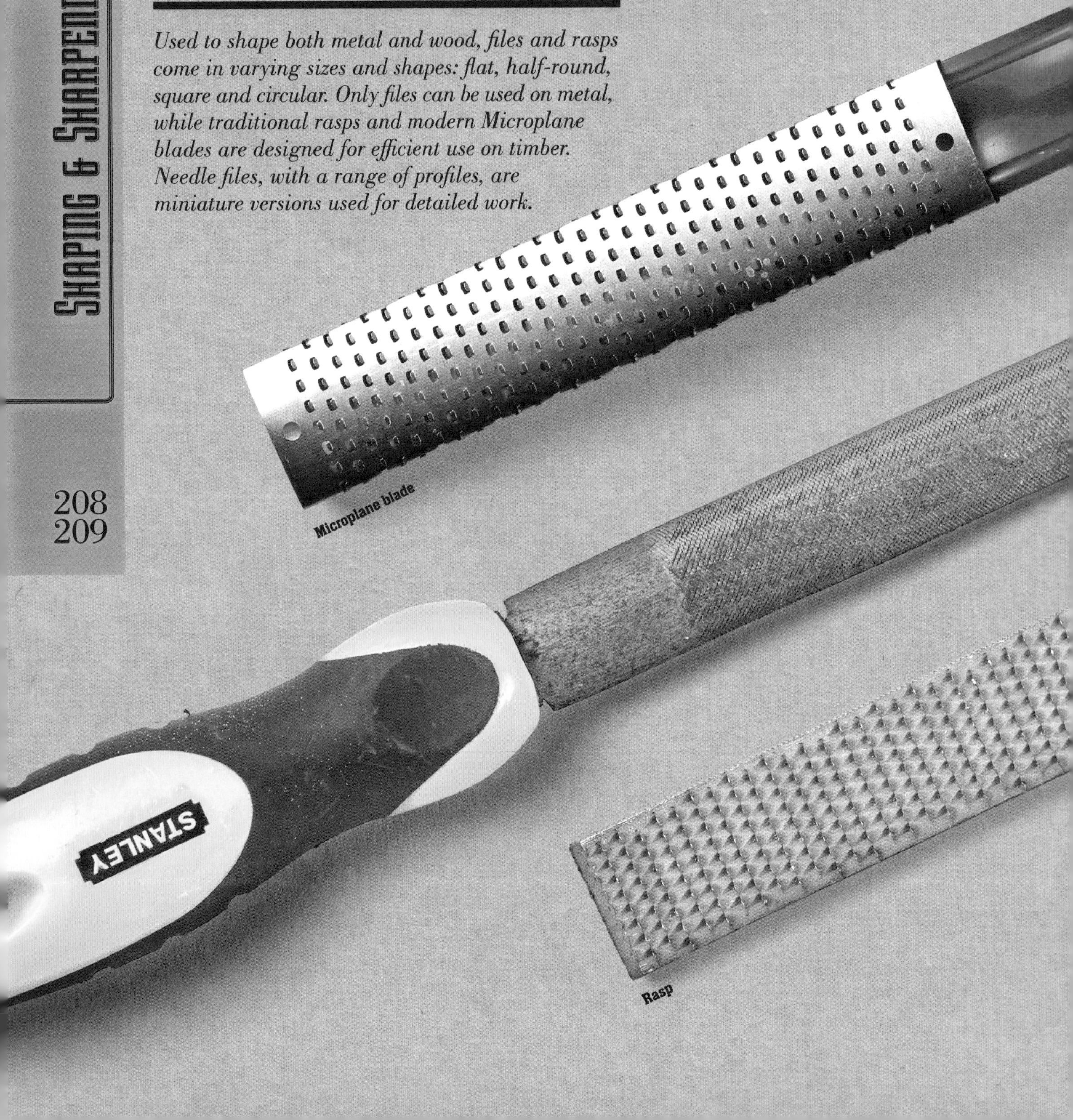

Microplane blade

Rasp

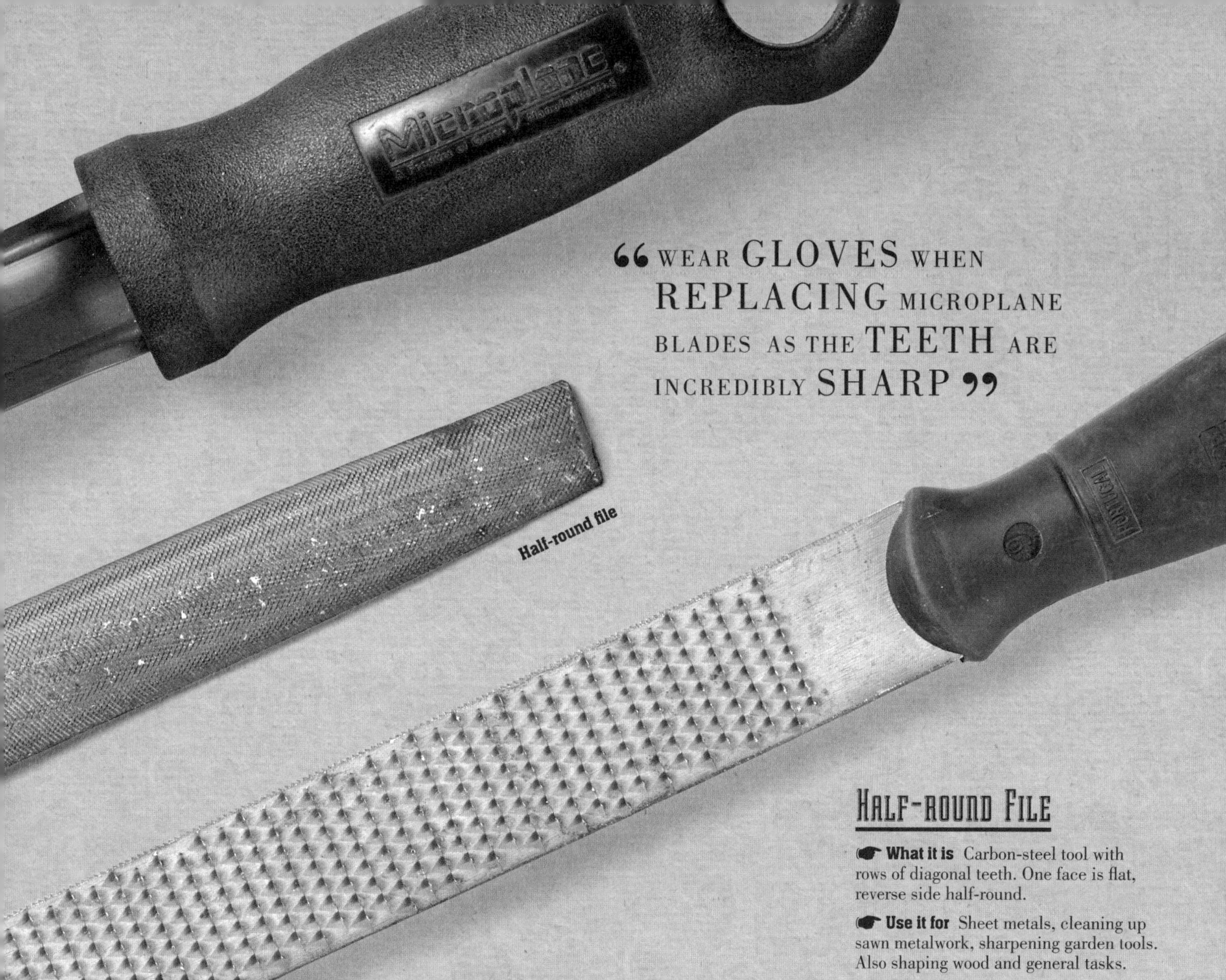

“WEAR GLOVES WHEN REPLACING MICROPLANE BLADES AS THE TEETH ARE INCREDIBLY SHARP”

“USE A FILE CARD TO CLEAN THE CLOGGED TEETH ON A FILE OR RASP”

Microplane Blade

What it is A stainless-steel blade with plastic handle and rows of razor-sharp, chemically formed teeth that create shavings when pushed on surface.

Use it for Shaping plasterboard, timber, and plastics rapidly and cleanly. Several profiles are available, including angled blades.

How to use Fit blade to holder, move tool back and forth over surface to be shaped. The tool cuts on the push stroke, but blade can be reversed to cut on the pull stroke.

Look for Some blades can also be fitted to hacksaw frames. Tools may have either fixed or snap-in handles.

Half-round File

What it is Carbon-steel tool with rows of diagonal teeth. One face is flat, reverse side half-round.

Use it for Sheet metals, cleaning up sawn metalwork, sharpening garden tools. Also shaping wood and general tasks.

How to use Grip end with thumb and forefinger, grasp handle with other hand and push tool forwards, keeping it level.

Look for Appropriate cut/coarseness type for job. Individual files often sold without handles, so check when buying.

Rasp

What it is Tool with coarser teeth than a file. These may be punched by machine or stitched by hand.

Use it for Removing waste areas of wood rapidly. Preliminary shaping work before filing, but leaves coarse marks.

How to use Grip end with thumb and forefinger, grasp handle with other hand and push tool forwards, keeping it level.

Look for Handmade rasps are efficient but expensive to buy. A 250mm (10in) size is a useful all-rounder.

Choosing a
Gouge or Spokeshave

Often a curved rather than straight cutting tool is required whenever you're working with wood, whether the project is carving a bowl or shaping a chair leg. This is the realm of gouges and spokeshaves. A gouge is basically a chisel with a curved profile blade, which is used for carving and scribing cuts. The twin-handled spokeshave works in a similar way to a bench plane, but its design allows it to be used to create convex or concave cuts.

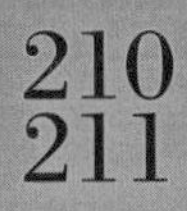

Out-cannel gouge

Flat-sole spokeshave

In-cannel gouge

Rounded-sole spokeshave

“ MAKE SURE YOUR **SLIPSTONE** MATCHES THE **GOUGE** CURVATURE. IF IT **DOESN'T**, YOU WILL STRUGGLE TO GET A KEEN **EDGE** WHEN SHARPENING ”

Out-cannel Gouge

What it is A convex-profile steel blade with cutting edge ground on the outside. Hardwood handle.

Use it for Carving and hollowing out hard- and softwoods.

How to use Select suitable blade width. Position blade on timber and push gouge handle or strike with mallet.

Look for Handles can vary in size and shape, so check for comfort and balance. Sharpen with a slipstone.

Flat-sole Spokeshave

What it is Cast-iron body with twin handles. Blade is ground at 25 degrees and secured with a wing nut and cap iron.

Use it for Creating convex curves in timber, particularly narrow edges; shaping spindles and similar.

How to use With the blade barely out, grip tool with both hands, push forwards, following grain. Adjust depth as required.

Look for Elaborate tools have twin thumbscrews to adjust blade depth. More basic tools are trickier to adjust.

In-cannel Gouge

What it is Convex profile steel blade with cutting edge ground on inside. Hardwood handle.

Use it for Trimming timber to match adjacent items, such as curved mortise and tenon joints on chairs.

How to use Select the blade to match curve of profile. Position blade on timber and strike with mallet.

Look for Handles can vary in size and shape, so check for comfort and balance. Sharpen with a slipstone.

Rounded-sole Spokeshave

What it is Cast-iron body with twin handles. Blade is ground at 25 degrees and secured with wing nut and cap iron.

Use it for Creating smooth, concave curves in hard- and softwoods, particularly on narrow edges.

How to use With blade barely out, grip tool with both hands, push forwards following grain. Adjust depth as required.

Look for Elaborate tools have twin thumbscrews to adjust blade depth. More basic tools are trickier to adjust.

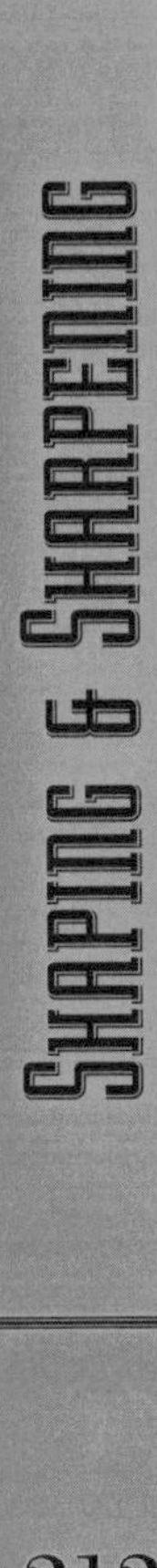

Choosing a Sharpening Stone

An edge tool needs sharpening with an abrasive stone. Whether it's a plane, chisel, or spokeshave, a consistent bevel must be maintained along the edge. Although natural sharpening stones are available, the most popular and cheapest are made from synthetic materials. A coarse stone will hone rapidly, but this is usually followed with a finer grade.

Oilstone

Oilstone (side view)

Diamond stone

“ADD THE CORRECT LUBRICANT WHEN USING A SHARPENING STONE TO PREVENT CLOGGING”

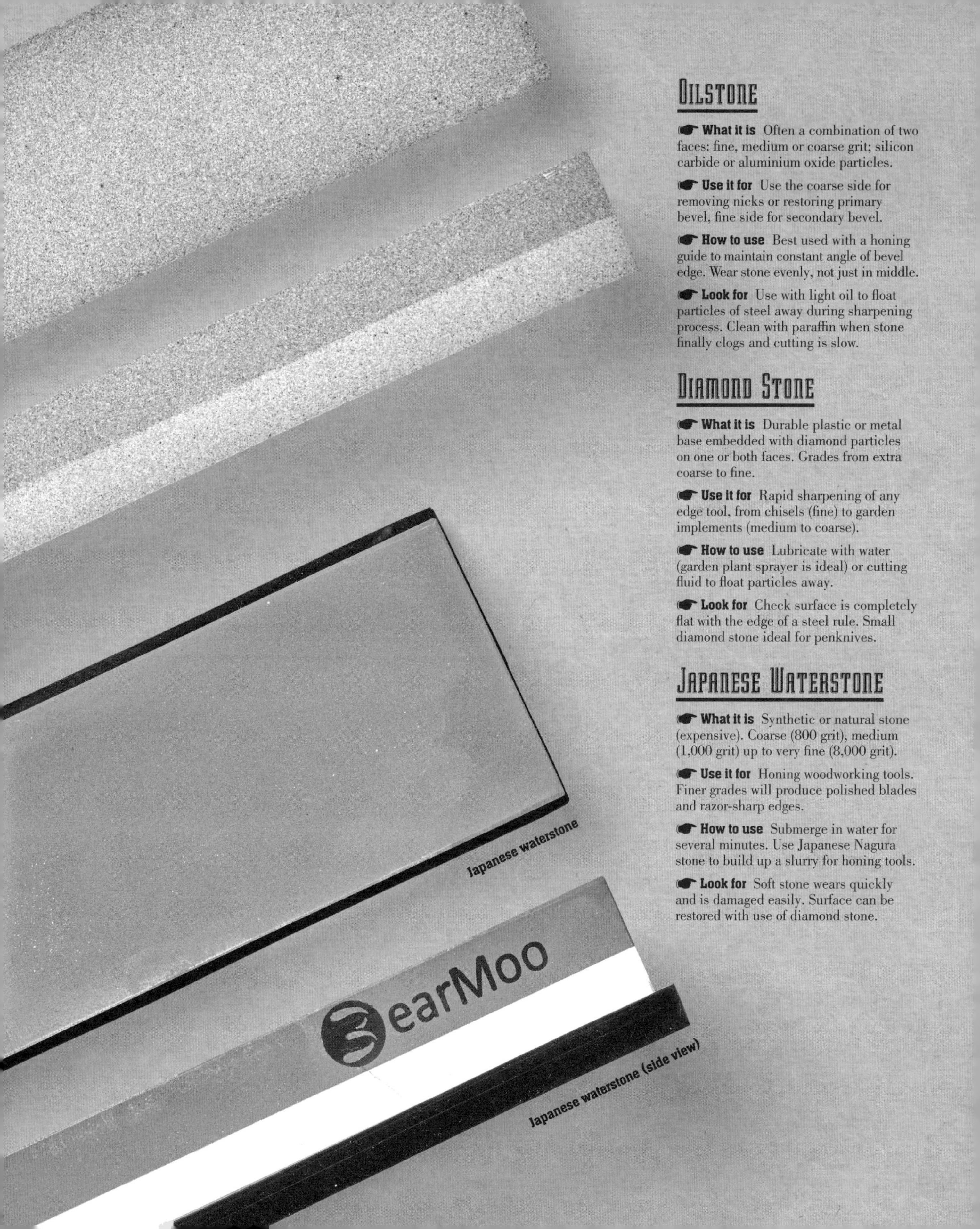

Japanese waterstone

Japanese waterstone (side view)

Oilstone

☛ **What it is** Often a combination of two faces: fine, medium or coarse grit; silicon carbide or aluminium oxide particles.

☛ **Use it for** Use the coarse side for removing nicks or restoring primary bevel, fine side for secondary bevel.

☛ **How to use** Best used with a honing guide to maintain constant angle of bevel edge. Wear stone evenly, not just in middle.

☛ **Look for** Use with light oil to float particles of steel away during sharpening process. Clean with paraffin when stone finally clogs and cutting is slow.

Diamond Stone

☛ **What it is** Durable plastic or metal base embedded with diamond particles on one or both faces. Grades from extra coarse to fine.

☛ **Use it for** Rapid sharpening of any edge tool, from chisels (fine) to garden implements (medium to coarse).

☛ **How to use** Lubricate with water (garden plant sprayer is ideal) or cutting fluid to float particles away.

☛ **Look for** Check surface is completely flat with the edge of a steel rule. Small diamond stone ideal for penknives.

Japanese Waterstone

☛ **What it is** Synthetic or natural stone (expensive). Coarse (800 grit), medium (1,000 grit) up to very fine (8,000 grit).

☛ **Use it for** Honing woodworking tools. Finer grades will produce polished blades and razor-sharp edges.

☛ **How to use** Submerge in water for several minutes. Use Japanese Nagura stone to build up a slurry for honing tools.

☛ **Look for** Soft stone wears quickly and is damaged easily. Surface can be restored with use of diamond stone.

“THE EXPECTATIONS OF LIFE DEPEND UPON DILIGENCE; THE MECHANIC THAT WOULD PERFECT HIS WORK MUST FIRST SHARPEN HIS TOOLS.”

CONFUCIUS

Maintain tools for
Shaping & Sharpening

For accuracy and reliability, shaping tools need to be maintained properly. Look after them and you can expect years of work in return.

SHARPENING EDGE TOOLS

Edge tools must be sharp for efficient cutting and safety. Blunt tools are more likely to slip in use and require greater effort for often mediocre results.

1 Flatten back
With the back of the blade flat on the sharpening stone, run the blade across the surface to remove any slight burrs. Use oil or water (to suit the stone) to float the steel particles away and prevent clogging.

Chisel blade *must be kept sharp for accurate cuts.*

2 Hone secondary bevel
With the blade held at 30 degrees, move the blade in a figure-of-eight pattern across the stone. It's important to keep the angle constant, so use a honing guide here.

3 Remove burr
Lay the blade flat on the stone again and slide it over the surface a couple of times. This will remove the burr and leave a sharp edge. Wipe the stone clean when you have finished.

ADJUSTING DEPTH OF CUT

With a sharpened blade fitted, make sure it is seated correctly in the tool. The blade should protrude evenly through the mouth on each side.

Planes
Hold the plane upside down to check the blade is even. If not, it will cut to one side, so use the lateral lever for adjustment. Set the cutting depth by rotating the knurled thumbwheel adjuster.

Spokeshaves
Simple spokeshaves do not have blade adjusters but are simply tightened with a wingnut. On adjustable spokeshaves, slacken off the cap iron, rotate the twin thumbscrews evenly, then retighten the front locking screw.

Tools	Inspection	
Chisels	▪ Check cutting edge for nicks or damage	
Planes	▪ Check for rust if tool is stored in unheated workshop – wipe oil from sole before use	
Files & Rasps	▪ Check teeth are not clogged with debris from previous job	
Gouges & Spokeshaves	▪ Check cutting edge for nicks or damage	
Sharpening Stones	▪ Check for cracks or blemishes –particularly with Japanese waterstones ▪ Stone should be dead flat – check with edge of steel rule	

	Sharpening	Cleaning	Adjustment	Storage
	▪ Primary bevel is 25 degrees, secondary bevel is 30 degrees – use honing guide to maintain consistent angle			▪ Either keep chisels in leather tool roll or fit plastic guards to ends of blades – different size guards to fit standard blade widths
	▪ Primary bevel is 25 degrees, secondary bevel is 30 degrees – use honing guide to maintain consistent angle		▪ Cap iron should be set about 2mm (0.08in) from end of blade – use lateral adjustment lever for even blade protrusion through mouth – set depth of cut with thumbwheel adjuster	▪ Wipe sole with camellia oil or oily rag (remember to wipe clean before using tool)
		▪ Remove clogged debris with file card		▪ Store in tool box or hang from hook
	▪ Usually just one bevel when sharpening gouges – hone spokeshave blade at around 30 degrees		▪ Spokeshaves may have two adjuster thumbscrews for setting depth of cut – adjust these for even blade	▪ Either keep gouges in leather tool roll or hang in tool rack ▪ Keep spokeshaves in toolbox or hang from hook
		▪ Oilstones should be cleaned with paraffin and abrasive pad – wash slurry from Japanese waterstone after use ▪ To keep stones flat, rub stone across sheet of medium grit silicon carbide paper taped to flat surface		▪ Keep rectangular oilstone in hardwood box made to fit ▪ Avoid storing wet Japanese waterstones in unheated workshop in winter – they may crack

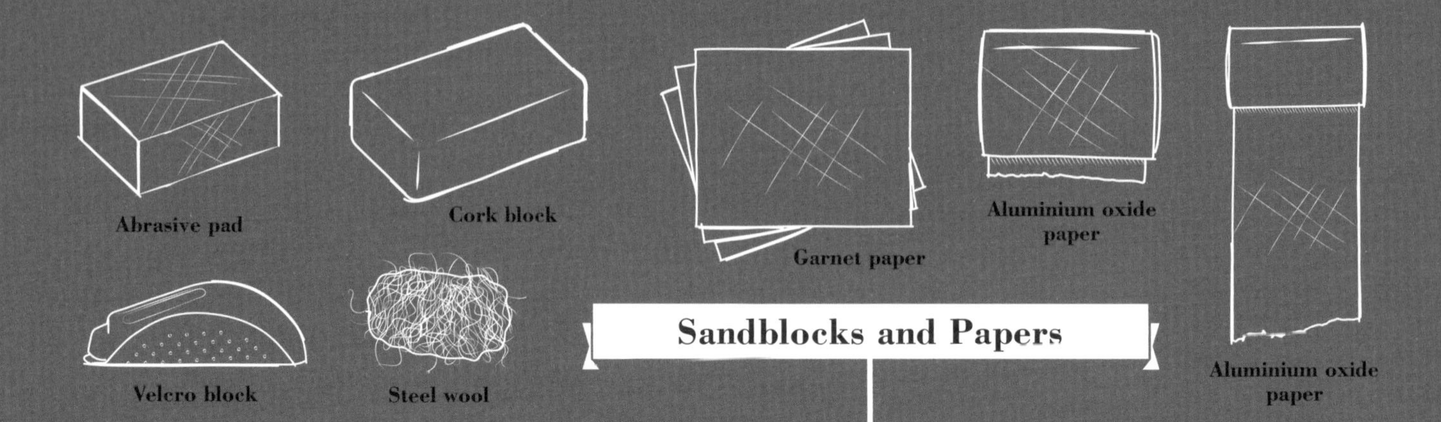

THE TOOLS for FINISHING & DECORATING

From achieving a satin-smooth surface on wood to putting on that final lick of paint, these tools will help you achieve the perfect finish to your projects.

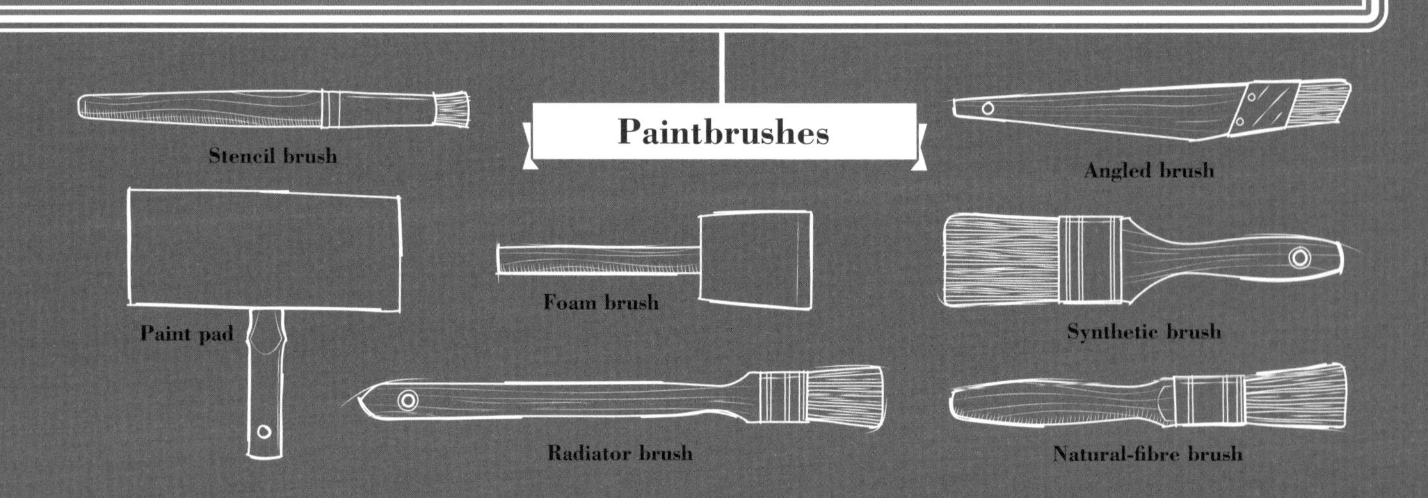

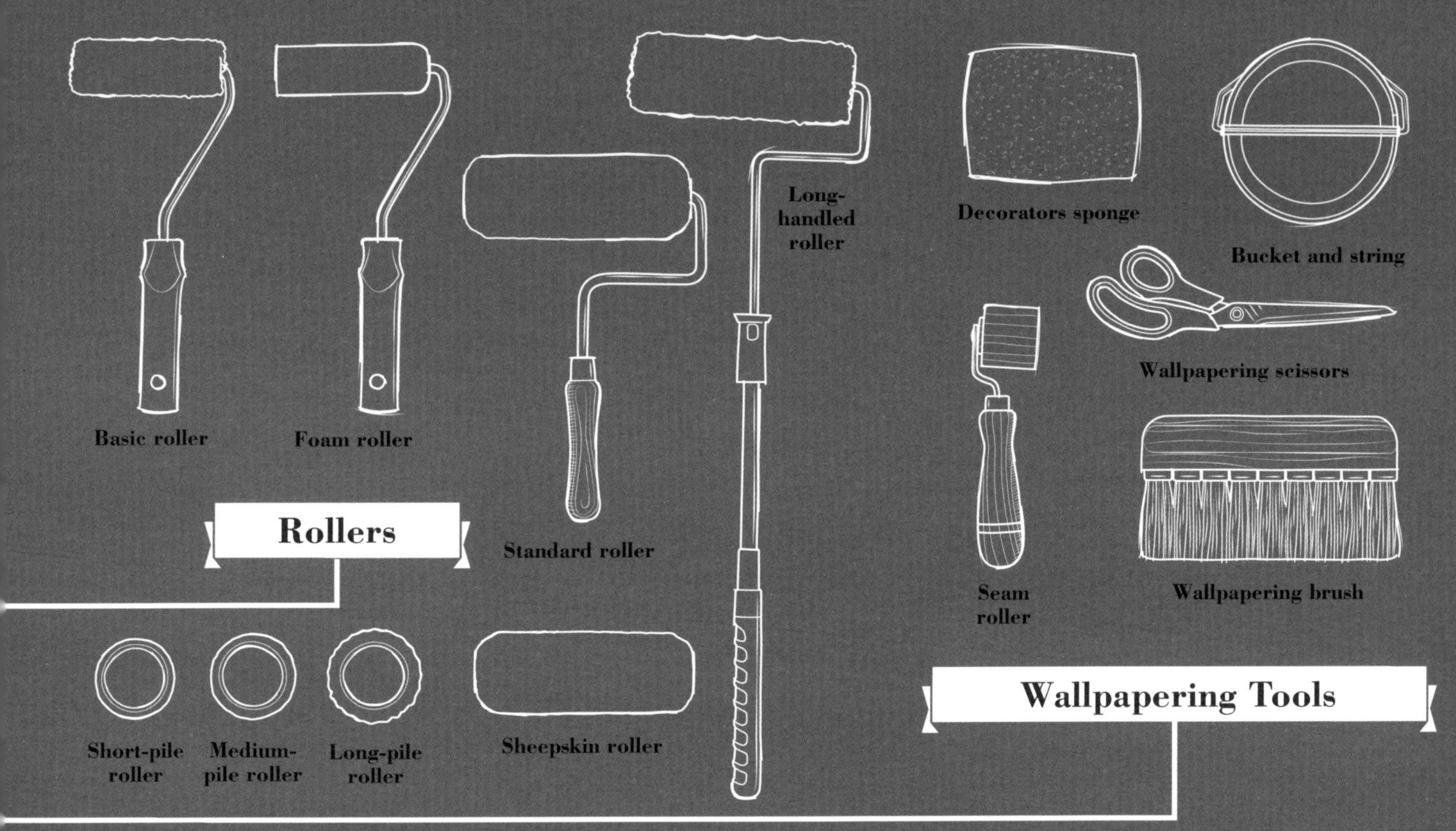

Basic roller
Foam roller
Standard roller
Long-handled roller
Rollers
Short-pile roller
Medium-pile roller
Long-pile roller
Sheepskin roller
Decorators sponge
Bucket and string
Wallpapering scissors
Seam roller
Wallpapering brush
Wallpapering Tools

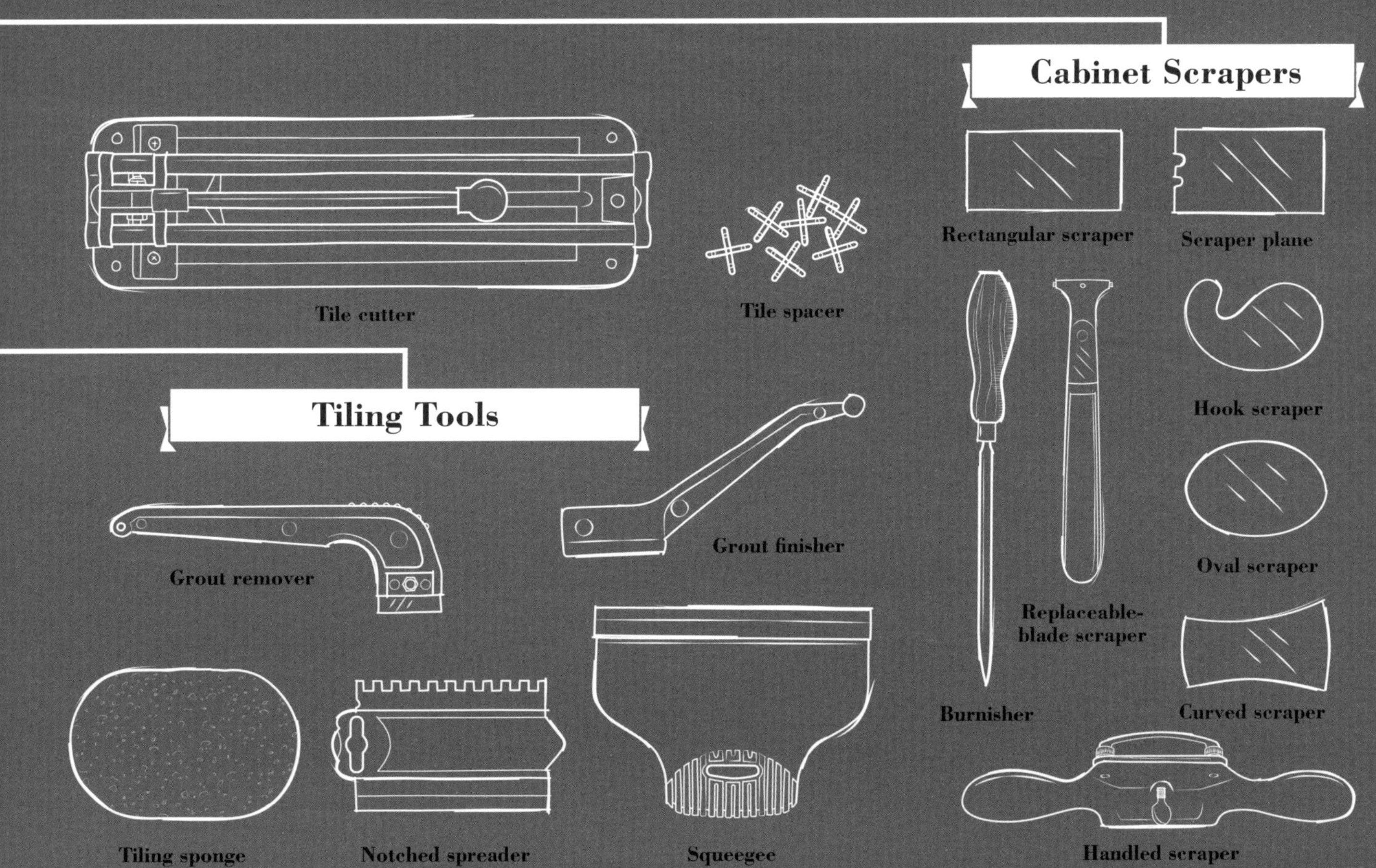

Cabinet Scrapers
Tile cutter
Tile spacer
Rectangular scraper
Scraper plane
Hook scraper
Tiling Tools
Grout remover
Grout finisher
Oval scraper
Replaceable-blade scraper
Burnisher
Curved scraper
Tiling sponge
Notched spreader
Squeegee
Handled scraper

HISTORY OF FINISHING & DECORATING

FIRST SCRAPERS

2.6–1.7 MYA

In Paleolithic times flat stone scrapers were used to carry out basic planing, such as smoothing out rough spots. The stone scraper is an ancestor of the metal cabinet scrapers used today.

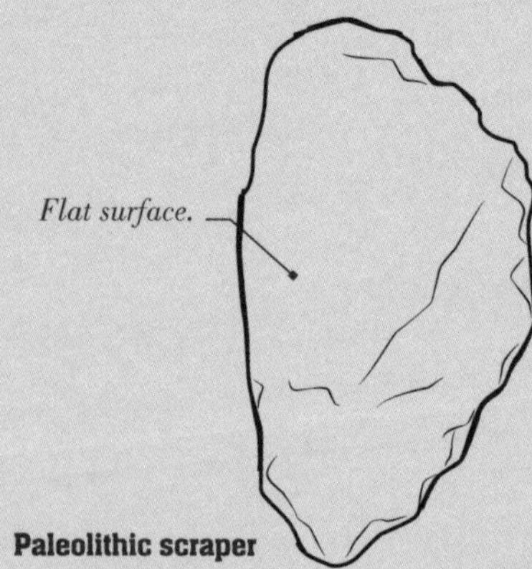

Paleolithic scraper

FIRST BRUSHES

2.5 MYA

Cave paintings found in the Périgord region of France and Altamira in Spain show that brushes were used to apply pigment to the cave walls during the Paleolithic period.

Paris

FRANCE

Moss or animal hair brushes used at cave sites in Périgord.

Reeds, bristles, twigs, or small bones were used at Altamira.

SPAIN

PORTUGAL

Madrid

Lisbon

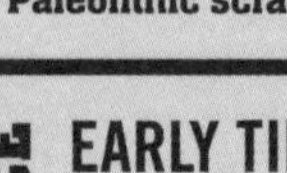

EARLY TILES

4000 BCE

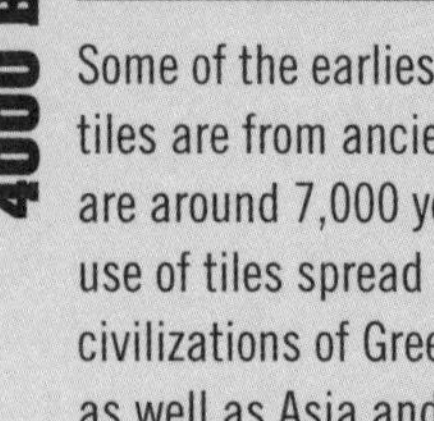

Some of the earliest decorative tiles are from ancient Egypt and are around 7,000 years old. The use of tiles spread through ancient civilizations of Greece and Rome, as well as Asia and North Africa.

ABRASIVE SAND

3000–1900 BCE

The Bronze Age saw the widespread use of sand as an abrasive to finish metal axe heads, while in Egypt during the same period, sandstone was used to smooth architectural stone.

TILED WALLS

Tile manufacture flourished in Mesopotamia, as evidenced by the glazed decorative block covering the Processional Wall and Ishtar Gate in Babylon (modern-day Iraq) built at this time.

THE PROCESSIONAL WAY was half a mile long and its 15m (49ft) high walls displayed

120 LIONS

> “DECORATING IS LIKE MATH, A GAME OF ADDING AND SUBTRACTING.”
>
> CHARLOTTE MOSS, AMERICAN INTERIOR DESIGNER

300B CE THE PAINTBRUSH

The paintbrush is thought to have been invented by Meng Tian, a Qin Dynasty general. Early brushes were designed for use in calligraphy but later models were used for painting pottery. The brushes consisted of a bamboo handle and animal hair, such as rabbit hair or longer hog bristles.

800s ISLAMIC TILE ART

Early examples of Islamic tile art at Tunisia's Great Mosque of Kairouan, dating from 836 CE, reveal intricate geometric patterns, such as eight-pointed stars, that would be perfected and elaborated over coming centuries.

> "GEOMETRY ENLIGHTENS THE INTELLECT AND SETS ONE'S MIND RIGHT."
>
> IBN KHALDUN, ARAB HISTORIAN

1200s EARLY SANDPAPER

In China, ground shells, sand, seeds, and natural gum were used to make sandpaper. Naturally rough sharkskin is also thought to have been used for the same purpose.

50 ROLLS

Many of the first wallpapers were hand-painted. In France, for example, Jean Bourdichon decorated 50 rolls of paper with angels on a blue background for Louis XI in 1481. The papers were mounted on panels so they could be moved easily.

1500s WALLPAPER REPLACES TAPESTRIES

In Europe, wallpaper was being manufactured and was particularly popular in England and France. In England, following Henry VIII's excommunication from the Catholic Church, wallpaper replaced the tapestries that had often been imported from France.

1500s TILES IN THE AMERICAS

Following the Spanish colonization of Central and South America, tile production evolved. Hand-made and brightly painted tiles from Mexico are prized to this day.

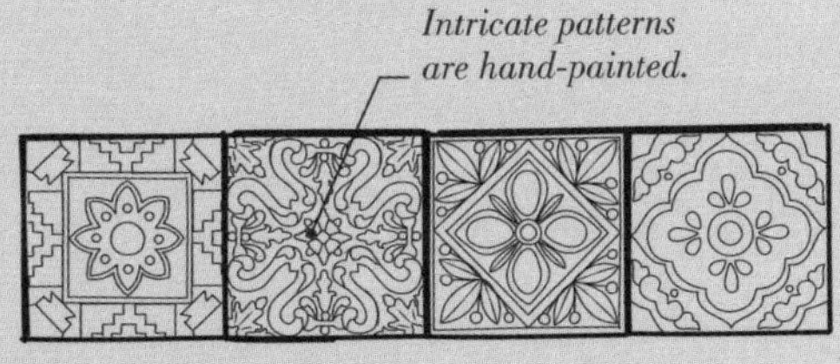

Mexican tiles

1785 WALLPAPER PRINTER

The first machine for printing wallpaper was invented by Christophe-Philippe Oberkampf. His machine printed coloured tints onto sheets of paper. In 1798, Frenchman Nicolas-Louis Robert invented a machine that could print unbroken rolls of paper, but this technology wasn't applied to wallpaper until the next century.

1800s MACHINE-MADE BRUSHES

Early paintbrushes were handmade, but in the 19th century, machines began being developed in various parts of the world for manufacturing the handles, mixing and tapering the bristles, and finally gluing them into place.

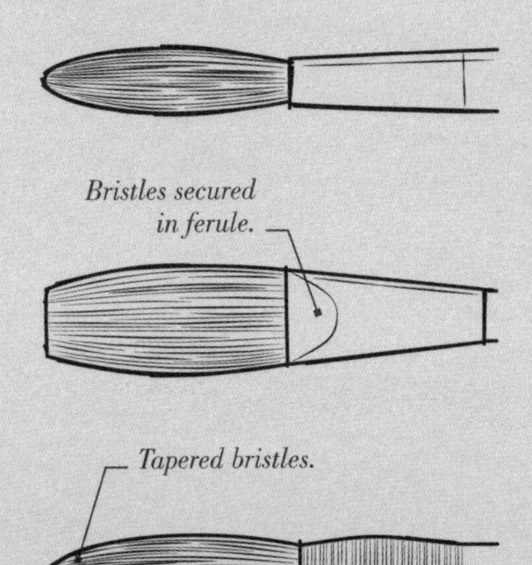

19th-century paintbrushes

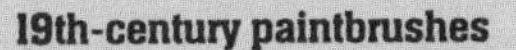

1833 GLASSPAPER

Particles of glass were used to make early forms of sandpaper, known as glasspaper. Around this time glasspaper started to be mass-produced, thanks to new adhesive techniques developed by John Oakey in London, UK.

1925 EARLY ROLLER

The New Yorker magazine made the first mention of the paint roller in one of its 1925 editions, where it praised the tool's virtues for decorating interiors, describing it as a "phenomenal success".

Choosing

Abrasives & Sanding Blocks

Sanding is a tedious but necessary task to get the best finish before varnishing or painting woodwork and other surfaces. Using the most suitable abrasive paper and a backing block helps to reduce the drudgery. Many abrasive grades (denoted by grit size) will be required on most jobs.

Velcro Block

☛ **What it is** Lightweight, rigid polyurethane foam block with a Velcro base for hook-and-loop abrasive discs.

☛ **Use it for** General sanding work where fast, convenient change of abrasive grits is required.

☛ **How to use** Attach abrasive paper disc and wrap it around sides. Grip finger moulds and sand with grain.

☛ **Look for** Make sure that abrasive disc diameter (125mm and 150mm/5in and 6in) matches size of block.

Aluminium Oxide Paper

☛ **What it is** Long-lasting particles (harder than garnet) resin-bonded to heavy paper backing. Standard sheet size (280 x 230mm/11 x 9in) or rolls 115mm (4.5in) wide. Grades from 40–320 grit.

☛ **Use it for** Painting and decorating preparation, coarser sanding of hard- and softwoods. Cut rolls to fit power sanders.

☛ **How to use** Tear to size and wrap around cork block or similar.

☛ **Look for** It's more economical to buy abrasive in roll form and cut to size, rather than individual sheets.

Garnet Paper

☛ **What it is** Crushed stone particles glued to paper backing. Less common, but lasts far longer than glass paper.

☛ **Use it for** Sanding hard- and softwoods in cabinetmaking, fine furniture, musical-instrument making.

☛ **How to use** Tear sheet to size and wrap around cork block. Sand in direction of grain, working through the grades.

☛ **Look for** Grades from 40 to 320 grit. Packs of 25 sheets is most economical way to buy.

Steel Wool

☛ **What it is** Fine carbon-steel strands meshed together and available in wad or roll form. Several grades, from 4 (coarse) to 0000 (very fine).

☛ **Use it for** Applying wax polish to wood, cleaning glass, marble, and delicate surfaces; removing rust, restoring metal surfaces to a bright finish.

☛ **How to use** Cut a piece with scissors. Wear thin work gloves if using steel wool with white spirit or meths.

☛ **Look for** Avoid using on oak surfaces as it may react and cause staining. Use stainless-steel wool if in doubt.

Cork Block

☛ **What it is** Compressed cork block to wrap abrasive paper around. Light weight is comfortable for long sanding sessions.

☛ **Use it for** Its size allows a standard sheet of abrasive paper to be torn into four equal pieces without wastage.

☛ **How to use** Fold paper around block and grip both sides. Apply light pressure and move across the surface.

☛ **Look for** Check that the surface of the block is flat and not damaged.

Sanding Pad

☛ **What it is** Double-sided, low-density foam sponge, faces coated with silicon carbide particles. Grades: 60 to 220 grit.

☛ **Use it for** Sanding curved and profiled surfaces. Finer grades ideal for preparing painted surfaces for recoating.

☛ **How to use** Use dry or wet. Dip into water. Wash under the tap to clean pad.

☛ **Look for** Deeper, high-density blocks are coated on all four sides, so ideal for sanding into corners.

USING A

Sanding Block

You can use abrasive paper by holding it in your fingers, but you'll obtain a crisper surface if it's wrapped around a sanding block. A traditional cork block can seem stiff to use, but it offers some resilience, unlike a piece of hardwood of similar size. When paired with suitable abrasive paper, this is an effective tool for sanding flat areas.

The Process

Before you start

- **Get the right grit** Make sure you have several grades of abrasive ready. The grit size will be printed on the back of paper abrasives.
- **Protect yourself** Always wear a dust mask when sanding, no matter what the material.
- **Prepare the work area** Try to sand timber outdoors, if possible. When sanding indoors, open windows but close doors to contain the dust.
- **Protect your hands** You may choose to wear flexible work gloves if sanding for long periods, to avoid skin abrasions due to long contact.

1 Select the right grade

Choose the most appropriate abrasive for the job. If you're uncertain about what grade to use, start with a finer grit (high number) and work back from there. It's harder to remove coarser sanding marks from wood than lighter ones.

Folding and creasing *the paper first allows it to tear more cleanly.*

2 Size up the paper

Prepare the abrasive paper to fit the sanding block. Allow enough excess to run up the sides and provide a grip for thumb and fingers. Creasing and tearing paper along the edge of a bench will give a fairly clean edge. Thin paper can be sliced easily with a steel rule. Avoid scissors as the abrasive will blunt these quickly. A standard sheet should be folded into four.

FOCUS ON...

Abrasive Particles

Abrasive sheets consist of particles of hard materials bonded to backing paper. Grit size refers to the density of particles per 6.4sq cm (1sq in) of abrasive paper. Coarser particles are larger and cut more rapidly than finer particles, which are more numerous. Glass particles (traditional glasspaper) are relatively soft, garnet is medium, while aluminium oxide is harder still. Hardest of all is silicon carbide.

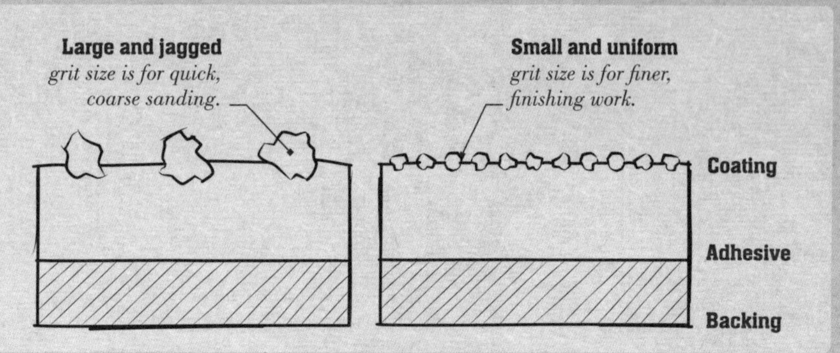

Overlap *enough paper on the sides to create a firm hold.*

3 Cover the block

Place the sanding block on the paper, then wrap the paper firmly around the block, creasing it along each corner so that the excess fits flat against the sides. When sanding timber, start with the coarsest grade first and finish with the finest (240 grit or higher). Never sand across the grain, as the resulting scratches will be tough to remove.

4 Sand it all over

Sand the edges of a workpiece, keeping the block flat. Rounded edges and profiles (such as traditional skirting or architrave) can be sanded by wrapping the paper around a hardwood dowel to suit the contour. Remove sharp corners and edges with a piece of worn abrasive paper.

After you finish

- **Clean it up** Brush all surfaces clean after sanding and vacuum up the dust if working indoors.
- **Store any excess** Keep sheets of abrasive paper in plastic bags if storing them in an unheated workshop.

Choosing a
Paintbrush

Choosing the right paintbrush can seem like an overwhelming task. Many different brush sizes are available to use for different tasks and you also need to choose between natural or synthetic fibres, depending on the type of paint you will be using. As a general rule, the smaller the area to paint, the smaller the brush should be.

Angled brush

Synthetic brush

Foam brush

Radiator brush

Stencil brush

Angled Brush

What it is A narrow brush with a slanted edge to the bristles.

Use it for Cutting in the join between walls and ceilings, or around door and window frames.

How to use Brush is best held in a pencil grip. Drag the brush down the join of the wall.

Look for Tight bristles that create a very sharp straight edge to them. Comfortable grip to hold.

Synthetic Brush

What it is Brush with bristles made from man-made materials such as nylon or polyester, or a mix of both.

Use it for Best used with water-based latex paints like emulsion as synthetic bristles don't absorb a lot of water.

How to use Dip brush into paint up to about one-third of bristle length. Scrape excess off before applying to wall.

Look for Quality, by flicking brush a few times to see if bristles come out. Cheap brushes lose bristles in paintwork.

Natural-fibre Brush

What it is Brush with animal-hair bristles, such as badger or hog.

Use it for Oil-based paints and varnishes due to their high absorbency.

How to use Load brush lightly with paint to create a sharp edge.

Look for A good quality brush to last a long time. Check for a sharp, even edge.

Foam Brush

What it is "Brush" made from sponge, usually chisel-shaped.

Use it for Creating a smooth finish with oil-based paints, varnishes, stains.

How to use Dip a third of sponge length into paint. Scrape off excess and paint in long strokes in one direction.

Look for A tight knit in the foam. Loose foam can leave bits in the paint.

Stencil Brush

What it is Specialist, usually round brush with short, tight, pile bristles.

Use it for Dabbing paint through a stencil.

How to use Load brush with tiny amount of paint, dab it over the stencil.

Look for Choose a head the right size for the stencil. Smaller heads are generally easier to control.

Radiator Brush

What it is Brush with a long handle, often with an angled head.

Use it for Painting behind radiators without removing them first. Also useful for applying wallpaper to a wall behind a radiator.

How to use Don't overload the brush. Work from the bottom up and paint in upward strokes.

Look for The correct bristle type for the paint being used. An angled head can make the job easier.

Paint Pad

What it is Rectangular pad with tightly packed foam. Comes in a range of sizes, some with adjustable handles.

Use it for Painting walls. Holds less paint than roller so needs loading more often. Good for edges, smooth surfaces. Creates less splatter then a roller.

How to use Dip into paint tray, scrape off excess. Drag paint in one direction along surface. Avoid back/forth motion.

Look for Good-quality foam on the pad. An adjustable handle is always a good idea.

Structure of a Paintbrush

Although they vary in size, shape, and purpose, all paintbrushes are essentially made the same way. The handle is connected to a bundle of filaments, which is normally referred to as the bristle, by a metal clamp known as a ferule. The differences arise in terms of the type of bristle used (natural or synthetic) and how the end of the bristle is finished, as well as the size of brush and type of handle it's mounted on.

Top end *of handle allows for directional action.*

Ferrule *is a metal clamp that holds the bristles into brush.*

Bristle edges *come in different shapes: tapered, chiselled, angled, or straight.*

Filaments *also known as bristles, made from either natural or synthetic fibres.*

Crimp *is where the ferrule presses onto brush handle.*

USING A

PAINTBRUSH

A paintbrush is ideal for intricate or detailed decorating where you need a smooth finish. For example, use a paintbrush to cover joins between walls and ceilings where rollers can't reach, or around light fittings and switches.

The Process

Before you start

- **Flick the brush** Flick the bristles of a new paintbrush back and forth a few times to release any loose bristles.
- **Use a kettle** A paint kettle (small paint bucket) makes it easier to keep the paintbrush handle clean and stops paint building up around the paint-tin rim.

1 Load the brush
Dip the paintbrush into the paint so that around one-third of the bristle length is covered. Brush the excess away on the side of the paint kettle or tin. Hold a wall brush by wrapping your entire hand around the handle; your thumb will be on one side of the ferrule, and your fingers on the other. If painting trim, hold the smaller trim brush as you would a pencil.

2 Paint in a stripe
Drag the brush in one long stripe along the area that needs to be painted. When the paintbrush starts to "dry wipe", stop dragging and reload it. Repeat until you've covered the desired area.

3 Hide your work
To ease away brush marks, feather the brush lightly on the wall in a back and forth motion.

After you finish

- **Wrap it up** If you're planning to use it for a second coat, wrap the brush in plastic wrap.
- **Wash it up** To clean, wash the brush under running water, brushing it against the bottom of the sink to work the paint out of the middle of the bristles. Wrap the bristle in a folded paper towel to help keep its shape.

FOCUS ON...

BRISTLES

A paintbrush functions because the filaments hold and channel paint all the way to the centre of the bristle. As the brush is dragged down the surface to be painted, the pressure exerted on the bristle forces the paint out from the centre of the bristle to the filament edges. This is why the edges of bristles need to be sharp in order to create a crisp, neat line of paint on whatever it is that is being painted.

Bottom end *of handle provides stability when painting.*

Hole *for hanging up and storing brushes after washing.*

TOP VIEW

"QUALITY DIRECTLY AFFECTS FINISH, SO INVEST IN A GOOD SET OF BRUSHES FOR LIFE AND LOOK AFTER THEM WELL"

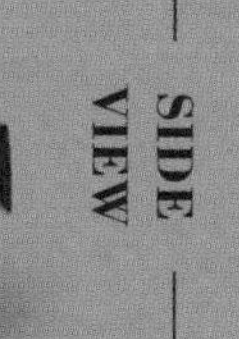

SIDE VIEW

Choosing a Roller

Like most decorating tools, the range and selection of paint rollers available can be very confusing. The type you need depends not only on the job you want to do, but also the type of paint you are using. As a rule of thumb, the more textured the wall, the thicker pile you need to have on your roller head.

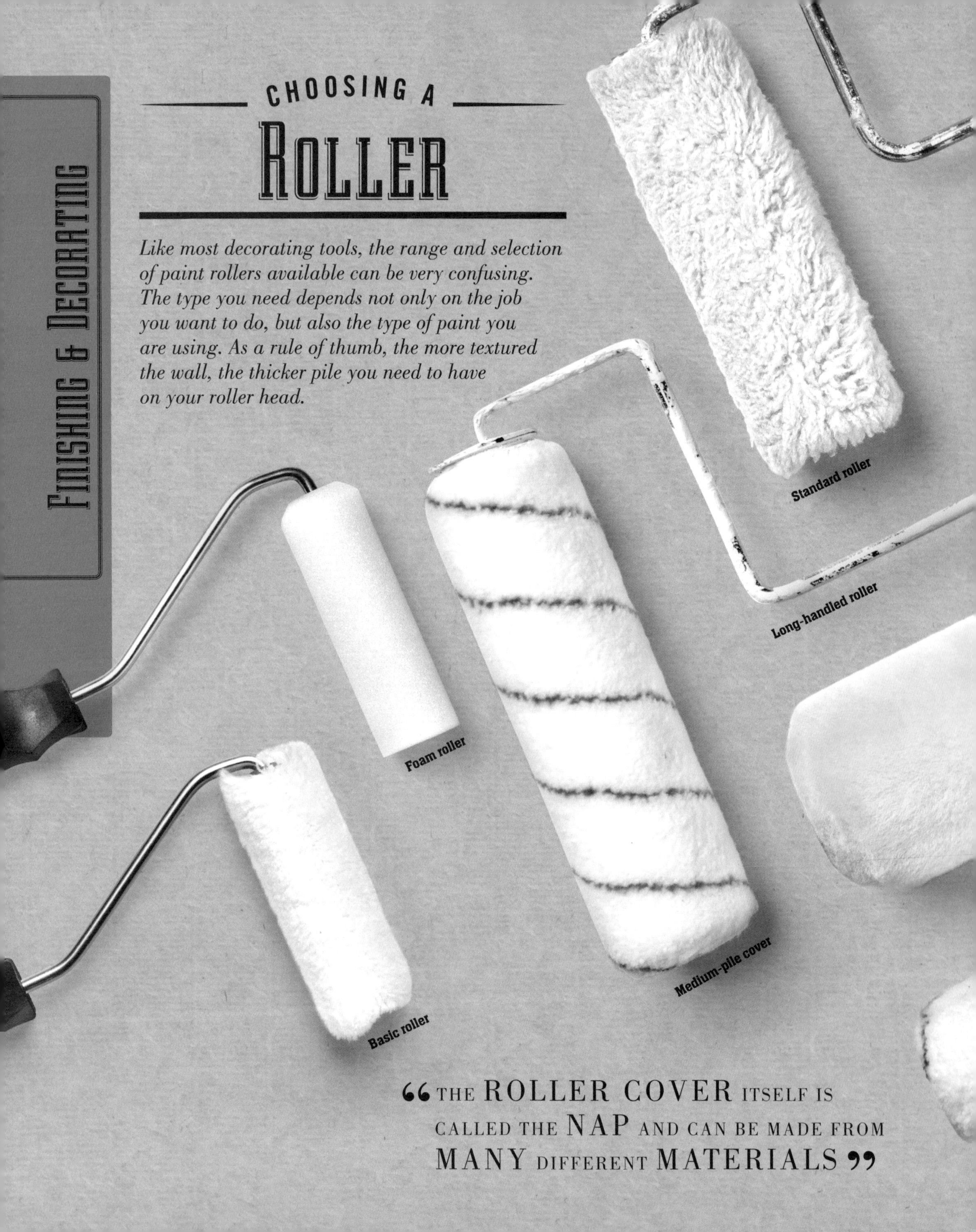

“THE ROLLER COVER ITSELF IS CALLED THE NAP AND CAN BE MADE FROM MANY DIFFERENT MATERIALS”

“COVER WET PAINT ROLLERS IN PLASTIC WRAP BETWEEN COATS TO PREVENT DRYING”

Standard Roller

What it is Medium-handled roller, usually in pack with a roller tray. Roller sleeve fits over a cage on handle end.

Use it for Water-based paints, like emulsion on large areas like walls.

How to use Pour some paint into the trough on the tray, dip in the roller, roll it back over the ridged tray slope to cover the head and remove excess.

Look for A comfortable handle and a medium pile to cope with various general painting jobs.

Long-handled Roller

What it is A telescopic extension pole that either fits the end of a normal roller handle or is an independent roller that extends.

Use it for Ceilings, tops of walls, painting floors without bending over.

How to use Attach to roller handle before loading with paint, then load paint in normal way. Extend to right length and roll paint onto surface.

Look for Ensure your regular roller handle has a hole so that the extension handle will fit it.

Foam Roller

What it is Budget alternative to traditional fibre or napped rollers. Absorbs paint very easily.

Use it for Very smooth surfaces, as the foam dispenses paint more evenly than fibre heads. Also good for thin paint.

How to use Load the roller head as normal, remove excess thoroughly.

Look for Value packs, as foam heads are mainly for one-time use.

Basic Roller

What it is Small roller to be used on a smaller roller base and handle.

Use it for Cutting in. Applying water-based paints, such as emulsion, to small areas, like window trims.

How to use Use with a small paint tray. Dip roller head into paint, roll back over flat part of the tray to remove excess.

Look for Normally comes in pack with foam rollers. Ensure hole for the roller head fits the roller handle.

Sheepskin Cover

What it is A natural-fibre roller head, made from sheepskin, lambswool or a lambswool mix. Sometimes called a lambswool roller cover.

Use it for Oil-based paint, varnish, stain. Can also be used with emulsion paint, but as these are more expensive rollers, standard is better choice there.

How to use Dip the roller head into the paint and roll back over the flat surface to remove excess. Apply paint on to the wall by rolling up and down evenly.

Look for Mohair is the best type to buy. Choose a long pile if you're painting a textured wall.

Short-, Medium-, and Long-pile Covers

What it is Roller heads or sleeves with different lengths of fibres (nap).

Use it for Different wall textures that need different lengths. The flatter the surface, the smoother the roller needed – so match a foam roller to flat wood, but a long-pile roller to textured ceilings.

How to use Load the roller head by dipping it into paint, then remove excess by rolling over the flat part of the tray.

Look for Choose the type and length of pile to match both surface type and paint type. For example, applying water-based paints on a smooth surface requires a short-pile roller.

Structure of a Roller

Rollers are great tools, designed for painting large, flat areas. They can come in different shapes and sizes, but they all work in the same way. A roller head is attached to a bar or cage that rolls around when pushed, and it is this mechanism and combination of parts that allow paint to be spread smoothly and evenly.

Plastic sleeve core *of roller head keeps textured sleeve rigid.*

Spring cage *holds the roller head, allowing it to rotate.*

Plastic end caps *on cage keeps the roller head in place.*

Metal handle core *usually made from steel.*

Attachment section *allows some handles to slot or screw into poles of various lengths.*

Roller head nap *comes in many lengths and materials.*

“YOU MAY HOLD A ROLLER FOR A LONG TIME, SO BE SURE TO USE A COMFORTABLE HANDLE TO PREVENT BLISTERS AND CALLOUSES”

Textured grip *of rubber or plastic for greater comfort when painting.*

EXTENDED VIEW

Screw threads *allow attachment to extension pole.*

COLLAPSED VIEW

FOCUS ON...

NAP LENGTHS

Different roller nap lengths work better on differently textured surfaces. As a general guideline, you need a smooth roller for a smooth surface. For rough surfaces such as masonry or artex, use a long-haired roller. The longer the nap or hair, the more paint the roller can hold, and the more bumps it can cover effectively.

USING A

ROLLER

Paint rollers are the most sensible choice for painting the body of walls, floors, and ceilings. For floors and ceilings, investing in an extension pole is essential if you want to make the job go quicker and more easily.

The Process

Before you start

- **Choose the right tool** Select the right nap length and size of roller for the type of surface you want to paint. Smooth rollers for smooth surfaces, thicker/longer naps for textured ones.
- **Line the tray** Line your roller tray with a plastic bag to avoid having to clean it out.

1 Pour the paint

Fill the trough in the roller tray about two-thirds full with paint. Ensure the roller head is pushed firmly onto the roller cage before you begin.

2 Load the roller

Dip the roller head into the paint. Pull it backwards and roll it over the flat surface of the tray a few times to spread the paint evenly over the roller head and remove any excess. This will help to prevent drips.

3 Paint the wall

To paint a wall, start by rolling the top half, so that any drips trail onto unpainted surface beneath. Work from bottom to top, and overlap roller widths each time, to ensure smooth coverage. Paint in sections, a few feet at a time. When you reach a corner, roll as closely to it as you can without touching, then start on the adjacent wall.

After you finish

- **Keep it covered** Wrap the roller in plastic wrap or a plastic bag if you take a break or are planning to use it again for a second coat.
- **Clean it up** Put the roller head under running water while scraping paint down its sides with a plastic scraper. Or use your hands (in rubber gloves) to squeeze paint out.

“ONLY THOSE WHO HAVE THE PATIENCE TO DO SIMPLE THINGS PERFECTLY EVER ACQUIRE THE SKILL TO DO DIFFICULT THINGS EASILY.”

FRIEDRICH VON SCHILLER

blue™

Choosing

Wallpapering Tools

Wallpapering can be a straightforward task when you use the right tools in the right order. From loading up with paste to rolling down the last seam, if you take time to get organized and learn how to use the different tools correctly, your project will run smoothly.

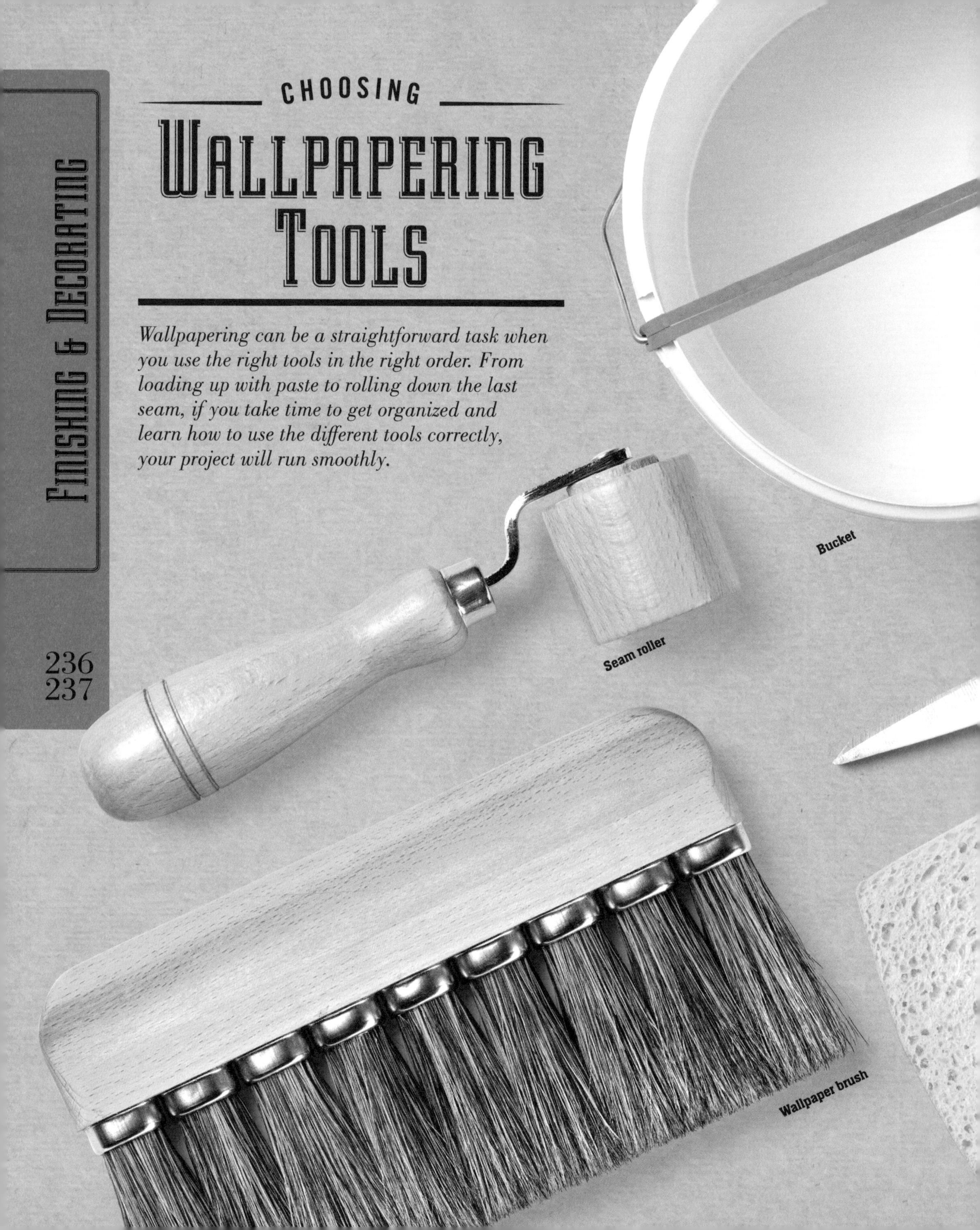

❝ KEEP THE DRY TOOLS AWAY FROM THE WET TOOLS OR ALL YOUR TOOLS WILL END UP DRIPPING IN STICKY GLUE ❞

Wallpaper scissors

Decorating sponge

Bucket

☛ **What it is** Large, wide-mouthed bucket, with handle.

☛ **Use it for** Mixing wallpaper adhesive. Tie string, or a rubber band, across bucket on handle hooks to wipe excess paste off glue brush.

☛ **How to use** Pour in warm water and add wallpaper adhesive slowly, stirring with a stick or wooden spoon as you go.

☛ **Look for** A sturdy handle is essential. Some buckets come with a strap so you can wipe excess glue off the brush.

Seam Roller

☛ **What it is** Small plastic roller with a smooth surface. Usually around 4–5cm (1.5–2in) in width.

☛ **Use it for** Rolling over the seams of two sections of wallpaper so they push down to the wall and meet together neatly.

☛ **How to use** Once two sections ("drops") of paper are on wall, gently roll down the join between them. Wipe off excess glue that pushes out of seam.

☛ **Look for** High-quality rollers are slightly soft to touch so they don't damage the paper by crushing it as it's rolled.

Wallpaper Scissors

☛ **What it is** Very long scissors with angled handles and very sharp blades.

☛ **Use it for** Cutting wallpaper to the correct length.

☛ **How to use** Mark how long your drop needs to be, then use the scissors to cut along that line. Can also be used to trim wet paper when it's on the wall.

☛ **Look for** A comfortable handle and long blade. The scissors need to be very sharp, so store wallpaper scissors separately from general-purpose scissors to avoid blunting them with everyday use.

Decorating Sponge

☛ **What it is** A medium-sized, thick cellulose sponge.

☛ **Use it for** Wiping and cleaning excess wallpaper adhesive off wallpaper after it has been hung.

☛ **How to use** Dip sponge in clean water, gently wringing it out before use. If it's too dry, it can tear the paper; too wet and it will drip and damage the paper.

☛ **Look for** A high-quality sponge to hold the right amount of water.

Wallpaper Brush

☛ **What it is** A long, wide brush with soft, medium-length bristles and a flat handle, usually made of wood.

☛ **Use it for** Smoothing out the surface of wallpaper when it has been hung to remove creases and bumps.

☛ **How to use** Once a section of wallpaper is on the wall, brush from the centre of the drop to the edge, moving from top to bottom and brushing gently as you go.

☛ **Look for** A little lip or dimple in the handle will make it easier to hold – very helpful if you have a lot of paper-hanging to do.

CHOOSING
Tiling Tools

Tiling is one home-improvement activity that requires a lot of different tools. While this can seem daunting at first, if you make sure that you use them in the right way, it's easy to achieve a professional-looking finish. Larger tools such as tile cutters are sold individually, but the most basic items can often be purchased in packs, making it very straightforward to find all you need to get the job done.

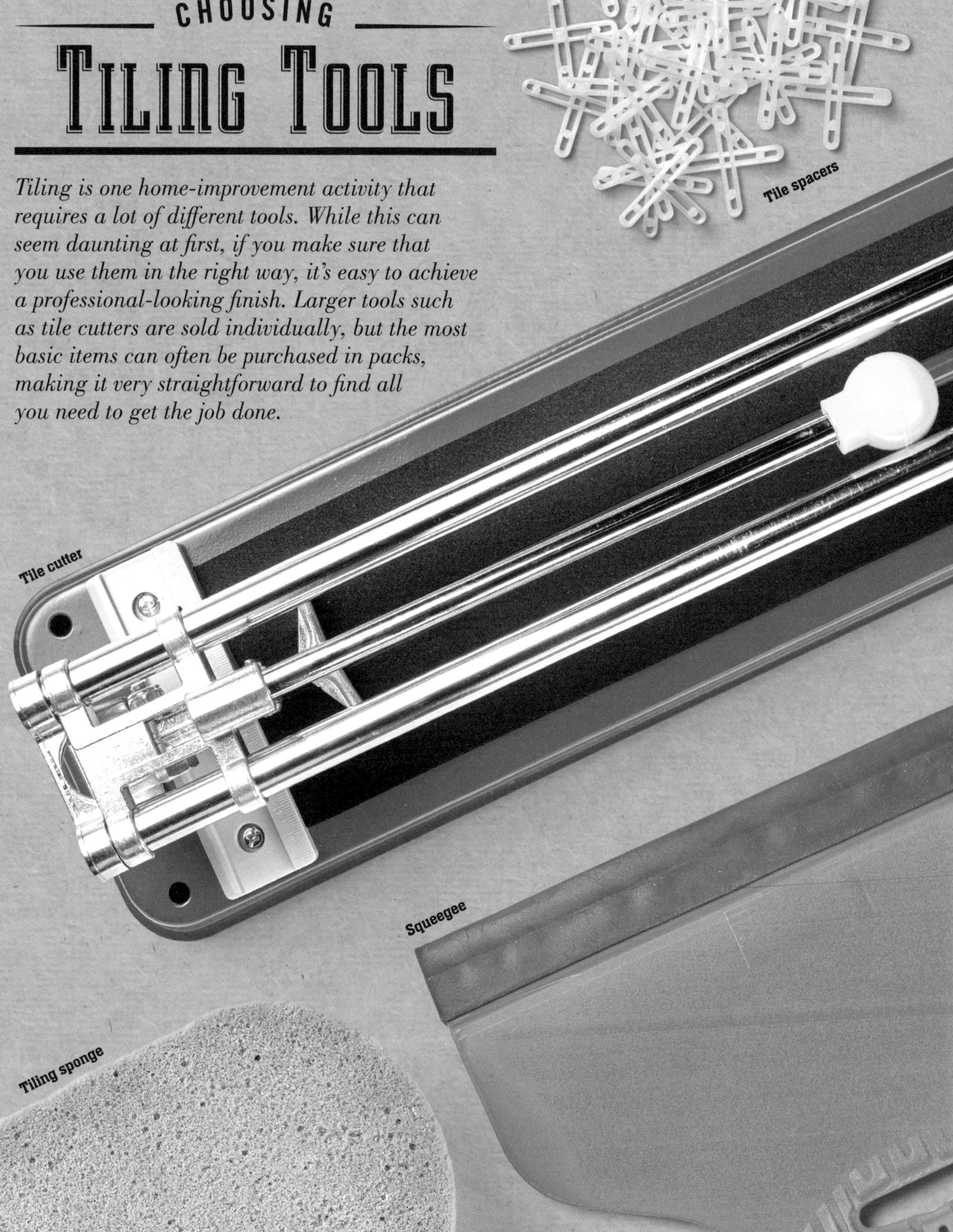

Squeegee

☛ What it is Long rubber tool, often found on other side of a notched spreader.

☛ Use it for Applying grout, removing the initial excess left as grout is spread.

☛ How to use Once grout has been applied, draw squeegee along tiles to force grout into the gaps between them, removing any excess at the same time.

☛ Look for A length that works with the size of tile.

Tile Spacers

☛ What it is Small plastic crosses in different sizes. Normally sold in packs.

☛ Use it for Ensuring tiles are evenly spaced so grout lines will be aligned.

☛ How to use Place in between each tile when applying. In larger tiles, place a few spacers between each tile and the next in line.

☛ Look for A spacer that is the same size as the desired grout lines. Larger grout lines need wider spacers.

Tile Cutter

☛ What it is A manual device that scores a tile, allowing the tile to be snapped apart by applying pressure.

☛ Use it for Making straight cuts in ceramic tiles. For harder tiles, a wet-wheel power cutter works best.

☛ How to use Mark where the tile should be cut, drag the blade along the mark. Line the push arm over the centre of the mark, push down to split the tile.

☛ Look for Ensure the cutter can handle the tile depth. Floor tiles are thicker than wall tiles and need a larger blade and push arm.

Notched Spreader

☛ What it is A flat tool 15–30cm (6–12in) long with notches along the side. Can be made of either plastic or metal.

☛ Use it for Spreading adhesive onto walls or floors for tiles to rest on. The notches create lines in the adhesive, which allows air in to dry it properly.

☛ How to use Apply even layer of adhesive to wall or floor, scrape along it with the notched side to create grooves. Press tile into adhesive with even pressure.

☛ Look for Small plastic spreaders are adequate for basic jobs. For larger jobs, a large metal spreader is better.

Tiling Sponge

☛ What it is A large sponge, sometimes with a plastic handle on the back.

☛ Use it for Cleaning excess grout off tiles before it hardens.

☛ How to use Soak sponge in water, wipe over the tiles' surface. Rinse and repeat often to remove excess grout.

☛ Look for Good quality. A cheap sponge will break up, leaving bits on the surface. Ensure that you can hold the sponge comfortably in your hands.

Grout Remover

☛ What it is A tool with a plastic handle and a thin, jagged metal blade.

☛ Use it for Breaking up and removing old or broken grout.

☛ How to use Drag the blade down the grout line repeatedly to break it up.

☛ Look for A good grip will make it easier to apply the necessary pressure.

Grout Finisher

☛ What it is A two-edged tool, normally plastic, with a small thin blade at one end and a ball on the other.

☛ Use it for Applying grout to corners and edges. The ball end is used to make grout lines look neat and even.

☛ How to use Apply grout to the blade and push it into the tiles. Drag the ball along the line to create a neat finish.

☛ Look for A good grip and feel in the hand. Inexpensive price, as this is a very basic tool.

CHOOSING A

Cabinet Scraper

Scrapers are simple tools used for the final preparation of timber surfaces before applying a finish. They have one or more cutting edges that are held at a set angle to the surface. Traditional cabinet scrapers are sharpened with a file and burnisher, while more modern tools have disposable blades that are suitable for use on paintwork.

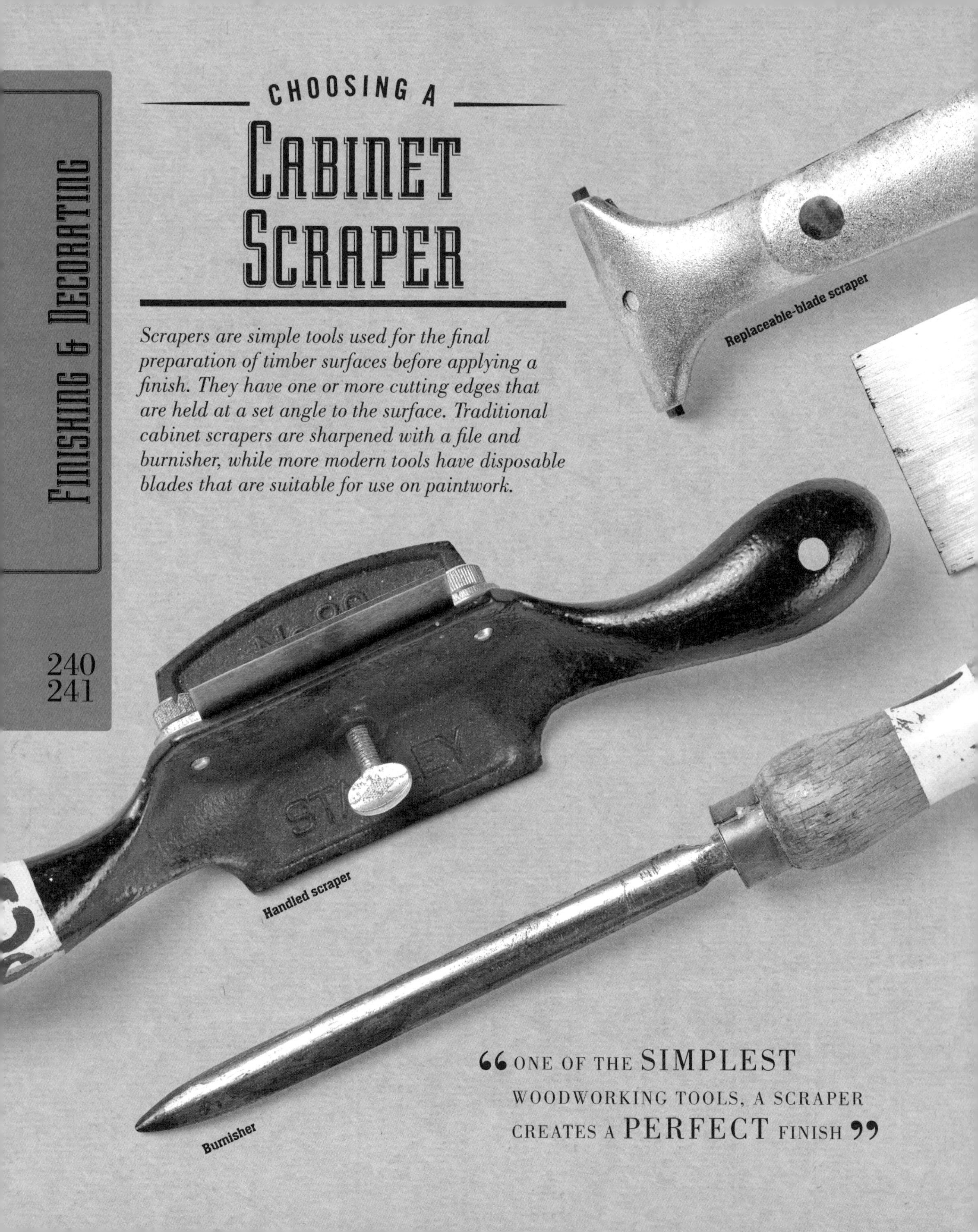

“ONE OF THE SIMPLEST WOODWORKING TOOLS, A SCRAPER CREATES A PERFECT FINISH”

Mixed-shape Cabinet Scrapers

What they are Thin, flexible tempered-steel plates with cutting edges formed with burnisher. Rectangular and various convex and concave curves.

Use them for Fine scraping of flat timber to produce final finish. Shaped scrapers for profiled mouldings such as beading, skirting, and architrave.

How to use Hold scraper with both hands. Flex steel with your thumbs to make cut as you push tool across wood.

Look for Heat is generated when scraping, so stick tape to face of tool. A burnisher and file are needed to form cutting edges.

“A CORRECTLY SHARPENED CABINET SCRAPER WILL PRODUCE A PAPER-THIN SHAVING ON WOOD”

Replaceable-blade Scraper

What it is Aluminium and soft-grip polypropylene handle with reversible tungsten carbide blade at one end.

Use it for Removing dried glue, varnish, rust, old paintwork prior to decorating. Cleaning floors and removing defects from surfaces generally.

How to use Grip in one hand and pull scraper backwards. Cuts on pull stroke rather than push stroke.

Look for Check width of tool when buying replacement blades.

Handled Scraper

What it is Twin-handled cast-steel body with flat sole holds scraper blade at angle. Adjustable tension and depth.

Use it for Fine scraping of planed wood and veneered surfaces. Ideal tool for wild hardwood grain that tears with normal planing.

How to use Secure blade in tool so it's touching benchtop. Adjust tension and push tool forwards across timber.

Look for The blade is sharpened like a normal cabinet scraper, so you will need a burnisher.

Burnisher

What it is Hardened steel blade, which may be oval or circular in section. Fitted with hardwood handle.

Use it for Forming a burr or hooked edge on steel cabinet scrapers.

How to use File edge square with scraper held in vice. Lay scraper flat on bench. Hold burnisher flat, draw tool along the edge. Put scraper in vice again and repeat with burnisher horizontal, then at slight angle to form hooked edge.

Look for Keep the blade clean with steel wool or fine abrasive paper.

"EACH PLANK...CAN HAVE ONLY ONE IDEAL USE. THE WOODWORKER MUST FIND THIS IDEAL USE AND CREATE AN OBJECT OF UTILITY TO MAN, AND IF NATURE SMILES, AN OBJECT OF LASTING BEAUTY."

GEORGE NAKASHIMA

MAINTAIN TOOLS FOR

FINISHING & DECORATING

These tools get dirty by nature, so factor in some cleaning time when using them. If paint or glue dry onto the tools they can no longer be used.

CLEAN AS YOU GO

It's great to have a dedicated place to clean tools, like a utility room or outdoor sink as they can cause a mess when cleaning. To avoid excess material drying on the tool while it's in use keep a rag to hand and wipe the tool regularly.

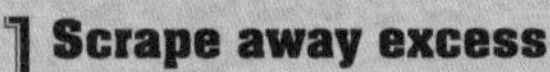

Rinse brushes *under running soapy water.*

1 Scrape away excess
Use a scraper to scrape excess paint, glue, or adhesive off used paintbrushes or rollers.

2 Soap and water
Run the tool under warm water and use regular washing-up liquid to loosen any materials. For stubborn dried-on paint, leave brushes to soak in warm water for up to two hours. For solvent-based paints you may need to use a little solvent-based cleaner, such as white spirit, to soak off residue in a secure container.

3 Clean and dry
Ensure tool is thoroughly clean before wiping over with a dry cloth and leaving to dry in a well-ventilated area. Do not dispose of any solvents down the sink. Cover the container you used and leave it until any paint has sunk to the bottom. Store any remaining solvent for use in future. Allow paint residue to dry out before sealing and throwing away in a bin.

TOOLS	INSPECTION
PAINTBRUSHES	▪ Check bristles on brushes – when you see lots of bristles start to come out in paint it shows paintbrush is nearing the end of its useable life
ROLLERS	
WALLPAPERING TOOLS	
TILING TOOLS	
CABINET SCRAPERS	▪ Check cutting edges for any damage

	Cleaning	Repair	Tips	Storage
	▪ For solvent-based paints use white spirit to clean paint off and then rinse through with warm water. For water-based paints use regular washing-up liquid and water	▪ Only a very expensive brush is worth repairing and this should be rare as expensive brushes don't break very often. Bristles can be re-glued into ferule but this is a tricky job. Sometimes a manufacturer will offer a repair service	▪ If you don't have access to a cleaning area straight away, wrap brushes in plastic so they don't dry out as this makes them harder to clean	▪ When drying brushes out after cleaning, hang bristles down so water doesn't run into ferule, loosening glue
	▪ Some rollers are inexpensive and therefore can be thrown away after a job – for others scrape paint off with plastic scraper under warm water and rinse through until water runs clear	▪ Not worth repair as price point is relatively low	▪ If your paint tray cracks mid-way through a job, tape it up with gaffer tape and cover tray with a bag to allow you to continue	▪ Store roller heads upright to dry them
	▪ Have a rag to hand when wallpapering to clean glue tools as you go	▪ Most wallpapering tools are not worth repairing; they won't break often but are inexpensive, so if they break it's worth just buying new ones ▪ Scissors can become blunt over time but can be sharpened using a sharpening stone		▪ Store all wallpapering tools together so they don't get mixed up with other tools and they are always ready to use when you come to wallpaper
	▪ Wipe all adhesive and grout off tools as you go. Then clean with soap and warm water			▪ Store all tiling tools together so they don't get mixed up with other tools
	▪ Wipe to remove any debris after each use		▪ Use a burnisher to rapidly restore the cutting edge	▪ Keep wrapped up and in toolbox, to protect sharp edges ▪ Store in dry area to prevent corrosion

POCKET REF
SINCE 1899
BAG BALM

TOOLSFORWORKINGWOOD.COM MERCHANTS & MANUFACTURERS BROOKLYN NEW

A

B

C

D

E

F

G

H

I

J

K

L

M

N

O

P, Q

R

S

T

U

V

W, Y

DK would like to thank the following people for their assistance in the publication of this book: Mandy Earey, Simon Murrell, and Charlotte Johnson for design assistance; Victoria Pyke for proofreading; Jamie Ambrose for research; Brian Lawrence and Gary Wade for hand-modelling; John Spence for advice and assistance on set; MMS Marketing Services for photography and logistics. Special thanks to the following for the kind donation of tools or assistance in sourcing images: Phil Davy and John Read.

MMS Photography
www.mms-ww.com

Axminster Tools & Machinery
www.axminster.co.uk

Niwaki Ltd
www.niwaki.com

Stanley Black & Decker
www.stanleyblackanddecker.com

Timeless Tools
www.timelesstools.co.uk

PICTURE CREDITS

The publisher would like to thank the following for their kind permission to reproduce their photographs:

(Key: a-above; b-below/bottom; c-centre; f-far; l-left; r-right; t-top)

8 123RF.com: donatas1205 (tr). 8-9 Alex Rosa: (cb). 9 123RF.com: Darius Dzinnik / Dar1930 (t). 84-85 Niwaki Ltd - www.niwaki.com: (Japanese shears). 88 Niwaki Ltd - www.niwaki.com: (cla). 89 Niwaki Ltd - www.niwaki.com: (ca). 200 Niwaki Ltd - www.niwaki.com: (b). 246-247 Alex Rosa.

ABOUT THE AUTHORS

Nick Offerman Actor, Humorist, Author, Woodworker. A native of Minooka, Illinois, Nick Offerman learned basic carpentry from his father, and the further use of tools from his farmer uncles and grandfathers. In Chicago, Offerman launched his professional acting career and also built scenery and props to supplement his meagre acting income. After earning a couple of feature film roles, Offerman relocated to Hollywood to further pursue his acting career where he also continued to work as a carpenter building cabins and decks while landing film and television work. Since 2000, Offerman owns and operates Offerman Woodshop in California where he and a team of redoubtable American woodworkers create good times, fine furniture, and other sundry items. Last October, Offerman released his third New York Times Bestseller, ***Good Clean Fun: Misadventures in Sawdust at Offerman Woodshop.*** His previous books are ***Paddle Your Own Canoe: One Man's Fundamentals for Delicious Living*** (2013), followed by ***Gumption: Relighting the Torch of Freedom with America's Gutsiest Troublemakers*** (2015).

Phil Davy has been crafting things from wood since he was a boy. His broad expertise includes making musical instruments, running woodwork training workshops, teaching carpentry and joinery, as well as being a qualified wood machinist. Phil joined "Good Woodworking" magazine as Technical Editor when it was launched in 1992, going on to edit Britain's biggest-selling woodwork magazine for nine years. He remains as a Consultant Editor.

Jo Behari set up the UK's first all-female home improvement and property maintenance company, Home Jane, which has won many awards. She co-authored ***The Girls Guide to DIY*** and co-presented the Channel 4 television show Make, Do and Mend. She is also the DIY expert columnist for "House Beautiful" magazine.

Luke Edwardes-Evans is a journalist and former editor of "Land Rover World", "Winning", "Cycle Sport", "Cycling Active" and "Tour" magazines. Other books he has contributed to include: ***The Advanced Cyclist's Training Manual*** and DK's ***The Complete Bike Owner's Manual.***

Matt Jackson is a landscape consultant, has worked in horticulture for more than 20 years, and now designs and restores heritage gardens. He has written about gardening in The Telegraph, and a book on ***Biodynamic & Lunar Gardening.***